Biological and Medical Physics, Biomedical Engineering

BIOLOGICAL AND MEDICAL PHYSICS, BIOMEDICAL ENGINEERING

This series is intended to be comprehensive, covering a broad range of topics important to the study of the physical, chemical and biological sciences. Its goal is to provide scientists and engineers with textbooks, monographs, and reference works to address the growing need for information. The fields of biological and medical physics and biomedical engineering are broad, multidisciplinary and dynamic. They lie at the crossroads of frontier research in physics, biology, chemistry, and medicine.

Books in the series emphasize established and emergent areas of science including molecular, membrane, and mathematical biophysics; photosynthetic energy harvesting and conversion; information processing; physical principles of genetics; sensory communications; automata networks, neural networks, and cellular automata. Equally important is coverage of applied aspects of biological and medical physics and biomedical engineering such as molecular electronic components and devices, biosensors, medicine, imaging, physical principles of renewable energy production, advanced prostheses, and environmental control and engineering.

More information about this series at http://www.springer.com/series/3740

Chirlmin Joo · David Rueda
Editors

Biophysics of RNA-Protein Interactions

A Mechanistic View

Springer

Editors
Chirlmin Joo
Kavli Institute of NanoScience
Delft University of Technology
Delft, Zuid-Holland, The Netherlands

David Rueda
Department of Medicine
Imperial College London
London, UK

ISSN 1618-7210 ISSN 2197-5647 (electronic)
Biological and Medical Physics, Biomedical Engineering
ISBN 978-1-4939-9724-4 ISBN 978-1-4939-9726-8 (eBook)
https://doi.org/10.1007/978-1-4939-9726-8

This Springer imprint is published by the registered company Springer Science+Business Media, LLC part of Springer Nature.
The registered company address is: 233 Spring Street, New York, NY 10013, U.S.A.

Contents

Part I
RNA Binding Proteins

Chapter 1
How Proteins Recognize RNA

Rajan Lamichhane

1.1 Introduction

According to the central dogma of molecular biology, genetic information is transformed from DNA to RNA during a process called transcription [1]. In eukaryotes, after transcription, the pre-mRNA undergoes several processing events including 5′ end capping, splicing, editing, and 3′ end polyadenylation before entering the ribosome for protein synthesis (Fig. 1.1). RNA has various structural, catalytic, and regulatory roles in the cell [2]. Perhaps in the cell, most functional RNAs interact with proteins to carry out functions, such as processing, nuclear export, transport, translation, modification, RNA stabilization, and localization [2–7]. For example, during posttranscriptional regulation of gene expression, RNA interacts directly with proteins to form ribonucleoprotein particles (RNPs) [8–10]. These RNPs are important for recognition of specific sequence elements present in RNA to control the function of the RNA molecule [9, 11]. Since there are many RNAs and a vast number of RNA-binding proteins, the biogenesis of RNPs must be performed with high fidelity. Incorrect formation of RNP complexes or aberrant expression of RNA-binding proteins can cause genetic disorders that may lead to diseases, such as neuromuscular and neurodegenerative disorders and cancers [12–18]. Therefore, understanding the molecular mechanism of protein–RNA interactions and their applications to function is an important aspect of structural and biological research [17, 19].

RNA molecules can adopt different secondary and tertiary structures from standard Watson–Crick base pairs to non-canonical base pairs, creating a platform that

---

This chapter is based on a dissertation submitted in 2011 (Department of Chemistry, Wayne State University, Detroit MI).

R. Lamichhane (✉)
Department of Biochemistry & Cellular and Molecular Biology, University of Tennessee, Knoxville, TN 37996, USA
e-mail: rajan@utk.edu

C. Joo and D. Rueda (eds.), *Biophysics of RNA-Protein Interactions*,
Biological and Medical Physics, Biomedical Engineering,
https://doi.org/10.1007/978-1-4939-9726-8_1

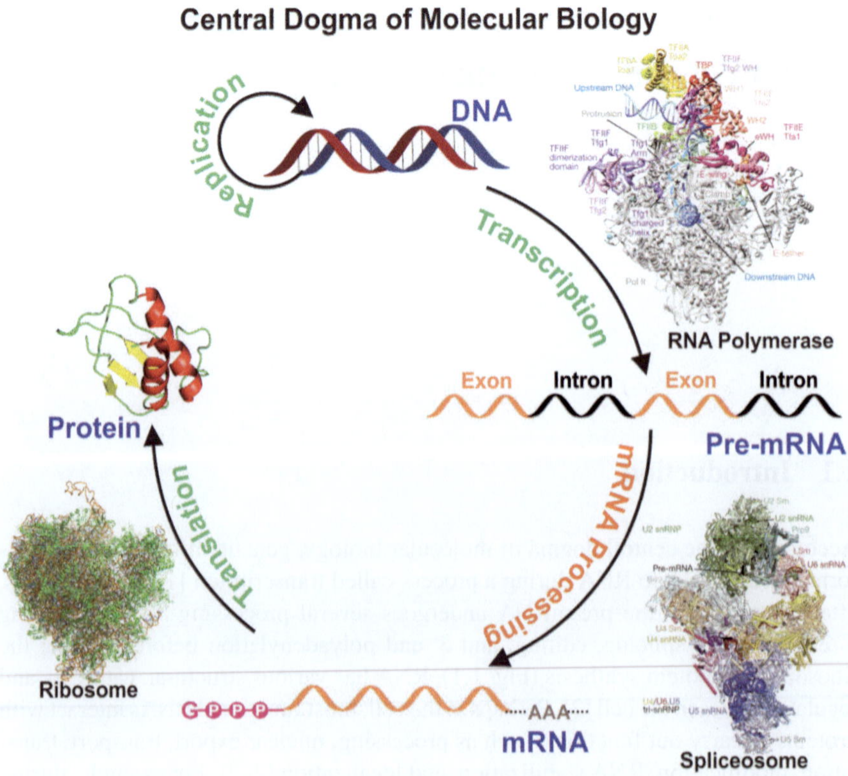

Fig. 1.1 Central dogma of molecular biology representing the general cellular processes in eukaryotic cells. DNA replicates its information and creates new copies of DNA during the process of replication. In eukaryotes, RNA polymerase transcribes DNA information into pre-mRNA, which undergoes RNA processing with the help of spliceosomes. Finally, ribosome translates RNA information into a protein. Structures are reprinted from the following: RNA polymerase II initiation complex is adapted from Plaschka et al. [20], with permission from Springer Nature; structure of a pre-catalytic spliceosome is adapted from Plaschka et al. [21], with permission from Springer Nature; structure of the human 80S ribosome is generated using PyMOL (PDB:4UG0) from Khatter et al. [22]

allows for interaction with a wide variety of ligands. These structures include single-stranded RNA (ssRNA), double-stranded RNA (dsRNA), hairpin loops, bulge loops, internal loops, junction loops, kink turn, and pseudoknots and are recognized by various proteins to form protein–RNA complexes (Fig. 1.2) [23]. These protein–RNA complexes have a wide variety of structural and functional roles in the cell [5, 7].

Despite their functional importance in biology, the actual mechanisms of protein–RNA interactions are poorly understood. Over the last decades, much work has been done to understand the structural and functional relationships of different types of protein–RNA interactions [4, 5, 19, 24, 25, 26]. Several biophysical methods have been used to characterize protein–RNA interactions. For example, X-ray

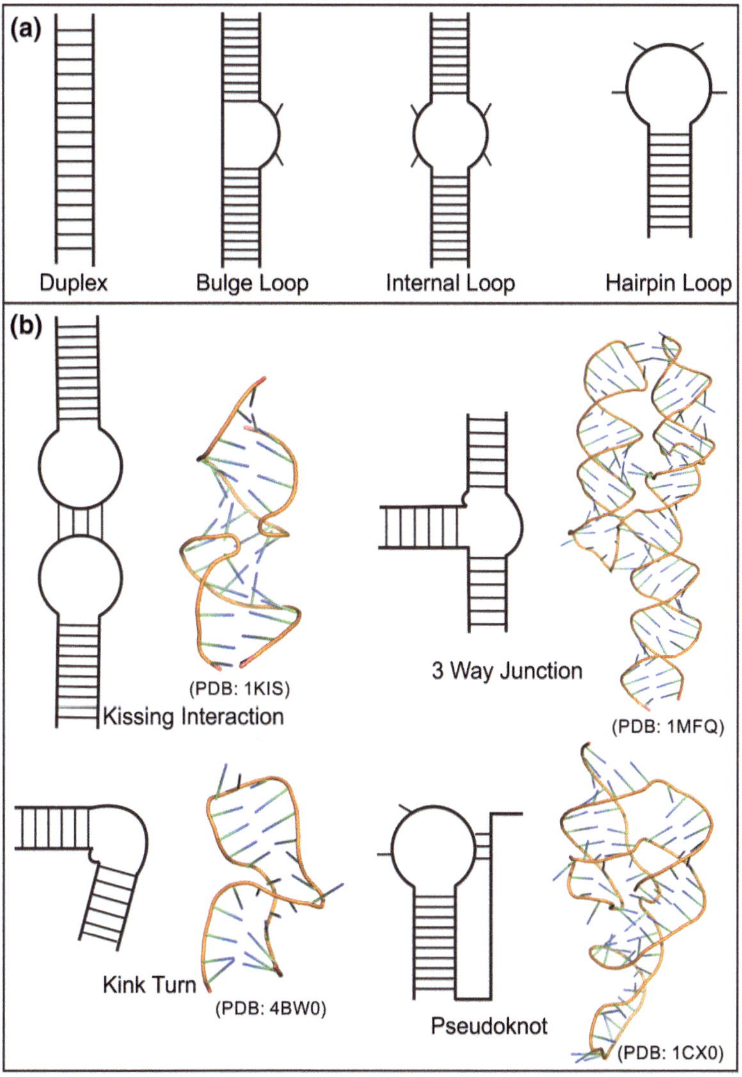

Fig. 1.2 Common RNA secondary structures and tertiary interactions. **a** Two-dimensional representation of common RNA secondary structural motifs (duplex RNA, bulge loop, internal loop, and hairpin loop). **b** Common RNA tertiary structural motifs and interactions with examples. Three-dimensional examples are generated using PyMOL and PDB files as mentioned: [kissing interaction (PDB: 1KIS); three-way junction (PDB: 1MFQ); kink turn (PDB: 4BW0); and pseudoknot (PDB: 1CX0)]

crystallography can be useful to obtain information concerning the detailed molecular interactions of a structured system, while cryo-electron microscopy (cryo-EM) can provide the overall shape of a protein–RNA complex. However, both of these methods have certain restrictions for a system with conformational flexibility and structural heterogeneity [19, 27, 28]. Recent advances have made nuclear magnetic resonance (NMR) one of the best techniques to study protein–RNA interactions in solution by using specific isotope labeling strategies. Coupling of NMR with complimentary small-angle X-ray scattering (SAXS) and electron paramagnetic resonance (EPR) is very helpful to solve larger protein–RNA complexes [27, 29, 30, 31]. Several solution-based protein–RNA structures have been reported in the Protein Database (PDB) [27]. Furthermore, computational modeling has also added insight into the structural analysis of protein–RNA complexes on the basis of different experimental interpretations [27, 32, 33]. The recent advancements of single-molecule spectroscopic techniques have added an effort to understand both the structural and the dynamic behaviors of protein–RNA interactions [34–37].

In this review, a comparison of structural and functional aspects of important known RNA-binding proteins will be discussed. Some important examples of common RNA-binding domains are summarized in Table 1.1 with their PDB entry numbers as an example.

Table 1.1 General properties and examples of common RNA-binding domains. The table is modified from [4], with permission from Springer Nature

Domain	Topology	RNA recognition motif	Protein interaction	Examples (PDB ID)
RRM	βαββαβ	β sheet makes a flat, solvent-exposed RNA-binding surface	Interacts with ssRNA through stacking, electrostatic interactions, and hydrogen bonding	PTB (2ADC) [38] Fox-1 (2ERR) [39]
KH	βααββα αββααβ	A cleft formed by GXXG loop and variable loop	Recognizes at least four nucleotides of ssRNA through hydrophobic interactions, backbone contacts from the loop, and hydrogen bonding with bases	Nova-1 (1EC6) [40] NusA (2ATW) [41]
TRAP	β-sandwich	Edges of β-strand	Bind GAG triplet through protein–base interactions, stacking, or hydrogen bonding	TRAP (1C9S) [42]

(continued)

Table 1.1 (continued)

Domain	Topology	RNA recognition motif	Protein interaction	Examples (PDB ID)
Sm/LSm proteins	αβββββ	Loops formed by β2-β3 and β4-β5	Recognizes poly U of ssRNA through stacking and hydrogen bonding	Sm core protein (1M8 V) [43], Hfq (1KQ2) [44]
Pumilio homology	α	Helix α2 provides the RNA interacting pocket	Stacking interactions and two amino acids in α2 make hydrogen bonds with Watson–Crick edge of a base	Pumilio 1 (1M8Y) [45], Nop9 (5WTY) [46]
Zinc finger	αβ	Amino acid residues in α helices	Sequence-specific (UAUU-TIS11d [47, 48]), hydrogen bonding to the protein backbone, and shape determine the specificity	TIS11D (1RGO) [48] MBNL (5U6H and 5U6L) [49]
PAZ	αβ (β-barrel)	Hydrophobic pocket formed by β-barrel and inserted αβ motif	Single-stranded RNA (ssRNA), and the 5′-phosphate and 3′-OH contribute to specificity	PAZ (1SI3) [50] Argonaute 2 (4OLA)[51]
dsRBM	αβββα	α1 helix and β1-β2 loop	Shape-specific recognition of RNA minor groove of A-form helix (stem-loop), and sequence-specific (G-X_n-A/G) contact with the 2′-OH of sugar and phosphate backbone	ADAR2 (2L3C) [52] Staufen (1EKZ) [53]
SAM	αααααα	Hydrophobic core packed with electropositive regions	Shape-specific recognition of RNA stem-loop, and interaction with phosphate backbone and a single nucleotide G at position 3 of the pentaloop	Vts1p (2ESE) [54]

1.2 RNA-Binding Proteins Are Modular

Most RNA-binding proteins have a modular structure formed by RNA-binding domains. These RNA-binding domains are encoded by sequences of 70–150 amino acids that are important for RNA recognition and interaction [4, 55, 56]. Most of the RNA-binding proteins (RBPs) consist of one or more RNA-binding domains (Fig. 1.3). These include the RNA-binding domain (RBD), most abundant and often called RNA recognition motif (RRM); K-homology (KH) domain; zinc finger (ZnF); Pumilio/FBF (PUF) domain; Piwi/Argonaute/Zwille (PAZ); sterile alpha motif (SAM) domain; double-stranded RNA-binding domain (dsRBD); DEAD box

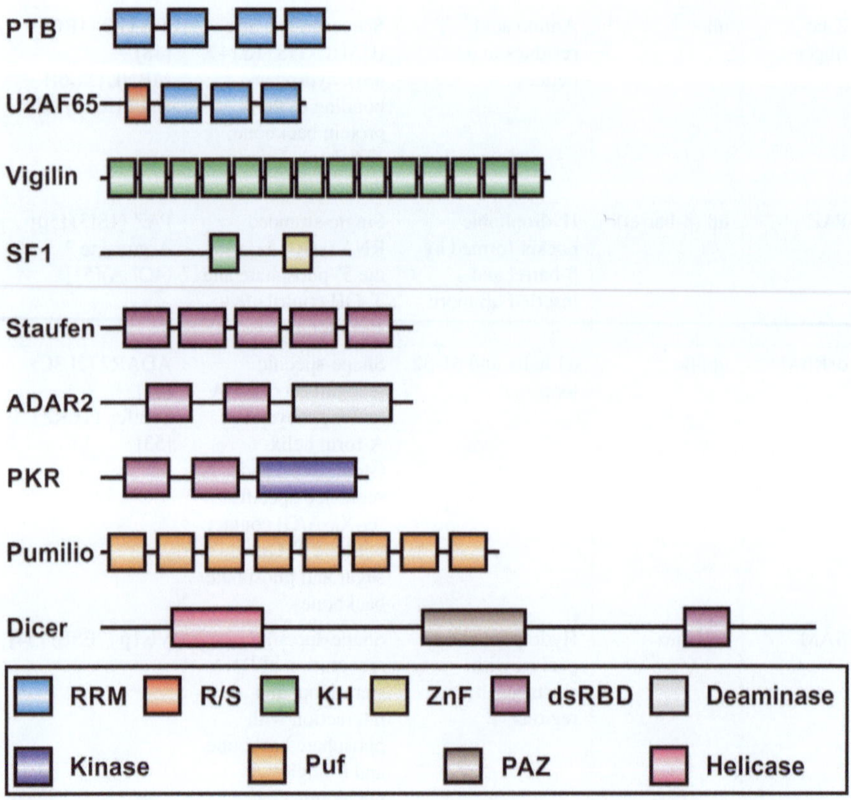

Fig. 1.3 Different modular structures of RNA-binding proteins (RBPs). Examples are taken from the most common RBPs. Each RBP contains many domains as shown by the colored boxes. These include RNA recognition motif (RRM), K-homology (KH) domain, RNA-binding zinc finger (ZnF), double-stranded RNA-binding domain (dsRBD), Puf RNA-binding repeats (Puf), and Piwi/Argonaute/Zwille (PAZ) domain. PTB, polypyrimidine tract binding; R/S, arginine/serine-rich domain; SF1, splicing factor-1; PKR, protein kinase R; U2AF, U2 auxiliary factor; and ADAR, adenosine deaminase. The figure is modified from [4], with permission from Springer Nature (the figure is not drawn to scale)

helicase domain (DDX); and the Sm domain. These modular architectures allow RBPs to recognize RNA with high specificity and affinity, as well as create functional diversity within the RBPs [2, 4, 57]. Proteins with multiple domains can bind long RNA strands or also interact with multiple RNAs; furthermore, modulation of RNA-binding domains with other auxiliary functional domains helps to recognize RNA as well as perform the enzymatic activity. For example, adenosine deaminases that act on RNA 2 (ADAR2) and protein kinase R (PKR) have similar dsRBD but different auxiliary functional domains. ADAR2 converts adenosine to inosine, while PKR has a kinase activity in its target RNA [58, 59].

Frequently, RNA-binding domains are connected with interdomain linkers of variable length. The importance of these linkers is in recognition of the discrete target, and they may act as spacers to regulate the catalytic action of each domain [4]. In some cases, linkers can interact with the RNA-binding domains to allow two domains to function synergistically as observed in polypyrimidine tract-binding protein domains 3 and 4 (PTB34) [35, 38]. Eukaryotic genomes have been shown to have higher numbers of modular RBPs, which might reflect the evolution of highly specific gene expression and modification patterns [2, 7, 9, 60].

1.3 Single-Stranded RNA Recognition

In most cases, RNA-binding proteins (RBPs) recognize ssRNA as their target. Many ssRNA-binding domains have been identified and have been shown to recognize RNA by conserved RNA-binding domains (RRM and KH) and by repeats of RNA-binding domains (TRAP and Sm). The oligonucleotide-/oligosaccharide-binding protein (OB-fold) domains recognize structured RNAs [61]. Many of ssRBPs are sequence-specific RNA-binding proteins with a hydrophobic binding surface to maximize intermolecular contacts with the RNA bases. The most common ssRBPs and their structures are discussed in detail.

1.3.1 RNA Recognition Motifs (RRMs)

The RNA recognition motif (RRM) domain is the most abundant and the best-characterized RNA-binding domain in higher eukaryotes. These domains, also known as ribonucleoprotein (RNP) domain or RNA-binding domain (RBD), consist of 80–100 amino acid residues [57, 62] and are often found in multiple copies. Single RRMs recognize a minimum of two to a maximum of eight nucleotides in the RNA [63, 64]. RRM has four antiparallel β-sheets packed against two α-helices with a topology of βαββαβ (Fig. 1.4a, b). An unusual fifth β-strand is present in RRM3 of polypyrimidine tract-binding protein (PTB) (Fig. 1.4c) [38, 65]. Most of the studied structures of RRM protein in complex with RNA have led to two proposed

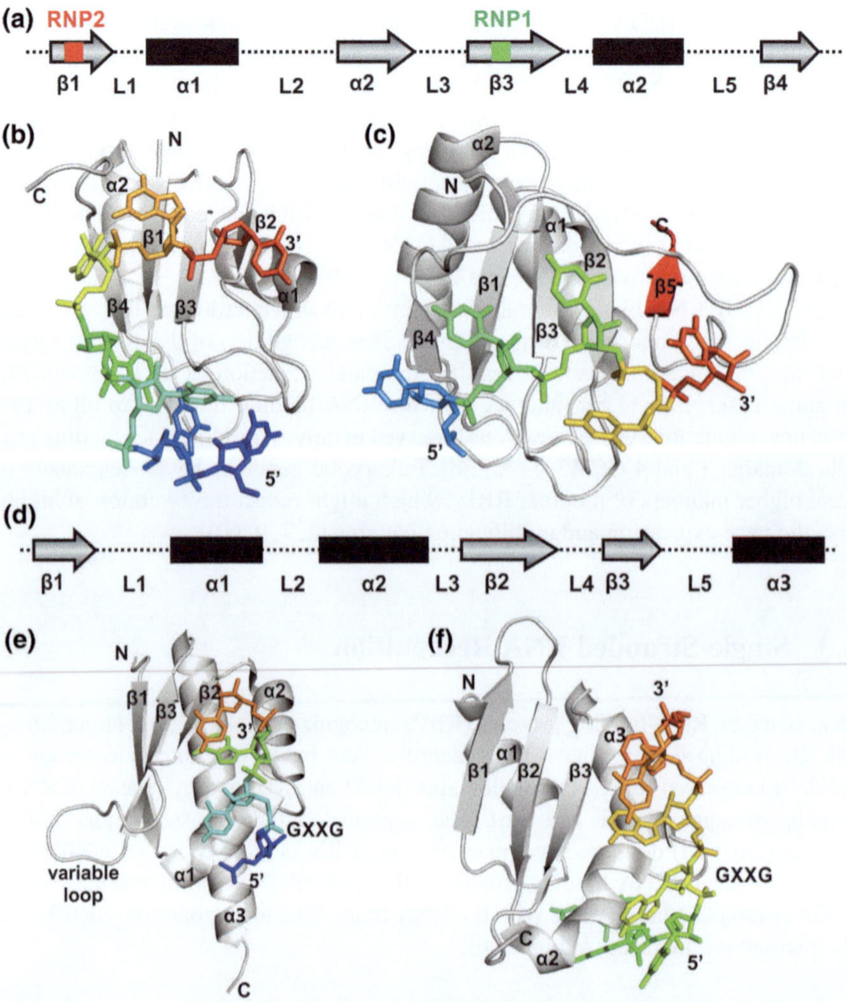

Fig. 1.4 Structures for common single-stranded RNA-binding protein RRM and KH domains. **a** The secondary structure for RRM domain with conserved sequences RNP2 (red) and RNP1 (green). **b** The RRM for Fox-1 domains (PDB: 2ERR). **c** The RRM domain 3 of PTB (PDB: 2ADC) showing the extra β-strand (red). **d** The secondary structure for type I KH domain. **e** Type I KH domain of Nova-1 (PDB: 1EC6) with GXXG conserved loop. **f** Type II KH domain in NusA (PDB: 2ATW). RNA nucleotides are represented in color, and protein secondary structures are shown in gray. The figures are generated with PyMOL

primary conserved sequence stretches that contribute to the RNA binding known as RNP1 ([R/K]-G-[F/Y]-[G/A]-[F/Y]-[I/L/V]-X-[F/Y]) and RNP2 ([I/L/V]-[F/Y]-[I/L/V]-X-N/L) (Fig. 1.4a) [62]. These RNA-binding sequences often rely on the surface of the central β-strands: β1 and β3 [38, 66, 67, 68]. To form these RRM–RNA complexes, solvent-exposed charged residues (Arg or Lys) form a salt bridge to the phosphodiester backbone of the RNA and two aromatic residues can form a ring-stacking interaction or hydrogen bonds with the RNA nucleobases [12, 62]. The wide range of RNA structures and recognition sequence elements has associated RRM proteins with diverse biological functions. These motifs in eukaryotes are implicated in posttranscriptional gene regulation, like pre-mRNA splicing, alternative splicing, capping, mRNA stability and export, RNA editing, and poly(A) recognition [19, 57]. During alternative splicing, many ssRBPs associate with pre-mRNA (RNPA1, $U2AF^{65}$, $U2AF^{35}$, PTB, Fox-1, sex-lethal) to regulate splicing [69]. For example, SR proteins recognize exonic splicing sites to promote alternative splicing whereas Fox-1 does the same activity by interaction with intronic splicing elements [70, 71]. Recent studies have shown that RRMs are also involved in protein–protein interactions for the recognition and interaction with RNA with very distinct mechanisms from protein–RNA interactions [57].

1.3.2 KH-Homology Domain

The heterogeneous nuclear ribonucleoprotein K-homology (KH) domain is highly expressed and most abundant in gene expression and regulatory systems in bacteria, archaea, and eukaryotes [72]. The KH domain consists of nearly 70 amino acid residues with a signature sequence of (I/L/V)IGXXGXX(I/L/V) at the center of the domain [72, 73]. All KH domains are composed of three β-sheets packed against three α-helices. KH domains are divided into two subfamilies: Type I has βααββα topology (Fig. 1.4d, e) (Nova), whereas type II has αββααβ topology (Fig. 1.4f) (NusA) [73]. An important feature of the KH domain is the presence of a variable length loop that connects $\beta 2$ and $\beta 3$ in type I and $\beta 3$ and $\alpha 2$ in type II [74]. In both type I and II, the consensus sequence is formed by a GXXG loop recognized four nucleotides. Hydrophobic interactions between bases and non-aromatic residues, backbone contacts with the GXXG loop, as well as hydrogen bonding with bases are the prevalent interactions observed between protein and RNA [4]. This ssRNA-binding protein domain can also be found in multiple copies (14 copies in chicken vigilin, three KH domains in hnRNP K) that can increase the RNA-binding affinity and cooperativity of this protein [75].

The KH domain is the most abundant RNA-binding domain in eubacteria and eukaryotes, suggesting the evolutionary importance of this ancient RNA-binding domain. Like RRM, KH protein domains are also involved in a myriad of biological processes like splicing (splicing factor 1, SF1) [76], alternative splicing (Nova family protein) [77], transcriptional and translational gene control (hnRNPK) [78], and mRNA stability, transport, and localization [19]. Unusual expression of this protein has been linked to many diseases, such as human fragile X mental retardation syndrome which is caused by a loss of FMR-1 expression where a mutation on the conserved KH motif has an RNA-binding defect [79].

1.3.3 RNA Recognition by Modular RNA-Binding Repeats

In some cases, RNA-binding domains oligomerize to form modular RNA-binding repeats. The numbers of modular repeats vary; for example, eleven repeats are observed in TRAP proteins, seven in Sm core proteins, and six in Lsm proteins Hfq [42, 43, 44, 80].

The tryptophan RNA-binding attenuation protein (TRAP) is comprised of 70 amino acids in each of the eleven monomers that fold into four antiparallel β-strands to form a β-sandwich-like structure. Tryptophan is inserted between the interfaces of two β-strands. Each monomer oligomerizes into an 11-mer symmetric ring as observed in the crystal structure of *Bacillus subtilis* TRAP bound with a 53-nucleotide ssRNA containing GAG triplets (Fig. 1.5a) [42]. Each monomer contains an RNA-binding pocket created by two β-strands to allow for binding to the GAG triplet through protein–base interactions [42].

The outer edge of the 11-mer oligomeric structure has a symmetric ring with an 80-Å diameter. TRAP regulates the expression of L-tryptophan biosynthesis genes in several bacilli, which is activated by bound L-tryptophan. For regulation, TRAP binds

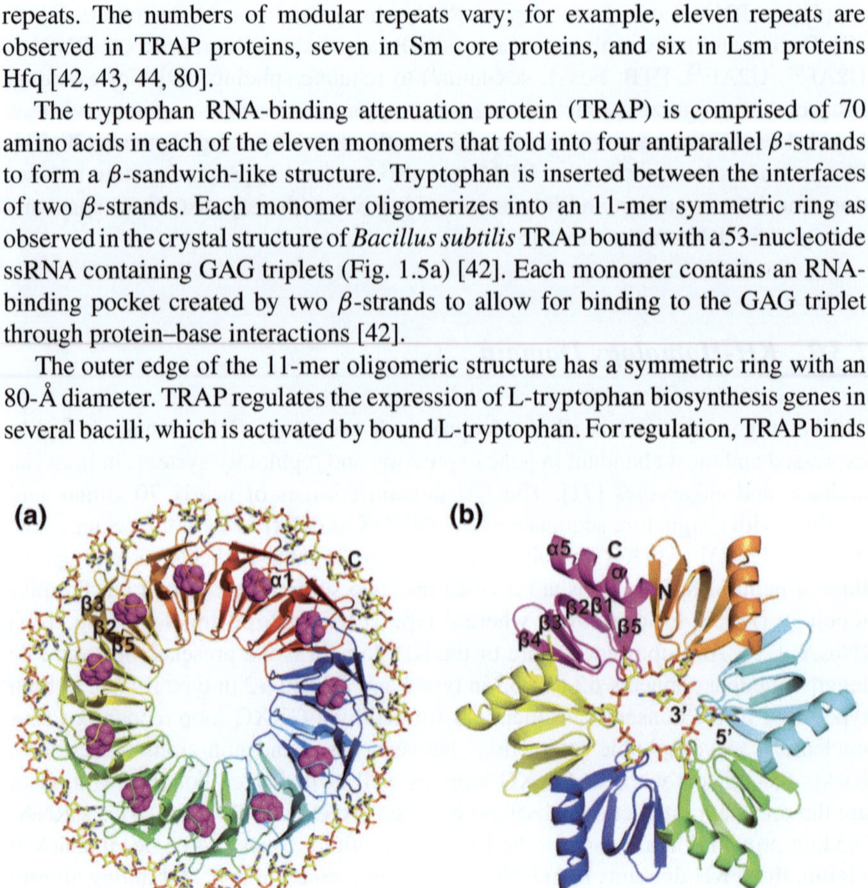

Fig. 1.5 RNA recognition by modular RNA-binding repeats. **a** The crystal structure of the 11-mer TRAP (PDB: 1C9S) protein with GAUGU ssRNA repeats. The surface in magenta is an L-tryptophan inserted in the β-sandwich. **b** Structure of Hfq (PDB: 1KQ2) showing the hexameric ring from *S. aureus*. The central core contains a bound 5′-AU_5G-3′ RNA. For clarity, each protein subunit is colored differently and RNA is in yellow sticks. The figures are generated from PyMOL

to the 5′ ssRNA leader sequence of an mRNA operon and terminates transcription by preventing the formation of the antiterminator stem-loop structure [19, 81].

The classical Sm fold is characterized by an N-terminal α-helix followed by five β-strands with a topology of αβββββ (Fig. 1.5b) [82]. The Sm proteins consist of nearly 80 residues and recognize the uridine-rich site (Sm site) present in small nuclear RNAs (snRNAs). Each Sm protein oligomerizes to form a heptameric ring (~70-Å diameter) structure around the poly(U) RNA [82]. The central hole of this ring can accommodate the U small nuclear RNP (UsnRNP) during pre-mRNA splicing [83, 84]. It has been proposed that the inter-subunit interaction during oligomerization is manifested by hydrophobic contacts between adjacent β-strands and each U-rich RNA is recognized by three conserved residues in the loops of *β2-β3* and *β4-β5* [43]. The interactions between the Sm protein domains and the RNA include stacking and hydrogen bonding. Unlike Sm proteins, LSm proteins, such as bacterial host factor for Q-β bacteriophage (Hfq), form a hexameric doughnut shape with a 12Å central cavity in the absence of RNA [44, 85, 86]. The crystal structure of *S. aureus* Hfq with a short RNA (5′-AU$_5$G-3′) showed that the RNA is bound around the basic central pore (Fig. 1.5b) [44]. Hfq is known to play a role in posttranscriptional gene regulation where it helps small noncoding RNAs (ncRNAs) to identify its target mRNA [87–90]. Recent studies have shown that an intrinsically disordered C-terminal domain (CTD) of Hfq acts as chaperone that auto-regulates RNA binding in bacteria [91, 92].

1.3.4 Other SsRNA-Binding Proteins

Several recent studies have shown other proteins that can bind RNA through different structural arrangements than the traditional RRM and KH domains. These protein domains include zinc fingers, Pumilio homology domain (PUF), PAZ domain, and OB-fold. Their structures, RNA recognition motifs, and protein interactions are summarized in Table 1.1 and are mentioned in many research and review articles [61, 62, 93, 94, 95].

1.4 Double-Stranded RNA Recognition

Double-stranded RNA-binding motifs (dsRBMs) recognize perfectly duplexed RNA and are distributed in eukaryotes, and bacterial and viral proteins [96]. This motif adopts an α/β sandwich global fold with an αβββα topology that contains 70–90 amino acid residues (Fig. 1.6a) [4, 23, 97, 98, 99]. Previous structural studies of dsRBM protein–RNA complexes proposed that these proteins bind in a shape-specific rather than sequence-specific [96, 99]. Many of the solved structures suggested that dsRBM recognizes the A-form helix of dsRNA, and intermolecular interactions involve the direct contact with the 2′-OH sugar and phosphate backbone [4, 53, 100, 101, 102]. But the recent solution NMR structure of an adenosine deaminase

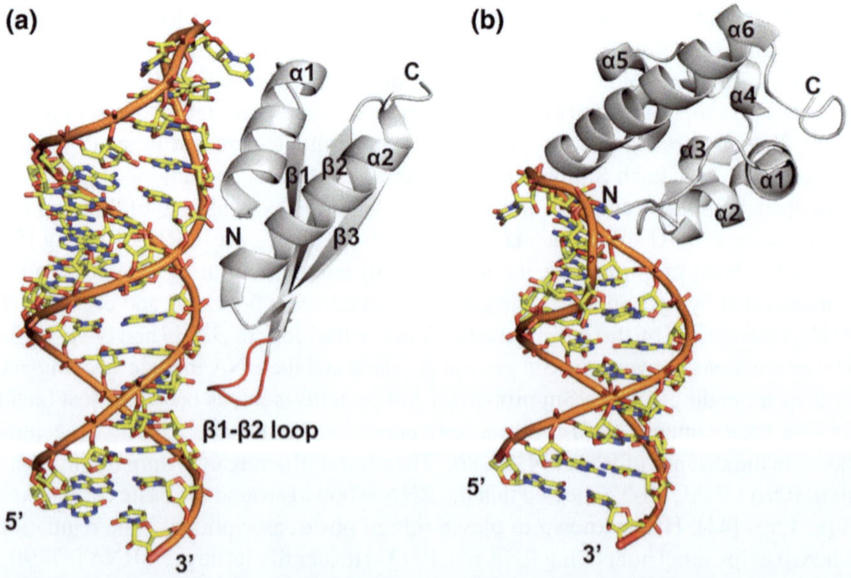

Fig. 1.6 Structure of RNA (yellow sticks) bound with dsRBM and SAM proteins (gray). **a** Upper stem-loop (USL of GluR-2 R/G) RNA recognition by dsRBM1 of ADAR2 (PDB: 2L3C). Shown in red is a β1-β2 loop that is important for sequence-specific recognition of RNA [52]. **b** The structure of Vts1p-SAM (PDB: 2ESE) domain in complex with SRE RNA. The figures are generated from PyMOL

(ADAR2) in complex with a stem-loop pre-mRNA encoding the R/G editing site of GluR-2 has revealed that dsRBM recognizes the shape as well as the sequence of the RNA [52]. The minor groove of the A-form helix in the stem-loop is specifically recognized by the N-terminal helix (α1) and β1-β2 loop of ADAR2 (Fig. 1.6a). The two domains of ADAR2, dsRBM1, and dsRBM2 preferentially recognize G-X_9-A and G-X_8-A RNA sequences, respectively, in a long stem-loop pre-mRNA. The sequence specificity of ADAR2 dsRBM is important for the proper editing function of the enzyme [52].

The double-stranded RBM is involved in several biological processes from RNA editing to protein phosphorylation in translational control [96]. For example, the RNase III domain is involved in RNA processing in the RNA interference (RNAi)/microRNA (miRNA) pathway [103–105]. *Drosophila melanogaster* staufen contains multiple copies of dsRBM domains that control RNP localization [105]. Furthermore, ADAR1 and ADAR2 are RNA editing proteins that regulate gene expression at the RNA level [106] by converting adenosine to inosine (A to I) by hydrolytic deamination in many mRNA and pre-mRNA transcripts [52, 107].

1.5 SAM-Binding Domain

The sterile alpha motif (SAM) domain is the most copious of the eukaryotic protein motifs, initially identified as a protein–protein interaction module involved in transcription regulation and signal transduction [54, 108]. Later, it was reported that the SAM domain also interacts with RNA to control posttranscriptional gene expression [109]. The SAM domain from *Saccharomyces cerevisiae* (Vts1p) and its homolog from *Drosophila melanogaster* (Smaug) specifically interact with the RNA stem-loop [109]. The RNA stem-loop recognized by Smaug contains a CNGGN pentaloop in the Smaug recognition element (SRE) present at the 3′ untranslated region (UTR) of the *nos* transcript [109, 110]. The solution NMR structure of Vts1p-SAM in complex with a 23-nucleotide SRE stem-loop RNA with a CUGGC pentaloop was recently solved (Fig. 1.6b). This study revealed that the SAM domain recognizes RNA in a shape-specific rather than sequence-specific manner specifically recognizing the G in position three of the pentaloop [54]. Two intermolecular hydrogen bonds specifically recognize the identity of the third G in the pentaloop, which also occupies the hydrophobic cavity formed by Leu465 and Ala495 [54]. This protein consists of six α-helices that adopt a globular protein fold and recognize the major groove of the RNA pentaloop through contacts with the RNA sugar phosphate backbone [54].

1.6 Protein–RNA Interactions in the Ribosome

The ribosome is a protein–RNA complex with a catalytic role in protein synthesis. This complex macromolecule consists of more than 50 different ribosomal proteins that interact with RNA. How all of these proteins interact with RNA to form an active structure of the ribosome was a question that proved elusive. The recent X-ray crystal structures of the ribosomal subunits offered a clear picture to explain the interactions between the ribosomal proteins and the RNA [111, 112]. The majority of the ribosomal proteins recognize ribosomal RNA by shape rather than by sequence. Hydrogen bonding, stacking, hydrophobic interactions, as well as interactions with the phosphate backbone were also observed among the characterized protein–RNA interactions.

Ribosomal proteins contain globular domains with similar α/β sandwich folds [111, 113]. The topologies of some of the ribosomal proteins are similar to other RNA-binding proteins as described before, reflecting the similar RNA-binding properties among them. Most of these proteins have extended structures like extended α-hairpin (S2), β-hairpin (S5, S10), N-terminal extension (S3), and C-terminal tail (S6) [112, 113]. These extensions are associated with basic amino acid side chains and have extensive contacts with ribosomal RNA that stabilize the tertiary structure of the ribosome and also participate in protein–protein interactions [113]. In the crystal structure, most of the primary binders are globular and surface-oriented and have direct interaction with RNA helices during assembly. For example, S15 is a primary binder with four α-helices and without any extensions that recognizes the junction of helices h20, h21, and h22 as well as helix h23a in the 16S ribosomal RNA [114]. Proteins with multiple extensions are buried in the RNA and are secondary or tertiary binders. Except for very few (h10, h14, and h33a), most of the RNA helices in the 16S RNA contact proteins and many proteins can recognize a single RNA helix. Most of the proteins in the large subunit, except L12, have direct interaction with RNA [111]. Therefore, it can be theorized that RNA-binding proteins may function in the proper folding of RNA. But some of the ribosomal proteins from large subunit (L1, L10, and L11) are directly involved in protein synthesis. Ribosomal proteins also have significant protein–protein interactions that influence the proper assembly of the ribosomal subunits [113].

1.7 Conclusions

RNA molecules can adopt different secondary and tertiary structures that not only allow it to perform structural, catalytic, and regulatory roles but also create a platform to interact with many proteins to form protein–RNA complexes. These protein–RNA complexes have a wide variety of structural and functional roles in the cell. Most of the RNA-binding proteins are modular, and their mode of RNA recognition is also different. We have discussed the common RNA-binding proteins and how they recognize target RNA based on available information from structural biology. Future works need to focus more on exploring the dynamics and mechanistic importance of protein–RNA interactions and their roles in cellular functions. The experimental approaches like single-molecule techniques in combination with computational biology might help to gain insight into the molecular mechanism and dynamics of protein–RNA interactions and their function.

Acknowledgements RL would like to thank Dr. Amanda Solem and Dr. Sharla Wood for their help and Dr. David Rueda for comments and suggestions on the manuscripts. RL is supported by the start-up funds from the University of Tennessee, Knoxville.

References

1. Crick, F. H. (1958). On protein synthesis. *Symposia of the Society for Experimental Biology, 12,* 138–163.
2. Glisovic, T., Bachorik, J. L., Yong, J., & Dreyfuss, G. (2008). RNA-binding proteins and post-transcriptional gene regulation. *FEBS Letters, 582,* 1977–1986.
3. Dreyfuss, G., Kim, V. N., & Kataoka, N. (2002). Messenger-RNA-binding proteins and the messages they carry. *Nat Rev Mol Cell Biol, 3,* 195–205.
4. Lunde, B. M., Moore, C., & Varani, G. (2007). RNA-binding proteins: Modular design for efficient function. *Nature Reviews Molecular Cell Biology, 8,* 479–490.
5. Foley SW, Kramer MC, Gregory BD (2017) RNA structure, binding, and coordination in Arabidopsis. Wiley Interdiscip Rev RNA.

6. Keene, J. D. (2007). RNA regulons: Coordination of post-transcriptional events. *Nature Reviews Genetics, 8,* 533–543.
7. Gerstberger, S., Hafner, M., & Tuschl, T. (2014). A census of human RNA-binding proteins. *Nature Reviews Genetics, 15,* 829–845.
8. Jones, S., Daley, D. T., Luscombe, N. M., Berman, H. M., & Thornton, J. M. (2001). Protein-RNA interactions: A structural analysis. *Nucleic Acids Research, 29,* 943–954.
9. Ray, D., Kazan, H., Cook, K. B., Weirauch, M. T., Najafabadi, H. S., Li, X., et al. (2013). A compendium of RNA-binding motifs for decoding gene regulation. *Nature, 499,* 172–177.
10. Singh, G., Pratt, G., Yeo, G. W., & Moore, M. J. (2015). The clothes make the mRNA: Past and present trends in mRNP fashion. *Annual Review of Biochemistry, 84,* 325–354.
11. Siomi, H., & Dreyfuss, G. (1997). RNA-binding proteins as regulators of gene expression. *Current Opinion in Genetics & Development, 7,* 345–353.
12. Burd, C. G., & Dreyfuss, G. (1994). Conserved structures and diversity of functions of RNA-binding proteins. *Science, 265,* 615–621.
13. Cooper, T. A., Wan, L., & Dreyfuss, G. (2009). RNA and disease. *Cell, 136,* 777–793.
14. Bekenstein, U., & Soreq, H. (2013). Heterogeneous nuclear ribonucleoprotein A1 in health and neurodegenerative disease: From structural insights to post-transcriptional regulatory roles. *Molecular and Cellular Neuroscience, 56,* 436–446.
15. Mannoor, K., Liao, J. P., & Jiang, F. (2012). Small nucleolar RNAs in cancer. *Biochimica et Biophysica Acta—Reviews on Cancer, 1826,* 121–128.
16. Ferreira, H. J., Heyn, H., Moutinho, C., & Esteller, M. (2012). CpG island hypermethylation-associated silencing of small nucleolar RNAs in human cancer. *RNA Biology, 9,* 881–890.
17. Mihailovic, M. K., Chen, A., Gonzalez-Rivera, J. C., & Contreras, L. M. (2017). Defective ribonucleoproteins, mistakes in RNA processing, and diseases. *Biochemistry-Us, 56,* 1367–1382.
18. Ramaswami, M., Taylor, J. P., & Parker, R. (2013). Altered ribostasis: RNA-protein granules in degenerative disorders. *Cell, 154,* 727–736.
19. Messias, A. C., & Sattler, M. (2004). Structural basis of single-stranded RNA recognition. *Accounts of Chemical Research, 37,* 279–287.
20. Plaschka, C., Hantsche, M., Dienemann, C., Burzinski, C., Plitzko, J., & Cramer, P. (2016). Transcription initiation complex structures elucidate DNA opening. *Nature, 533,* 353–358.
21. Plaschka, C., Lin, P. C., & Nagai, K. (2017). Structure of a pre-catalytic spliceosome. *Nature, 546,* 617–621.
22. Khatter, H., Myasnikov, A. G., Natchiar, S. K., & Klaholz, B. P. (2015). Structure of the human 80S ribosome. *Nature, 520,* 640–645.
23. Tian, B., Bevilacqua, P. C., Diegelman-Parente, A., & Mathews, M. B. (2004). The double-stranded-RNA-binding motif: Interference and much more. *Nature Reviews Molecular Cell Biology, 5,* 1013–1023.
24. Auweter, S. D., Oberstrass, F. C., & Allain, F. H. (2007). Solving the structure of PTB in complex with pyrimidine tracts: An NMR study of protein-RNA complexes of weak affinities. *Journal of Molecular Biology, 367,* 174–186.
25. Clery, A., Blatter, M., & Allain, F. H. (2008). RNA recognition motifs: Boring? Not quite. *Current Opinion in Structural Biology, 18,* 290–298.
26. Auweter, S. D., & Allain, F. H. (2008). Structure-function relationships of the polypyrimidine tract binding protein. *Cellular and Molecular Life Sciences, 65,* 516–527.
27. Jones, S. (2016). Protein-RNA interactions: Structural biology and computational modeling techniques. *Biophysical Reviews, 8,* 359–367.
28. Mackereth, C. D., Simon, B., & Sattler, M. (2005). Extending the size of protein-RNA complexes studied by nuclear magnetic resonance spectroscopy. *ChemBioChem, 6,* 1578–1584.
29. Carlomagno, T. (2014). Present and future of NMR for RNA-protein complexes: A perspective of integrated structural biology. *Journal of Magnetic Resonance, 241,* 126–136.
30. Lapinaite, A., Simon, B., Skjaerven, L., Rakwalska-Bange, M., Gabel, F., & Carlomagno, T. (2013). The structure of the box C/D enzyme reveals regulation of RNA methylation. *Nature, 502,* 519.

31. Duss, O., Michel, E., Yulikov, M., Schubert, M., Jeschke, G., & Allain, F. H. T. (2014). Structural basis of the non-coding RNA RsmZ acting as a protein sponge. *Nature, 509,* 588.
32. Ren, H., & Shen, Y. (2015). RNA-binding residues prediction using structural features. *BMC Bioinformatics, 16,* 249.
33. Mackereth, C. D., & Sattler, M. (2012). Dynamics in multi-domain protein recognition of RNA. *Current Opinion in Structural Biology, 22,* 287–296.
34. Lamichhane, R., Hammond, J. A., Pauszek, R. F., Anderson, R. M., Pedron, I., van der Schans, E., et al. (2017). A DEAD-box protein acts through RNA to promote HIV-1 Rev-RRE assembly. *Nucleic Acids Research, 45,* 4632–4641.
35. Lamichhane, R., Daubner, G. M., Thomas-Crusells, J., Auweter, S. D., Manatschal, C., Austin, K. S., et al. (2010). RNA looping by PTB: Evidence using FRET and NMR spectroscopy for a role in splicing repression. *Proceedings of the National Academy of Sciences of the United States of America, 107,* 4105–4110.
36. Karunatilaka, K. S., Solem, A., Pyle, A. M., & Rueda, D. (2010). Single-molecule analysis of Mss116-mediated group II intron folding. *Nature, 467,* 935–U975.
37. Bonilla, S., Limouse, C., Bisaria, N., Gebala, M., Mabuchi, H., & Herschlag, D. (2017). Single-molecule fluorescence reveals commonalities and distinctions among natural and in vitro-selected RNA tertiary motifs in a multistep folding pathway. *Journal of the American Chemical Society, 139,* 18576–18589.
38. Oberstrass, F. C., Auweter, S. D., Erat, M., Hargous, Y., Henning, A., Wenter, P., et al. (2005). Structure of PTB bound to RNA: Specific binding and implications for splicing regulation. *Science, 309,* 2054–2057.
39. Auweter SD (2006) Structure and function of PTB and fox, two regulators of alternative splicing. Swiss Federal Institute of Technology, Zurich.
40. Lewis, H. A., Musunuru, K., Jensen, K. B., Edo, C., Chen, H., Darnell, R. B., et al. (2000). Sequence-specific RNA binding by a Nova KH domain: Implications for paraneoplastic disease and the fragile X syndrome. *Cell, 100,* 323–332.
41. Beuth, B., Pennell, S., Arnvig, K. B., Martin, S. R., & Taylor, I. A. (2005). Structure of a Mycobacterium tuberculosis NusA-RNA complex. *EMBO Journal, 24,* 3576–3587.
42. Antson, A. A., Dodson, E. J., Dodson, G., Greaves, R. B., Chen, X., & Gollnick, P. (1999). Structure of the trp RNA-binding attenuation protein, TRAP, bound to RNA. *Nature, 401,* 235–242.
43. Thore, S., Mayer, C., Sauter, C., Weeks, S., & Suck, D. (2003). Crystal structures of the Pyrococcus abyssi Sm core and its complex with RNA. Common features of RNA binding in archaea and eukarya. *Journal of Biological Chemistry, 278,* 1239–1247.
44. Schumacher, M. A., Pearson, R. F., Moller, T., Valentin-Hansen, P., & Brennan, R. G. (2002). Structures of the pleiotropic translational regulator Hfq and an Hfq-RNA complex: A bacterial Sm-like protein. *EMBO Journal, 21,* 3546–3556.
45. Wang, X., McLachlan, J., Zamore, P. D., & Hall, T. M. (2002). Modular recognition of RNA by a human pumilio-homology domain. *Cell, 110,* 501–512.
46. Wang, B., & Ye, K. (2017). Nop9 binds the central pseudoknot region of 18S rRNA. *Nucleic Acids Research, 45,* 3559–3567.
47. Wang, X., Zamore, P. D., & Hall, T. M. (2001). Crystal structure of a Pumilio homology domain. *Molecular Cell, 7,* 855–865.
48. Hudson, B. P., Martinez-Yamout, M. A., Dyson, H. J., & Wright, P. E. (2004). Recognition of the mRNA AU-rich element by the zinc finger domain of TIS11d. *Nature Structural and Molecular Biology, 11,* 257–264.
49. Park, S., Phukan, P. D., Zeeb, M., Martinez-Yamout, M. A., Dyson, H. J., & Wright, P. E. (2017). Structural basis for interaction of the tandem zinc finger domains of human muscle-blind with cognate RNA from human cardiac troponin T. *Biochemistry-Us, 56,* 4154–4168.
50. Ma, J. B., Ye, K., & Patel, D. J. (2004). Structural basis for overhang-specific small interfering RNA recognition by the PAZ domain. *Nature, 429,* 318–322.
51. Schirle, N. T., & MacRae, I. J. (2012). The crystal structure of human Argonaute2. *Science, 336,* 1037–1040.

52. Stefl, R., Oberstrass, F. C., Hood, J. L., Jourdan, M., Zimmermann, M., Skrisovska, L., et al. (2010). The solution structure of the ADAR2 dsRBM-RNA complex reveals a sequence-specific readout of the minor groove. *Cell, 143,* 225–237.
53. Ramos, A., Grunert, S., Adams, J., Micklem, D. R., Proctor, M. R., Freund, S., et al. (2000). RNA recognition by a Staufen double-stranded RNA-binding domain. *EMBO Journal, 19,* 997–1009.
54. Oberstrass, F. C., Lee, A., Stefl, R., Janis, M., Chanfreau, G., & Allain, F. H. (2006). Shape-specific recognition in the structure of the Vts1p SAM domain with RNA. *Nature Structural and Molecular Biology, 13,* 160–167.
55. Varani, G. (1995). Exceptionally stable nucleic acid hairpins. *Annual review of biophysics and biomolecular structure, 24,* 379–404.
56. Daelemans, D., Costes, S. V., Cho, E. H., Erwin-Cohen, R. A., Lockett, S., & Pavlakis, G. N. (2004). In vivo HIV-1 Rev multimerization in the nucleolus and cytoplasm identified by fluorescence resonance energy transfer. *Journal of Biological Chemistry, 279,* 50167–50175.
57. Maris, C., Dominguez, C., & Allain, F. H. (2005). The RNA recognition motif, a plastic RNA-binding platform to regulate post-transcriptional gene expression. *The FEBS Journal, 272,* 2118–2131.
58. Valente, L., & Nishikura, K. (2005). ADAR gene family and A-to-I RNA editing: Diverse roles in posttranscriptional gene regulation. *Progress in Nucleic Acid Research and Molecular Biology, 79,* 299–338.
59. Garcia, M. A., Meurs, E. F., & Esteban, M. (2007). The dsRNA protein kinase PKR: Virus and cell control. *Biochimie, 89,* 799–811.
60. Keene, J. D. (2001). Ribonucleoprotein infrastructure regulating the flow of genetic information between the genome and the proteome. *Proceedings of the National Academy of Sciences of the United States of America, 98,* 7018–7024.
61. Theobald, D. L., Mitton-Fry, R. M., & Wuttke, D. S. (2003). Nucleic acid recognition by OB-fold proteins. *Annual review of biophysics and biomolecular structure, 32,* 115–133.
62. Auweter, S. D., Oberstrass, F. C., & Allain, F. H. (2006). Sequence-specific binding of single-stranded RNA: Is there a code for recognition? *Nucleic Acids Research, 34,* 4943–4959.
63. Calero, G., Wilson, K. F., Ly, T., Rios-Steiner, J. L., Clardy, J. C., & Cerione, R. A. (2002). Structural basis of m7G pppG binding to the nuclear cap-binding protein complex. *Nature Structural and Molecular Biology, 9,* 912–917.
64. Price, S. R., Evans, P. R., & Nagai, K. (1998). Crystal structure of the spliceosomal U2B"-U2A' protein complex bound to a fragment of U2 small nuclear RNA. *Nature, 394,* 645–650.
65. Conte, M. R., Grune, T., Ghuman, J., Kelly, G., Ladas, A., Matthews, S., et al. (2000). Structure of tandem RNA recognition motifs from polypyrimidine tract binding protein reveals novel features of the RRM fold. *The EMBO Journal, 19,* 3132–3141.
66. Handa, N., Nureki, O., Kurimoto, K., Kim, I., Sakamoto, H., Shimura, Y., et al. (1999). Structural basis for recognition of the tra mRNA precursor by the Sex-lethal protein. *Nature, 398,* 579–585.
67. Allain, F. H., Bouvet, P., Dieckmann, T., & Feigon, J. (2000). Molecular basis of sequence-specific recognition of pre-ribosomal RNA by nucleolin. *EMBO Journal, 19,* 6870–6881.
68. Auweter, S. D., Fasan, R., Reymond, L., Underwood, J. G., Black, D. L., Pitsch, S., et al. (2006). Molecular basis of RNA recognition by the human alternative splicing factor Fox-1. *EMBO Journal, 25,* 163–173.
69. Black, D. L., & Grabowski, P. J. (2003). Alternative pre-mRNA splicing and neuronal function. *Progress in Molecular and Subcellular Biology, 31,* 187–216.
70. Caceres, J. F., Misteli, T., Screaton, G. R., Spector, D. L., & Krainer, A. R. (1997). Role of the modular domains of SR proteins in subnuclear localization and alternative splicing specificity. *Journal of Cell Biology, 138,* 225–238.
71. Huh, G. S., & Hynes, R. O. (1994). Regulation of alternative pre-mRNA splicing by a novel repeated hexanucleotide element. *Genes and Development, 8,* 1561–1574.
72. Siomi, H., Matunis, M. J., Michael, W. M., & Dreyfuss, G. (1993). The pre-mRNA binding K protein contains a novel evolutionarily conserved motif. *Nucleic Acids Research, 21,* 1193–1198.

73. Grishin, N. V. (2001). KH domain: One motif, two folds. *Nucleic Acids Research, 29,* 638–643.
74. Backe, P. H., Messias, A. C., Ravelli, R. B., Sattler, M., & Cusack, S. (2005). X-ray crystallographic and NMR studies of the third KH domain of hnRNP K in complex with single-stranded nucleic acids. *Structure, 13,* 1055–1067.
75. Gibson, T. J., Thompson, J. D., & Heringa, J. (1993). The KH domain occurs in a diverse set of RNA-binding proteins that include the antiterminator NusA and is probably involved in binding to nucleic acid. *FEBS Letters, 324,* 361–366.
76. Liu, Z., Luyten, I., Bottomley, M. J., Messias, A. C., Houngninou-Molango, S., Sprangers, R., et al. (2001). Structural basis for recognition of the intron branch site RNA by splicing factor 1. *Science, 294,* 1098–1102.
77. Jensen, K. B., Dredge, B. K., Stefani, G., Zhong, R., Buckanovich, R. J., Okano, H. J., et al. (2000). Nova-1 regulates neuron-specific alternative splicing and is essential for neuronal viability. *Neuron, 25,* 359–371.
78. Ostareck-Lederer, A., Ostareck, D. H., & Hentze, M. W. (1998). Cytoplasmic regulatory functions of the KH-domain proteins hnRNPs K and E1/E2. *Trends in Biochemical Sciences, 23,* 409–411.
79. De Boulle, K., Verkerk, A. J., Reyniers, E., Vits, L., Hendrickx, J., Van Roy, B., et al. (1993). A point mutation in the FMR-1 gene associated with fragile X mental retardation. *Nature Genetics, 3,* 31–35.
80. Link, T. M., Valentin-Hansen, P., & Brennan, R. G. (2009). Structure of *Escherichia coli* Hfq bound to polyriboadenylate RNA. *Proceedings of the National Academy of Sciences of the United States of America, 106,* 19292–19297.
81. Babitzke, P. (1997). Regulation of tryptophan biosynthesis: Trp-ing the TRAP or how Bacillus subtilis reinvented the wheel. *Molecular Microbiology, 26,* 1–9.
82. Kambach, C., Walke, S., Young, R., Avis, J. M., de la Fortelle, E., Raker, V. A., et al. (1999). Crystal structures of two Sm protein complexes and their implications for the assembly of the spliceosomal snRNPs. *Cell, 96,* 375–387.
83. Stark, H., Dube, P., Luhrmann, R., & Kastner, B. (2001). Arrangement of RNA and proteins in the spliceosomal U1 small nuclear ribonucleoprotein particle. *Nature, 409,* 539–542.
84. Pomeranz Krummel, D. A., Oubridge, C., Leung, A. K., Li, J., & Nagai, K. (2009). Crystal structure of human spliceosomal U1 snRNP at 5.5 A resolution. *Nature, 458,* 475–480.
85. Mikulecky, P. J., Kaw, M. K., Brescia, C. C., Takach, J. C., Sledjeski, D. D., & Feig, A. L. (2004). Escherichia coli Hfq has distinct interaction surfaces for DsrA, rpoS and poly(A) RNAs. *Nature Structural and Molecular Biology, 11,* 1206–1214.
86. Sauter, C., Basquin, J., & Suck, D. (2003). Sm-like proteins in Eubacteria: The crystal structure of the Hfq protein from Escherichia coli. *Nucleic Acids Research, 31,* 4091–4098.
87. Majdalani, N., Cunning, C., Sledjeski, D., Elliott, T., & Gottesman, S. (1998). DsrA RNA regulates translation of RpoS message by an anti-antisense mechanism, independent of its action as an antisilencer of transcription. *Proceedings of the National Academy of Sciences of the United States of America, 95,* 12462–12467.
88. Masse, E., & Gottesman, S. (2002). A small RNA regulates the expression of genes involved in iron metabolism in Escherichia coli. *Proceedings of the National Academy of Sciences of the United States of America, 99,* 4620–4625.
89. Lee, T., & Feig, A. L. (2008). The RNA binding protein Hfq interacts specifically with tRNAs. *RNA, 14,* 514–523.
90. Santiago-Frangos A, Woodson SA (2018) Hfq chaperone brings speed dating to bacterial sRNA. Wiley Interdiscip Rev RNA:e1475.
91. Santiago-Frangos A, Jeliazkov JR, Gray JJ, Woodson SA (2017) Acidic C-terminal domains autoregulate the RNA chaperone Hfq. Elife 6.
92. Santiago-Frangos, A., Kavita, K., Schu, D. J., Gottesman, S., & Woodson, S. A. (2016). C-terminal domain of the RNA chaperone Hfq drives sRNA competition and release of target RNA. *Proceedings of the National Academy of Sciences of the United States of America, 113,* E6089–E6096.

93. Friesen, W. J., & Darby, M. K. (1998). Specific RNA binding proteins constructed from zinc fingers. *Nature Structural and Molecular Biology, 5,* 543–546.
94. Spassov, D. S., & Jurecic, R. (2003). The PUF family of RNA-binding proteins: Does evolutionarily conserved structure equal conserved function? *IUBMB Life, 55,* 359–366.
95. Yan, K. S., Yan, S., Farooq, A., Han, A., Zeng, L., & Zhou, M. M. (2003). Structure and conserved RNA binding of the PAZ domain. *Nature, 426,* 468–474.
96. Chang, K. Y., & Ramos, A. (2005). The double-stranded RNA-binding motif, a versatile macromolecular docking platform. *The FEBS Journal, 272,* 2109–2117.
97. Bycroft, M., Grunert, S., Murzin, A. G., Proctor, M., & St Johnston, D. (1995). NMR solution structure of a dsRNA binding domain from Drosophila staufen protein reveals homology to the N-terminal domain of ribosomal protein S5. *EMBO Journal, 14,* 3563–3571.
98. Kharrat, A., Macias, M. J., Gibson, T. J., Nilges, M., & Pastore, A. (1995). Structure of the dsRNA binding domain of E. coli RNase III. *EMBO Journal, 14,* 3572–3584.
99. Masliah, G., Barraud, P., & Allain, F. H. T. (2013). RNA recognition by double-stranded RNA binding domains: A matter of shape and sequence. *Cellular and Molecular Life Sciences, 70,* 1875–1895.
100. Ryter, J. M., & Schultz, S. C. (1998). Molecular basis of double-stranded RNA-protein interactions: Structure of a dsRNA-binding domain complexed with dsRNA. *EMBO Journal, 17,* 7505–7513.
101. Gan, J., Tropea, J. E., Austin, B. P., Court, D. L., Waugh, D. S., & Ji, X. (2006). Structural insight into the mechanism of double-stranded RNA processing by ribonuclease III. *Cell, 124,* 355–366.
102. Wu, H., Henras, A., Chanfreau, G., & Feigon, J. (2004). Structural basis for recognition of the AGNN tetraloop RNA fold by the double-stranded RNA-binding domain of Rnt1p RNase III. *Proceedings of the National Academy of Sciences of the United States of America, 101,* 8307–8312.
103. Robertson, H. D. (1982). Escherichia coli ribonuclease III cleavage sites. *Cell, 30,* 669–672.
104. Bernstein, E., Caudy, A. A., Hammond, S. M., & Hannon, G. J. (2001). Role for a bidentate ribonuclease in the initiation step of RNA interference. *Nature, 409,* 363–366.
105. Lee, Y., Ahn, C., Han, J., Choi, H., Kim, J., Yim, J., et al. (2003). The nuclear RNase III Drosha initiates microRNA processing. *Nature, 425,* 415–419.
106. Nishikura, K. (2006). Editor meets silencer: Crosstalk between RNA editing and RNA interference. *Nature Reviews Molecular Cell Biology, 7,* 919–931.
107. Bass, B. L., & Weintraub, H. (1988). An unwinding activity that covalently modifies its double-stranded RNA substrate. *Cell, 55,* 1089–1098.
108. Qiao, F., & Bowie, J. U. (2005). The many faces of SAM. *Sci STKE, 2005,* 1–10.
109. Aviv, T., Lin, Z., Lau, S., Rendl, L. M., Sicheri, F., & Smibert, C. A. (2003). The RNA-binding SAM domain of Smaug defines a new family of post-transcriptional regulators. *Nature Structural and Molecular Biology, 10,* 614–621.
110. Dahanukar, A., Walker, J. A., & Wharton, R. P. (1999). Smaug, a novel RNA-binding protein that operates a translational switch in Drosophila. *Molecular Cell, 4,* 209–218.
111. Ban, N., Nissen, P., Hansen, J., Moore, P. B., & Steitz, T. A. (2000). The complete atomic structure of the large ribosomal subunit at 2.4 Å resolution. *Science, 289,* 905–920.
112. Wimberly, B. T., Brodersen, D. E., Clemons, W. M., Jr., Morgan-Warren, R. J., Carter, A. P., Vonrhein, C., et al. (2000). Structure of the 30S ribosomal subunit. *Nature, 407,* 327–339.
113. Brodersen, D. E., Clemons, W. M., Jr., Carter, A. P., Wimberly, B. T., & Ramakrishnan, V. (2002). Crystal structure of the 30 S ribosomal subunit from Thermus thermophilus: Structure of the proteins and their interactions with 16 S RNA. *Journal of Molecular Biology, 316,* 725–768.
114. Agalarov, S. C., Sridhar Prasad, G., Funke, P. M., Stout, C. D., & Williamson, J. R. (2000). Structure of the S15, S6, S18-rRNA complex: Assembly of the 30S ribosome central domain. *Science, 288,* 107–113.

Chapter 2
The Interaction Between L7Ae Family of Proteins and RNA Kink Turns

Lin Huang and David M. J. Lilley

2.1 Introduction

The kink turn (normally abbreviated to k-turn) is an extremely common structural motif in duplex RNA that kinks the helical axis with an included angle of 50° (Fig. 2.1). A standard k-turn comprises a three-nucleotide bulge, followed by tandem sheared G·A and A·G base pairs, and there is a systematic nomenclature to identify each nucleotide within the structure. The helix 5′ to the bulge is called the C-helix, while that 3′ to the bulge (i.e., the helix containing the G·A pairs) is termed the NC-helix. The G·A base pairs position the conserved adenine nucleotides such that they place the sugar edges of the two adenine nucleobases facing the minor groove of the C-helix. Two critical cross-strand hydrogen bonds are formed across the interface between the two minor grooves, discussed in the following section.

The majority of the known k-turns bind specific proteins and most also mediate tertiary contacts. For example, the very well-studied *H. marismortui* ribosomal Kt-7 (*Hm*Kt-7) binds the L24 protein, and the kink allows the terminal loop of the C-helix to make a loop–loop contact with another stem-loop related by a common three-way helical junction [1]. Multiple k-turns are found in the ribosome, in the spliceosomal U4 snRNA [2, 3], and in seven known riboswitches [4–9]. In addition, k-turns play a critical role in the assembly of the box C/D and H/ACA snoRNP apparatus [10–13] that carry out the site-specific 2′*O*-methylation and pseudouridylation, respectively, of RNA in the nucleoli of archaea and eukaryotes (discussed further below).

In some k-turns, the C-helix is replaced by a loop of typically ~8 nucleotides. These are called k-loops [11]. The k-turns are also subject to significant variation. The standard k-turn comprises a three-nucleotide bulge followed by the tandem G·A and A·G base pairs. However, this basic motif can be elaborated in a variety of

L. Huang · D. M. J. Lilley (✉)
Cancer Research UK Nucleic Acid Structure Research Group, MSI/WTB Complex, The University of Dundee, Dow Street, Dundee DD1 5EH, UK
e-mail: d.m.j.lilley@dundee.ac.uk

C. Joo and D. Rueda (eds.), *Biophysics of RNA-Protein Interactions*, Biological and Medical Physics, Biomedical Engineering,
https://doi.org/10.1007/978-1-4939-9726-8_2

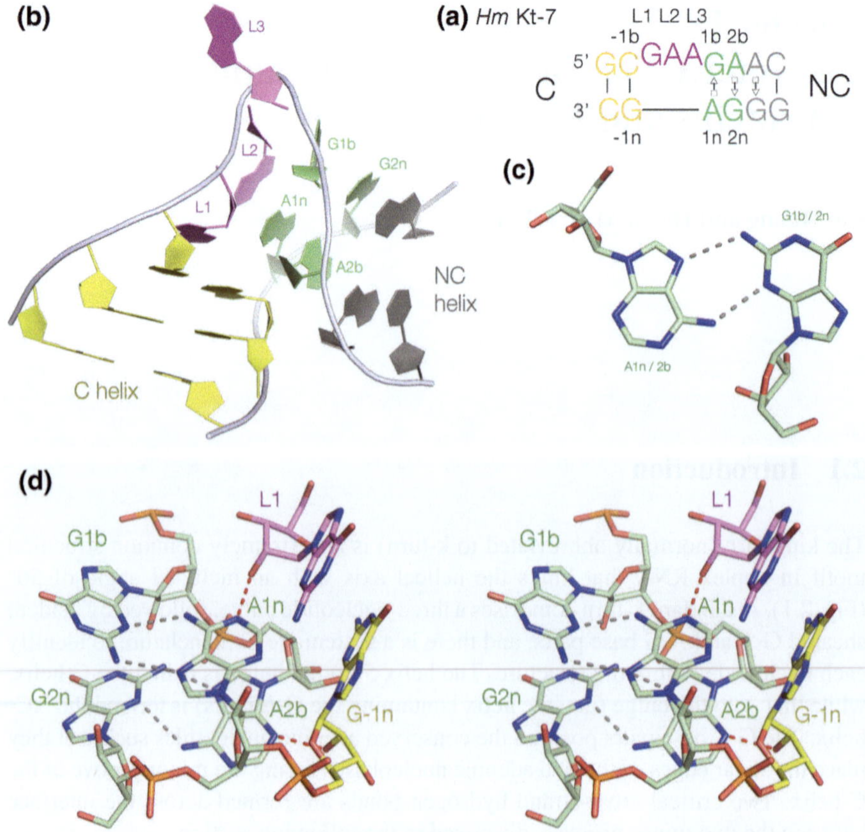

Fig. 2.1 The k-turn motif in RNA. **a** The sequence of *H. marismortui* Kt-7 (*Hm*Kt-7) with the standard nomenclature for the nucleotide positions indicated. **b** The structure of *Hm*Kt-7 folded into its k-turn conformation. This structure was determined by X-ray crystallography of a protein-free duplex RNA (PDE ID 4CS1). Under these conditions, *Hm*Kt-7 adopts an N3 conformation. **c** A *trans* G(sugar)·A(Hoogsteen) sheared base pair. Both G·A base pairs of the k-turn adopt this structure. **d**. The core of the structure of the *Hm*Kt-7 k-turn, showing the two G·A base pairs and the key cross-strand hydrogen bonds (drawn red) from L1 O2′ to A1n N1 and from G-1n O2′ to A2b N3. The structure is shown as a parallel-eye stereoscopic pair

ways. In nonstandard k-turns, there is some departure from the G·A sequences, yet these fold into recognizable k-turn structures [14, 15]. In the complex k-turns, the positioning of the key nucleotides in the structure may not map in a linear way onto the sequence, exemplified by Kt-15 in the *H. marismortui* ribosome [1]. The k-turn structure may even be elaborated into a three-way junction, termed the k-junction [16].

In free solution in the absence of added metal ions and proteins, the k-turns are predominantly unfolded, with a significantly larger included angle between the helical axes that is more typical of a normal three-base bulge. They can be induced to fold into the more-tightly kinked structure in one of three ways. Some k-turns (e.g., *Hm*Kt-7) will spontaneously fold on the addition of metal ions in a two-state process, requiring ~100 μM Mg^{2+} ions or ~30 mM Na^{+} ions. However, not all k-turns will respond to the addition of metal ions, and whether or not they do is in part determined by the 3b,3n sequence [17]. Second, most k-turns will undergo folding on binding proteins [18], exemplified by the L7Ae family discussed below. Lastly, the formation of tertiary contacts (e.g., that found in the SAM-I riboswitch) can stabilize the folded conformation of the k-turn [19].

2.2 The Structure of K-Turns in RNA

In the standard k-turn, the L1 nucleobase stacks onto the end of the C-helix, the L2 nucleobase stacks onto the NC-helix and L3 extends into solution. The tandem G·A, A·G sheared base pairs are the core of the k-turn structure (Fig. 2.1). Both are *trans* G(sugar)·A(Hoogsteen) base pairs, with hydrogen bonds between GN2 to AN7, and AN6 to GN3, although the latter does not form in the G2n·A2b base pair in one conformation of the k-turn (see below). In the kinked conformation, these base pairs stack the two adenine nucleobases with their glycosidic bonds on opposite sides, but with their minor groove edges oriented in the same direction, toward the minor groove of the C-helix. Thus, the minor grooves of the NC- and C-helices are juxtaposed in the core of the k-turn, allowing two critical cross-strand A-minor interactions [20] to form. These are donated by 2′-hydroxyl groups on the two strands and accepted by the two adenine nucleobases of the tandem G·A, A·G base pairs. One is donated by the O2′ of L1 (the first nucleotide of the loop) and accepted by A1n N1 [21]. This is invariant, and in some nonstandard k-turns (or even k-junctions [16]) where this is not possible, there is a surrogate hydrogen bond donor that takes the same role. The second is donated by O2′ of the nucleotide at -1n (the non-bulged strand nucleotide of the base pair of the C-helix adjacent to the bulge) and accepted by A2b [22].

All the k-turns (including nonstandard ones) can be equally divided into two classes that differ in which ring nitrogen atom accepts this bond [22]. In one group, the acceptor is A2b N3, and in the other, it is A2b N1. We therefore name these two groups the N3 and N1 class k-turns. In the N1 structure, the A2b N6 to G2n N3 distance is generally too long (typically >4.3 Å) to be considered a proper hydrogen bond, so that the G2n·A2b base pair interacts by a single hydrogen bond in the N1 structure. Using a systematic crystallographic analysis, we have deduced that the primary determinant of which structure is adopted preferentially is the 3b·3n sequence [23]. Interestingly, the same position also determines whether or not the k-turn will fold in response to the addition of metal ions, although all will fold on the addition of L7Ae [17].

The A-minor interactions in the core of the k-turn require that the axis of the helix is kinked with an included angle of ~50°. The two axes do not intersect, but are displaced relative to one another significantly. Comparison of the structures of *Hm*Kt-7 in the N3 and N1 conformations (it is one of the few k-turns that exists in both conformations) shows that switching between the structures results in an axial rotation of the C-helix that could potentially affect tertiary interactions or protein binding [22].

2.3 The L7Ae Family of Proteins and Their Cellular Roles

The L7Ae family of proteins [24, 25] include the ribosomal proteins L7Ae and L30e, human 15.5 k protein [26], and the yeast snu31p. Bacterial homologs including YbxF and YlxQ have also been identified [27]. Some functional substitution between these proteins is possible in some cases [28, 29].

In the ribosome, L7Ae and L30e bind to k-turns, stabilizing their folded conformation. For example, L7Ae binds to the complex k-turn Kt-15 in the large subunit of the archaeal ribosome [1], and we have determined a crystal structure of a complex of L7Ae bound to Kt-15 inserted into the SAM-I riboswitch (LH and DMJL unpublished data). 15.5 k binds a standard k-turn in the U4 snRNA in the pre-catalytic U4-U6.U5 tri-snRNP complex of the spliceosome cycle [2, 26]. In fact, this was probably the first k-turn complex crystal structure to be determined, although it required the identification of multiple k-turns within the ribosome to recognize it as a recurrent motif [30]. L7Ae and 15.5 k are important subunits within the archaeal and mammalian (respectively) box C/D and H/ACA snoRNPs that direct site-specific 2′O-methylation and pseudouridylation of target RNAs [28, 29, 31], and in the U3 snoRNP [32]. This will be discussed further below. L7Ae has also been identified as a subunit in the ribozyme ribonuclease P, where it also binds a k-turn [33].

2.4 The Molecular Recognition of K-Turns by L7Ae-Family Proteins

A number of crystal structures have been determined for L7Ae-family proteins bound to k-turns in different contexts, including the ribosome [1], box C/D [10, 34, 35] and H/ACA [11] and U4 snRNA [2] as well as the bacterial homologs [36].

We determined the crystal structure of *Archaeoglobus fulgidus* L7Ae (*Af*L7Ae) bound to the *H. marismortui* Kt-7 k-turn [37]. The diffraction extended to 2.3 Å with high-quality electron density, so this was the highest resolution structure of L7Ae bound to a standard k-turn. Comparing this to other complexes allows us to see the general principles of the recognition of k-turn structure by this class of proteins.

The protein is located on the outer face of the RNA, placing an α-helix into the major groove that runs around the structure (Fig. 2.2). This is strongly reminiscent of a bacterial repressor protein placing a recognition helix into the major groove of DNA. Normally, the major groove of A-form RNA is deep and narrow and thus inaccessible to a protein, but at the k-turn, it becomes opened by being splayed-out on the outside of the kinked structure. This helix makes three kinds of interaction with

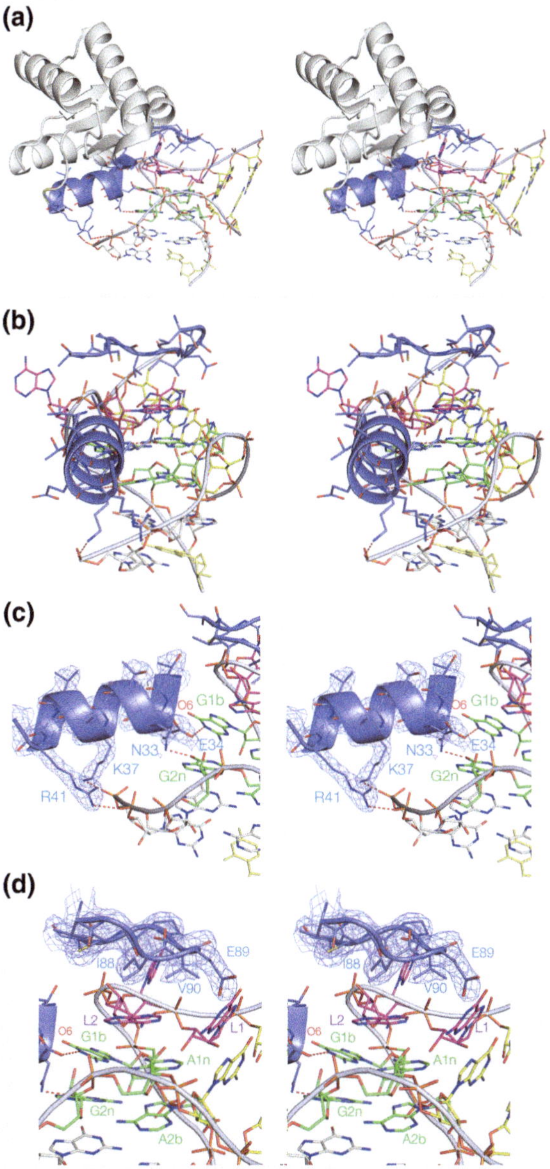

Fig. 2.2 Crystal structure of the complex of *A. fulgidus* L7Ae bound to *Hm*Kt-7 at 2.3 Å resolution (PDB ID 4BW0). Each view is shown as a parallel-eye stereoscopic pair. **a**. The overall structure of the Kt-7:L7Ae complex with the key α-helix and loop that interact with the RNA highlighted in blue. The view is from the side of the k-turn with the unbulged strand. **b**. A view into the major groove splayed around the outside of the k-turn, with just the key α-helix and loop shown for the protein. **c**. The α-helix, showing the electron density of the composite omit map contoured at 2σ. At the C-terminal end of the helix, the side chains of E37 and R41 make non-specific contacts with the RNA backbone, while at the N-terminal end N33 and E34 hydrogen bond to the nucleobases of G2n and G1b, respectively. The O6 carbonyl atom of G1b lies close to the axis of the α-helix, feeling the positive pole of the helix dipole. This α-helix can be considered the recognition helix for the L7Ae. **d**. The hydrophobic loop, with electron density shown contoured at 2σ. The loop contains a number of hydrophobic side chains (e.g., I88 and V90), enveloping the loop region of the k-turn These data were originally published in Huang and Lilley [37]

the RNA. At its C-terminal end, basic side chains make non-specific interactions with the polynucleotide backbone. At the N-terminal end, the helix is juxtaposed with the nucleobases of the conserved guanines forming the tandem G·A base pairs. The side chains of E34 and N33 are hydrogen bonded to G1b N1 and G2n O6, respectively. These can be regarded as the specificity interactions, recognizing the conserved guanine nucleotides in the core of the k-turn. By contrast, the conserved adenine bases are buried on the inner side of the k-turn and are not contacted by protein. In addition, the O6 atom of G1b is located close to the axis of the helix, exposed to the positive pole of the helix dipole. This should provide a degree of electrostatic stabilization of the complex. A second major point of contact is made by a hydrophobic loop of sequence VGIEVPC. This covers the L1 and L2 nucleobases of the k-turn loop burying 730 Å^2, with the hydrophobic side chains of I88 and V90 making close contact.

Most of these features are recapitulated in the other complexes between L7Ae family proteins and k-turns, and the above can be taken as a general description of the manner of the interaction. The specificity interactions are found in the L7Ae:box C/D [10] and 15.5 k:U4 [2] complexes. The location of G1b O6 on the α-helix positive pole is also universal. The bacterial YbxF protein binds k-turns but not k-loops. A structure of YbxF bound to the SAM-I riboswitch k-turn [36] reveals the same general structure and organization, but without the specific side chain interactions. The reported affinity of $K_d = 270$ nM is very much lower than that for AfL7Ae and may reflect in part this lack of specific interactions.

2.5 L7Ae-Family Proteins Bind k-Turns with High Affinity, Generating the Kinked Conformation

The addition of L7Ae to an RNA duplex containing a potential k-turn sequence leads to the formation of the kinked conformation. This is most easily demonstrated using an RNA with a central k-turn and arms of ~12 bp, and fluorophores at the 5′-termini that can act as donor and acceptor in a fluorescence resonance energy transfer (FRET) experiment. On adopting the kinked conformation, the fluorophores become closer (i.e., the end-to-end distance shortens) and thus FRET efficiency (E_{FRET}) increases [18, 21, 38] (Fig. 2.3). For the fluorescein-cyanine 3 FRET pair and a 27 bp RNA, E_{FRET} typically increases from 0.2 to 0.55 on the addition of L7Ae [18]. This has also been shown using fluorescent lifetime measurements [18] and X-ray scattering from k-turn-containing RNA with terminally attached gold nanoparticles [39].

The affinity of *Af*L7Ae is too high to measure from a simple binding experiment of this kind. Instead, we measured the rate of association (k_{on}) and the rate of dissociation (k_{off}) and calculated the affinity from their ratio ($K_d = k_{off}/k_{on}$) [40]. Using stopped-flow mixing, k_{on} was found to be only a little slower than diffusion controlled, and the calculated affinity was K_d = 10 pM. This is extremely high and thus cannot be measured by conventional means such as electrophoretic retardation analysis

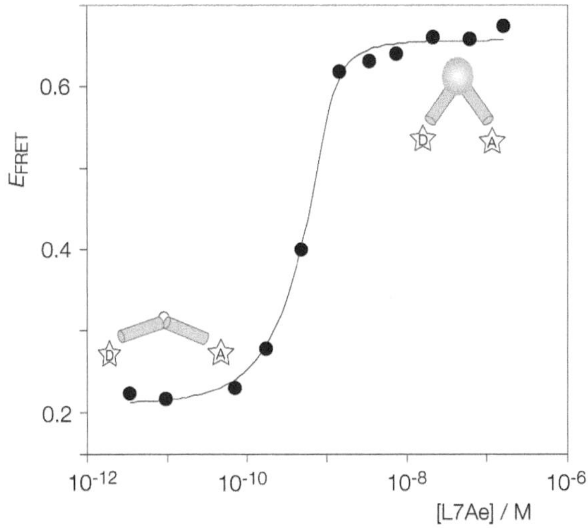

Fig. 2.3 Folding of *Hm*Kt-7 on binding *A. fulgidus* L7Ae. The k-turn folding upon the binding of L7Ae has been measured using steady-state FRET in bulk solution, using RNA terminally labeled with the fluorophores fluorescein and Cy3. Folding kinks the RNA, thus shortening the end-to-end distance leading to an increase in the efficiency of energy transfer (E_{FRET}) between the 5′-terminally attached fluorescein (D) and Cy3 (A) fluorophores. E_{FRET} is plotted as a function of L7Ae concentration, and the data have been fitted to a two-state model for L7Ae binding (line). These data were originally published in Turner et al. [40]

because it is not possible to detect the RNA at the low concentrations required to be in equilibrium even by fluorescence.

2.6 The Manner of K-Turn Folding Resulting from the Binding of L7Ae-Family Proteins

The L7Ae-family proteins bind to k-turns with considerable selectivity and extremely high affinity, forming a complex in which the RNA is in the kinked conformation. This represents protein-mediated RNA folding on an unusually large scale. This has often been referred to as ‘induced fit,’ but the term is used too loosely in this context. We could envisage two processes that could lead to this:

1. Conformational selection [41]. We know from fluorescence lifetime measurements that a small fraction of folded k-turn RNA always exists [38]. If the L7Ae-related proteins bind tightly to this component, they will drive the equilibrium toward a population of bound, folded molecules.
2. Induced fit [42]. An alternative is a more active process whereby the protein somehow mechanically coerces the conformation of the RNA to change on binding.

The fundamental difference between the two alternative mechanisms is whether or not the RNA folds before the protein binds, and this has been discussed previously for many macromolecular-ligand interactions [43–48]. We therefore used single-molecule FRET to analyze the change in conformation at the moment of L7Ae binding [49]. We developed a novel way to tether *Af*L7Ae protein as a fusion with U1A protein to the surface of a microscope slide, so that the only immobilized fluorescent k-turn-containing RNA molecules must be bound to protein. We found that such bound k-turns remained in a high FRET state (i.e., in the folded conformation) for as long as observed. No transitions to an unfolded state (low FRET) were observed in hundreds of trajectories. We then performed a real-time analysis of binding in an attempt to detect the transient formation of a bound RNA molecule in an unfolded state were they to exist. The fluorescent k-turn RNA (in the absence of divalent ions so that the RNA was predominantly unfolded in free solution) was introduced into the cell while simultaneously collecting the emitted fluorescent light. Thus, we could observe binding events in real time and measure E_{FRET} at the earliest time within the resolution of our data collection (down to 8-ms frame rate). An example is shown in Fig. 2.4.

At the start of the trajectory, no RNA is bound, so that the intensities of donor and acceptor (I_D magenta, and I_A blue, respectively) are at background levels. At

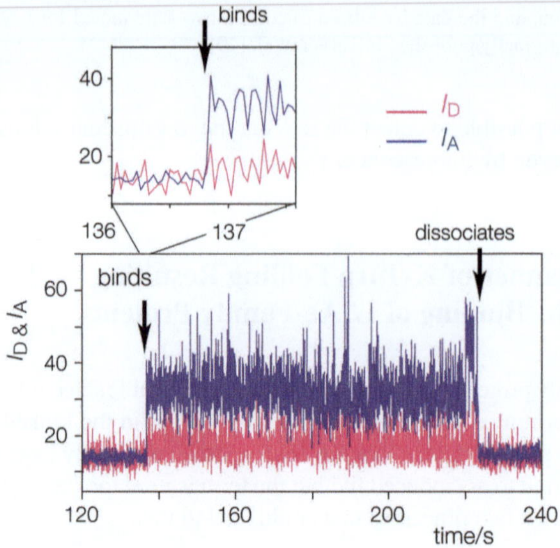

Fig. 2.4 Observation of a single molecule of *Hm*Kt-7 molecule binding to L7Ae in real time. A time trace of donor intensity (I_D, magenta) and acceptor intensity (I_A, blue) at 16-ms frame rate. I_A rises at 136.9 s upon binding of L7Ae. At 224 s, I_A falls back to its initial level; this is most probably due to dissociation of the complex. The expansion of the binding region shows that I_A rises fully within a single frame (corresponding to the higher FRET efficiency for the folded k-turn), with no evidence for bound RNA existing transiently in an unfolded conformation These data were originally published in Wang et al. [49]

the point at which binding occurs both I_A and I_D increase, with $I_A > I_D$ consistent with the kinked RNA conformation. The high FRET state is achieved within a single frame. We have not observed any transient states with lower E_{FRET} values in many trajectories even at the highest time resolution of our EM-CCD camera (8-ms frame). These data are consistent with conformational capture, providing no evidence for a less-kinked RNA conformation bound to the L7Ae protein. However, it remains possible that such a species could exist more transiently and thus not detected within the time frame of our detection.

2.7 Modulation of L7Ae-Family Protein Binding and k-Turn Folding by N^6-Methylation of Adenine

RNA is subject to site-specific covalent modifications [50, 51], the most frequent of which is methylation of the N6 group of adenine [52–54]. Using X-ray crystallography of short duplex RNA molecules, we have recently shown that that N^6-methyladenine (N^{6}mA) is tolerated at Watson–Crick *cis* A-U and A·G base pairs without affecting the base pairing. However, the *trans* Hoogsteen–sugar A·G base pair (sometimes referred to as a sheared base pair) is completely disrupted by the addition of a single methyl group at adenine N6 [55]. Tandem sheared G·A and A·G base pairs form the core of the k-turn, so perhaps unsurprisingly we found that inclusion of N^{6}mA at the 1n position of Kt-7 prevented its folding into the kinked conformation.

For this position to be methylated in the cell would require the A1n to be located within the context of a GAC sequence, which is the target for the METTL3–METTL14 methyltransferase complex [53, 54]. In other words, this position could only become methylated if the -1n position was C, i.e., the first base pair of the C-helix adjacent to the bulge was G-C. In the majority of k-turns, such as the human U4 k-turn and most ribosomal k-turns, the -1b, -1n base pair is C-G. However, we noted that some box C/D snoRNP sequences have G-C at this position and therefore performed a bioinformatic search of human snoRNA sequences to see how frequently G-C was found at the -1b, -1n position. We found that 27 human box C/D sequences had G-C at this location and so were potential targets for adenine methylation at the 1n position. Further bioinformatic analysis revealed that about half of these were actually methylated in the cell, the list including box C/D and C′/D′ k-turns, and some k-loops. Moreover, the C at the -1n position was strongly conserved in evolution in the vertebrates for the box C/D k-turns that were methylated in humans, but was less well conserved for those that were not methylated in humans.

Box C/D snoRNP complexes direct the site-specific 2′-*O*-methylation of rRNA and tRNA in archaea and the eukaryotes by providing complementary RNA (12 nt 'guide' RNA) for specificity and a SAM-dependent methyltransferase enzyme to modify the target ribose [25, 35, 56–61]. The assembly of the box C/D snoRNP is an ordered sequential process [31, 62–65], the first step of which is the binding of the

15.5 k protein to the box C/D and C′/D′ k-turns. Once this has occurred, then two further proteins Nop56 and Nop58 (Nop5 in archaea) bind and then finally fibrillarin (the methyltransferase enzyme that methylates O2′ on the target RNA) binds to yield the fully active snoRNP complex.

We have found that methylation of A-1n prevents the specific binding (electrophoretic analysis and microcalorimetry) and consequent RNA folding (FRET) of 15.5 k to human box C/D k-turns shown to be methylated *in vivo* [55]. An example of human box C/D U48 snoRNA is shown in Fig. 2.5. While 15.5 k binds to form a discrete complex with the unmodified RNA, the N^6mA-containing RNA exhibits non-specific binding (a smear on the gel, not a sharp band) and incomplete folding. Since the interaction with the 15.5 k protein occurs primarily in the major groove on the outer face of the k-turn, primarily with the guanine nucleotides at the 1b and 2n positions, while the A1n is on the inside of the structure [66], this effect is essentially indirect recognition of the methylation through its effect on the RNA structure.

Binding of 15.5 k protein stabilizes box C/D snoRNA [67], and if complex formation fails to occur, the RNA is unstable to degradation [68]. If 15.5 k fails to bind the box C/D k-turns, then the snoRNP assembly is blocked from proceeding further. Thus, these data indicate how modulation of k-turn folding can affect the assembly of the box C/D snoRNP and thus the O2′-methylation of the target RNA, and is quite plausibly an important regulatory mechanism in the cell.

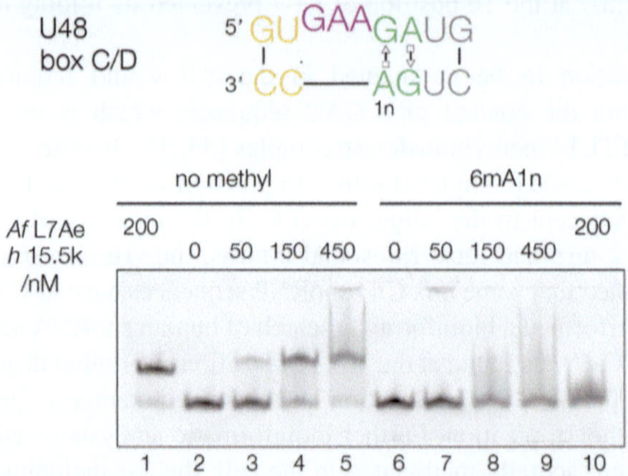

Fig. 2.5 Disruption of 15.5 k protein binding of k-turn conformation by N^6-methylation of adenine in human box C/D SNORD48 (U48) snoRNA, studied by gel electrophoretic retardation analysis. The U48 RNA was studied with and without N^6-methylation at the A1n position. 200 μM RNA was incubated with the indicated concentration of human 15.5 k, or *A. fulgidus* L7Ae proteins and applied to 10% polyacrylamide gels electrophoresed under non-denaturing conditions. Binding of either protein to the unmodified RNA (tracks 1 through 5) led to the formation of discrete retarded species. At higher concentrations of 15.5 k, a continuous smear of complexes ran up the gel suggesting non-specific binding beyond stoichiometric conditions. By contrast, no specific RNA-protein complexes were observed when N^6-methyladenine-containing RNA was used (tracks 6 through 10) Related data were published in Huang et al. [55]

2.8 L7Ae-Bound K-Turns in Nanoconstruction

The precise geometry and trajectory of the k-turn suggest its utility as a building block in the construction of molecular nanoscale objects. We have shown that a unit comprising two *Hm*Kt-7 k-turns on opposite strands of an RNA duplex (a two-k-turn unit, or 2K unit) is a horseshoe-shaped molecule that can associate in the crystal lattice *via* end-to-end stacking to form a variety of shapes including dumbbell (two 2K units) and triangle (three 2K units) [69]. We also found that a single duplex RNA containing six k-turns formed a quasi-cyclic triangular structure in the crystal where the ends were so perfectly stacked that the molecule had no preferred rotational setting. Extending this to complexes of the 2K unit with *Af*L7Ae bound, we obtained crystals containing three 2K-L7Ae complexes (triangular, with six bound *Af*L7Ae molecules) (Fig. 2.6) and four 2K-L7Ae complexes (square, with eight

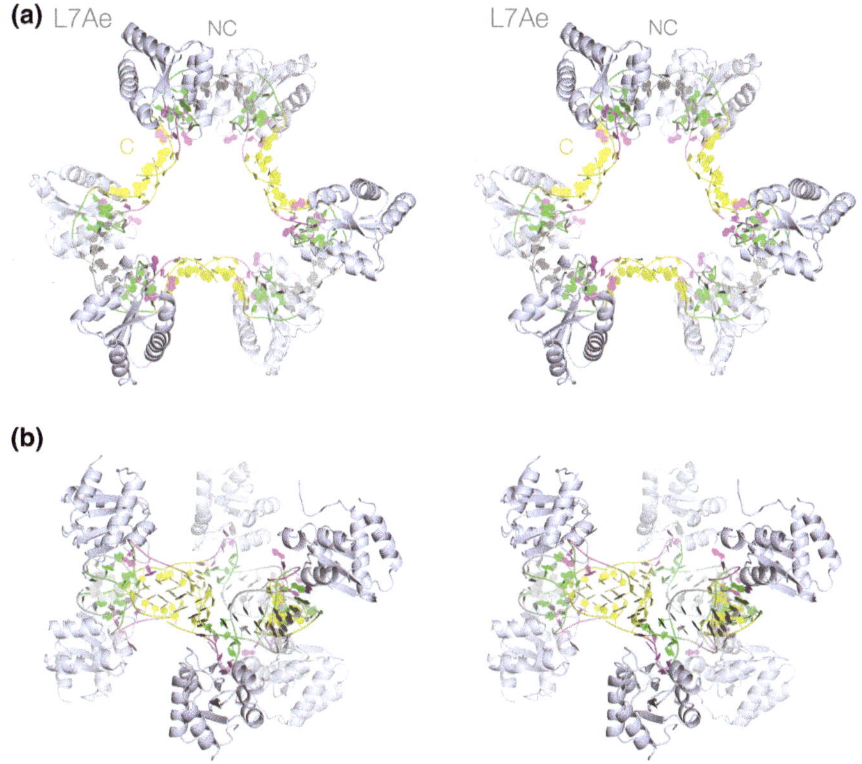

Fig. 2.6 Molecular nanoengineering using k-turn-L7Ae complexes. Two parallel-eye stereoscopic views of a trimeric assembly of 2 K units (each comprising two *Hm*Kt-7 motifs with a common C-helix and a 3b = 3n = U sequence, bound to *A. fulgidus* L7Ae). The structure was determined in space group P212121 at a resolution of 2.65 Å (PDB ID 5C4U). a. top view, down the threefold rotation axis of the structure. b. Side view along the plane of the triangular association of 2 K units These data were originally published in Huang and Lilley [69]

bound *Af*L7Ae molecules) [69]. These species have great potential in the future construction of functional nanoscale objects, perhaps in combination with other RNA motifs.

2.9 Summary

k-turns are extremely widespread in folded cellular RNA species that are involved in translation, splicing, and modification of RNA and in gene regulation. In general, most k-turns bind proteins, of which the most common are members of the L7Ae superfamily. L7Ae-related proteins are bound to k-turns in the ribosome, during the spliceosome cycle and to box C/D snoRNA species. We have found that specific protein binding can occur with extremely high affinity, to induce the formation of the kinked conformation most probably by conformational selection. The majority of k-turns mediate tertiary interactions in their RNA, due to the tight axial kink that forms in the RNA on folding. This also lends the k-turn motif as a building block for RNA nanoconstruction. This is really the role it has evolved to fill in the cell over millions of years, as a key element in nature's own nanoarchitecture of RNA.

References

1. Ban, N., Nissen, P., Hansen, J., Moore, P. B., & Steitz, T. A. (2000). The complete atomic structure of the large ribosomal subunit at 2.4 Å resolution. *Science, 289,* 905–920.
2. Vidovic, I., Nottrott, S., Hartmuth, K., Luhrmann, R., & Ficner, R. (2000). Crystal structure of the spliceosomal 15.5 kD protein bound to a U4 snRNA fragment. *Molecular Cell, 6,* 1331–1342.
3. Wozniak, A. K., Nottrott, S., Kuhn-Holsken, E., Schroder, G. F., Grubmuller, H., Luhrmann, R., et al. (2005). Detecting protein-induced folding of the U4 snRNA kink-turn by single-molecule multiparameter FRET measurements. *RNA, 11,* 1545–1554.
4. Montange, R. K., & Batey, R. T. (2006). Structure of the S-adenosylmethionine riboswitch regulatory mRNA element. *Nature, 441,* 1172–1175.
5. Blouin, S., & Lafontaine, D. A. (2007). A loop loop interaction and a K-turn motif located in the lysine aptamer domain are important for the riboswitch gene regulation control. *RNA, 13,* 1256–12567.
6. Smith, K. D., Lipchock, S. V., Ames, T. D., Wang, J., Breaker, R. R., & Strobel, S. A. (2009). Structural basis of ligand binding by a c-di-GMP riboswitch. *Nature Structural and Molecular Biology, 16,* 1218–1223.
7. Peselis, A., & Serganov, A. (2012). Structural insights into ligand binding and gene expression control by an adenosylcobalamin riboswitch. *Nature Structural and Molecular Biology, 19,* 1182–1184.
8. Baird, N. J., & Ferre-D'Amare, A. R. (2013). Modulation of quaternary structure and enhancement of ligand binding by the K-turn of tandem glycine riboswitches. *RNA, 19,* 167–176.
9. Zhang, J., & Ferre-D'Amare, A. R. (2013). Co-crystal structure of a T-box riboswitch stem I domain in complex with its cognate tRNA. *Nature, 500,* 363–366.
10. Moore, T., Zhang, Y., Fenley, M. O., & Li, H. (2004). Molecular basis of box C/D RNA-protein interactions; Cocrystal structure of archaeal L7Ae and a box C/D RNA. *Structure, 12,* 807–818.

11. Hamma, T., & Ferré-D'Amaré, A. R. (2004). Structure of protein L7Ae bound to a K-turn derived from an archaeal box H/ACA sRNA at 1.8 Å resolution. *Structure, 12,* 893–903.
12. Szewczak, L. B., Gabrielsen, J. S., Degregorio, S. J., Strobel, S. A., & Steitz, J. A. (2005). Molecular basis for RNA kink-turn recognition by the h15.5 K small RNP protein. *RNA, 11,* 1407–1419.
13. Youssef, O. A., Terns, R. M., & Terns, M. P. (2007). Dynamic interactions within sub-complexes of the H/ACA pseudouridylation guide RNP. *Nucleic Acids Research, 35,* 6196–6206.
14. Schroeder, K. T., & Lilley, D. M. (2009). Ion-induced folding of a kink turn that departs from the conventional sequence. *Nucleic Acids Research, 37,* 7281–7289.
15. Schroeder, K. T., Daldrop, P., McPhee, S. A., & Lilley, D. M. (2012). Structure and folding of a rare, natural kink turn in RNA with an A·A pair at the 2b·2n position. *RNA, 18,* 1257–1266.
16. Wang, J., Daldrop, P., Huang, L., & Lilley, D. M. (2014). The k-junction motif in RNA structure. *Nucleic Acids Research, 42,* 5322–5331.
17. McPhee, S. A., Huang, L., & Lilley, D. M. (2014). A critical base pair in k-turns that confers folding characteristics and correlates with biological function. *Nature comm., 5,* 5127.
18. Turner, B., Melcher, S. E., Wilson, T. J., Norman, D. G., & Lilley, D. M. J. (2005). Induced fit of RNA on binding the L7Ae protein to the kink-turn motif. *RNA, 11,* 1192–1200.
19. Schroeder, K. T., Daldrop, P., & Lilley, D. M. J. (2011). RNA tertiary interactions in a riboswitch stabilize the structure of a kink turn. *Structure, 19,* 1233–1240.
20. Nissen, P., Ippolito, J. A., Ban, N., Moore, P. B., & Steitz, T. A. (2001). RNA tertiary interactions in the large ribosomal subunit: The A-minor motif. *Proceedings of the National Academy of Sciences of the United States of America, 98,* 4899–4903.
21. Liu, J., & Lilley, D. M. J. (2007). The role of specific 2′-hydroxyl groups in the stabilization of the folded conformation of kink-turn RNA. *RNA, 13,* 200–210.
22. Daldrop, P., & Lilley, D. M. J. (2013). The plasticity of a structural motif in RNA: Structural polymorphism of a kink turn as a function of its environment. *RNA, 19,* 357–364.
23. Huang L, Wang J, Lilley DMJ (2016) A critical base pair in k-turns determines the conformational class adopted, and correlates with biological function. Nucleic Acids Research.
24. Koonin, E. V., Bork, P., & Sander, C. (1994). A novel RNA-binding motif in omnipotent suppressors of translation termination, ribosomal proteins and a ribosome modification enzyme? *Nucleic Acids Research, 22,* 2166–2167.
25. Watkins, N. J., Segault, V., Charpentier, B., Nottrott, S., Fabrizio, P., Bachi, A., et al. (2000). A common core RNP structure shared between the small nucleolar box C/D RNPs and the spliceosomal U4 snRNP. *Cell, 103,* 457–466.
26. Nottrott, S., Hartmuth, K., Fabrizio, P., Urlaub, H., Vidovic, I., Ficner, R., et al. (1999). Functional interaction of a novel 15.5kD [U4/U6.U5] tri-snRNP protein with the 5′ stem-loop of U4 snRNA. *The EMBO Journal, 18,* 6119–6133.
27. Sojka, L., Fucik, V., Krasny, L., Barvik, I., & Jonak, J. (2007). YbxF, a protein associated with exponential-phase ribosomes in Bacillus subtilis. *Journal of Bacteriology, 189,* 4809–4814.
28. Kuhn, J. F., Tran, E. J., & Maxwell, E. S. (2002). Archaeal ribosomal protein L7 is a functional homolog of the eukaryotic 15.5kD/Snu13p snoRNP core protein. *Nucleic Acids Research, 30,* 931–941.
29. Rozhdestvensky, T. S., Tang, T. H., Tchirkova, I. V., Brosius, J., Bachellerie, J.-P., & Hüttenhofer, A. (2003). Binding of L7Ae protein to the K-turn of archaeal snoRNAs: A shared RNA binding motif for C/D and H/ACA box snoRNAs in Archaea. *Nucleic Acids Research, 31,* 869–877.
30. Klein, D. J., Schmeing, T. M., Moore, P. B., & Steitz, T. A. (2001). The kink-turn: A new RNA secondary structure motif. *The EMBO Journal, 20,* 4214–4221.
31. Watkins, N. J., Dickmanns, A., & Luhrmann, R. (2002). Conserved stem II of the box C/D motif is essential for nucleolar localization and is required, along with the 15.5 K protein, for the hierarchical assembly of the box C/D snoRNP. *Molecular and Cellular Biology, 22,* 8342–8352.
32. Marmier-Gourrier, N., Clery, A., Senty-Segault, V., Charpentier, B., Schlotter, F., Leclerc, F., et al. (2003). A structural, phylogenetic, and functional study of 15.5-kD/Snu13 protein binding on U3 small nucleolar RNA. *RNA, 9,* 821–838.

33. Cho, I. M., Lai, L. B., Susanti, D., Mukhopadhyay, B., & Gopalan, V. (2010). Ribosomal protein L7Ae is a subunit of archaeal RNase P. *Proceedings of the National Academy of Sciences of the United States of America, 107,* 14573–14578.
34. Suryadi, J., Tran, E. J., Maxwell, E. S., & Brown, B. A. (2005). The crystal structure of the Methanocaldococcus jannaschii multifunctional L7Ae RNA-binding protein reveals an induced-fit interaction with the box C/D RNAs. *Biochemistry, 44,* 9657–9672.
35. Xue, S., Wang, R., Yang, F., Terns, R. M., Terns, M. P., Zhang, X., et al. (2010). Structural basis for substrate placement by an archaeal box C/D ribonucleoprotein particle. *Molecular Cell, 39,* 939–949.
36. Baird, N. J., Zhang, J., Hamma, T., & Ferré-D'Amaré, A. R. (2012). YbxF and YlxQ are bacterial homologs of L7Ae, and bind K-turns but not K-loops. *RNA, 18,* 759–770.
37. Huang, L., & Lilley, D. M. J. (2013). The molecular recognition of kink turn structure by the L7Ae class of proteins. *RNA, 19,* 1703–1710.
38. Goody, T. A., Melcher, S. E., Norman, D. G., & Lilley, D. M. J. (2004). The kink-turn motif in RNA is dimorphic, and metal ion dependent. *RNA, 10,* 254–264.
39. Shi, X., Huang, L., Lilley, D. M., Harbury, P. B., & Herschlag, D. (2016). The solution structural ensembles of RNA kink-turn motifs and their protein complexes. *Nature Chemical Biology, 12,* 146–152.
40. Turner, B., & Lilley, D. M. J. (2008). The importance of G.A hydrogen bonding in the metal ion- and protein-induced folding of a kink turn RNA. *J Molec Biol, 381,* 431–442.
41. Tsai, C. J., Ma, B., Sham, Y. Y., Kumar, S., & Nussinov, R. (2001). Structured disorder and conformational selection. *Proteins, 44,* 418–427.
42. Koshland, D. E. (1958). Application of a theory of enzyme specificity to protein synthesis. *Proceedings of the National Academy of Sciences of the United States of America, 44,* 98–104.
43. Pitici, F., Beveridge, D. L., & Baranger, A. M. (2002). Molecular dynamics simulation studies of induced fit and conformational capture in U1A-RNA binding: Do molecular substates code for specificity? *Biopolymers, 65,* 424–435.
44. Okazaki, K., & Takada, S. (2008). Dynamic energy landscape view of coupled binding and protein conformational change: Induced-fit versus population-shift mechanisms. *Proceedings of the National Academy of Sciences of the United States of America, 105,* 11182–11187.
45. Weikl, T. R., & von Deuster, C. (2008). Selected-fit versus induced-fit protein binding: Kinetic differences and mutational analysis. *Proteins, 75,* 104–110.
46. Hammes, G. G., Chang, Y. C., & Oas, T. G. (2009). Conformational selection or induced fit: A flux description of reaction mechanism. *Proceedings of the National Academy of Sciences of the United States of America, 106,* 13737–13741.
47. Csermely, P., Palotai, R., & Nussinov, R. (2010). Induced fit, conformational selection and independent dynamic segments: An extended view of binding events. *Trends in Biochemical Sciences, 35,* 539–546.
48. Zhou, H. X. (2010). From induced fit to conformational selection: A continuum of binding mechanism controlled by the timescale of conformational transitions. *Biophysical Journal, 98,* L1517.
49. Wang, J., Fessl, T., Schroeder, K. T., Ouellet, J., Liu, Y., Freeman, A. D., et al. (2012). Single-molecule observation of the induction of k-turn RNA structure on binding L7Ae protein. *Biophys J, 103,* 2541–2548.
50. He, C. (2010). Grand challenge commentary: RNA epigenetics? *Nature Chemical Biology, 6,* 863–865.
51. Cantara, W. A., Crain, P. F., Rozenski, J., McCloskey, J. A., Harris, K. A., Zhang, X., et al. (2011). The RNA modification database, RNAMDB: 2011 update. *Nucleic Acids Research, 39,* D195–D201.
52. Schibler, U., Kelley, D. E., & Perry, R. P. (1977). Comparison of methylated sequences in messenger RNA and heterogeneous nuclear RNA from mouse L cells. *Journal of Molecular Biology, 115,* 695–714.
53. Dominissini, D., Moshitch-Moshkovitz, S., Schwartz, S., Salmon-Divon, M., Ungar, L., Osenberg, S., et al. (2012). Topology of the human and mouse m6A RNA methylomes revealed by m6A-seq. *Nature, 485,* 201–206.

54. Meyer, K. D., Saletore, Y., Zumbo, P., Elemento, O., Mason, C. E., & Jaffrey, S. R. (2012). Comprehensive analysis of mRNA methylation reveals enrichment in 3′ UTRs and near stop codons. *Cell, 149,* 1635–1646.
55. Huang, L., Ashraf, S., Wang, J., & Lilley, D. M. (2017). Control of box C/D snoRNP assembly by N6-methylation of adenine. *EMBO Reports, 18,* 1631–1645. https://doi.org/10.15252/embr.201743967.
56. Kiss-Laszlo, Z., Henry, Y., Bachellerie, J. P., Caizergues-Ferrer, M., & Kiss, T. (1996). Site-specific ribose methylation of preribosomal RNA: A novel function for small nucleolar RNAs. *Cell, 85,* 1077–1088.
57. Tycowski, K. T., Smith, C. M., Shu, M. D., & Steitz, J. A. (1996). A small nucleolar RNA requirement for site-specific ribose methylation of rRNA in Xenopus. *Proceedings of the National Academy of Sciences of the United States of America, 93,* 14480–14485.
58. Tran, E. J., Zhang, X., & Maxwell, E. S. (2003). Efficient RNA 2′-O-methylation requires juxtaposed and symmetrically assembled archaeal box C/D and C′/D′ RNPs. *The EMBO Journal, 22,* 3930–3940.
59. Bleichert, F., Gagnon, K. T., Brown, B. A., Maxwell, E. S., Leschziner, A. E., Unger, V. M., et al. (2009). A dimeric structure for archaeal box C/D small ribonucleoproteins. *Science, 325,* 1384–1387.
60. Ye, K., Jia, R., Lin, J., Ju, M., Peng, J., Xu, A., et al. (2009). Structural organization of box C/D RNA-guided RNA methyltransferase. *Proceedings of the National Academy of Sciences of the United States of America, 106,* 13808–13813.
61. Lin, J., Lai, S., Jia, R., Xu, A., Zhang, L., Lu, J., et al. (2011). Structural basis for site-specific ribose methylation by box C/D RNA protein complexes. *Nature, 469,* 559–563.
62. Watkins, N. J., Newman, D. R., Kuhn, J. F., & Maxwell, E. S. (1998). In vitro assembly of the mouse U14 snoRNP core complex and identification of a 65-kDa box C/D-binding protein. *RNA, 4,* 582–593.
63. Omer, A. D., Ziesche, S., Ebhardt, H., & Dennis, P. P. (2002). In vitro reconstitution and activity of a C/D box methylation guide ribonucleoprotein complex. *Proceedings of the National Academy of Sciences of the United States of America, 99,* 5289–5294.
64. Schultz, A., Nottrott, S., Watkins, N. J., & Luhrmann, R. (2006). Protein-protein and protein-RNA contacts both contribute to the 15.5 K-mediated assembly of the U4/U6 snRNP and the box C/D snoRNPs. *Molecular and Cellular Biology, 26,* 5146–5154.
65. McKeegan, K. S., Debieux, C. M., Boulon, S., Bertrand, E., & Watkins, N. J. (2007). A dynamic scaffold of pre-snoRNP factors facilitates human box C/D snoRNP assembly. *Molec Cell Biol, 27,* 6782–6793.
66. Huang, L., & Lilley, D. M. J. (2013). The molecular recognition of kink-turn structure by the L7Ae class of proteins. *RNA, 19,* 1703–1710.
67. Szewczak, L. B., DeGregorio, S. J., Strobel, S. A., & Steitz, J. A. (2002). Exclusive interaction of the 15.5 kD protein with the terminal box C/D motif of a methylation guide snoRNP. *Chemistry and Biology, 9,* 1095–1107.
68. Caffarelli, E., Fatica, A., Prislei, S., De Gregorio, E., Fragapane, P., & Bozzoni, I. (1996). Processing of the intron-encoded U16 and U18 snoRNAs: The conserved C and D boxes control both the processing reaction and the stability of the mature snoRNA. *The EMBO Journal, 15,* 1121–1131.
69. Huang, L., & Lilley, D. M. J. (2016). A quasi-cyclic RNA nano-scale molecular object constructed using kink turns. *Nanoscale, 8,* 15189–15195.

Chapter 3
Evolving Methods in Defining the Role of RNA in RNP Assembly

Jaya Sarkar, Jong Chan Lee and Sua Myong

3.1 Introduction

The role of liquid–liquid phase separation (LLPS) in biology has received intense attention over the past few years. Its biological relevance continues to grow from being the basis behind the formation of ribonucleoprotein (RNP) granules [1], heterochromatin compaction [2, 3] to microtubule assembly [4]. Owing to their composition, certain types of RNP granules, such as stress granules (SGs), have the potential to act as the melting pot of misfolded proteins and protein aggregates that can lead to the formation of pathological bodies found in neurodegeneration. In fact, mutations in several SG proteins accelerate aberrant aging of these RNP bodies and are causative of neurodegeneration. Our focus, in this chapter, is on the potential role of RNA as an essential component of these RNP granules, more specifically: What is the molecular basis of RNA–protein interaction involved in the assembly, maintenance, and pathological progression of SGs? To address this question, here, we try to consolidate some of the myriad of recent findings in the field; discuss some current methodologies in their strengths and weaknesses; and finally put forth our methods and insights in an attempt to tackle some of the gaps and outstanding questions in the field. Together, these approaches may lead to a better understanding of disease pathogenesis and developing therapeutic interventions.

J. Sarkar · S. Myong (✉)
Biophysics Department, Johns Hopkins University, Baltimore, MD 21218, USA
e-mail: smyong@jhu.edu

J. C. Lee
Department of Biophysics and Biophysical Chemistry, Johns Hopkins University, Baltimore, MD 21218, USA

S. Myong
Center for Physics of Living Cells, University of Illinois, Urbana, IL 61801, USA

C. Joo and D. Rueda (eds.), *Biophysics of RNA-Protein Interactions*,
Biological and Medical Physics, Biomedical Engineering,
https://doi.org/10.1007/978-1-4939-9726-8_3

3.1.1 Composition of RNP Granules

Broadly, RNP granules are a general term used for membraneless phase-separated organelles containing a high local concentration of proteins and RNA. In eukaryotes, some of these are nuclear (such as Cajal bodies and PML bodies), while some are cytoplasmic (such as SGs and P bodies) [5]. A prominent example of RNP granules in other organisms includes germ cell granules (P granules) in *Caenorhabditis elegans*. In our discussion here, we focus on two widely used models in probing RNP granule mechanisms—SGs and P granules, using them as examples when appropriate. SGs are sites of RNA triage, formed from untranslated mRNAs and RNA-binding proteins (RBPs), when eukaryotic cells are under stress [6]. P granules play a key role in germ cell development in *C. elegans* [7].

RNP granules contain RNA, RBPs, and also non-RNA-binding proteins. The RBPs present in RNP granules contain signature motifs or domains—RNA recognition motifs (RRMs) and intrinsically disordered regions (IDRs). IDRs are also termed low complexity domains that are structurally disordered. While some IDRs feature uncharged polar amino acid residues infused with bulky aromatic residues (such as Gln-Gly-Ser-Tyr or Gly-rich patches), others may have charged residues (such as Arg-Gly-rich patches). Such amino acid composition renders the granule-forming RBPs interactive, thus making them ideal agents for nucleating (homotypic interaction) and recruiting others (heterotypic interaction) to promote large assemblies such as RNP granules (discussed in the next section). Taken together, the RNA-binding ability and self- and cross-interactive nature of RNP forming proteins enable them to establish multivalent yet dynamic RNP network (Fig. 3.1).

Under healthy conditions, assembly and disassembly of SGs are all a part of regular cellular dynamics, designed to protect untranslated mRNAs during stress. Components of SG identified by the earlier study include stalled preinitiation complex containing ribosomal subunits; translation associated factors, such as initiation factors eIF2, eIF3, PABP; and mRNA structure/function regulating proteins such as Staufen and G3BP [8]. Recent proteomic analyses of SGs isolated from yeast and mammalian cells have revealed a more diverse composition of these granules [9], identifying the presence of novel and conserved classes of proteins that include: ATP-dependent RNA and DNA helicases, and numerous DEAD-box proteins; ATP-dependent protein and nucleic acid remodeling factors, such as heat shock proteins and chaperones; ribosome biogenesis proteins; and housekeeping proteins such as aminoacyl-tRNA synthetases.

Neuronal inclusions from amyotrophic lateral sclerosis (ALS) and frontotemporal dementia (FTD) patients contain two RNA-processing nuclear proteins namely FUS and TDP-43 [10]. In vitro studies showed that disease-associated mutations in these proteins exhibited signs of accelerated aging or maturation of protein droplets characterized by loss of liquid-like property and subsequent appearance of solid-like crystals [11]. Such drastic change indicates aberrations in IDR–protein interaction that disrupts the inherent LLPS mechanism that is believed to control the assembly and dynamics of RNP granules. Interestingly, for mammalian cells, stress-specific

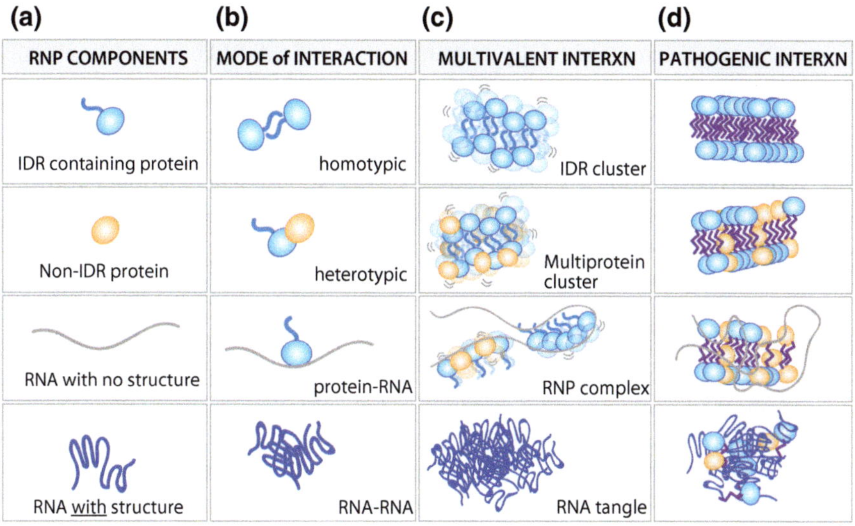

Fig. 3.1 RNP components and interaction. **a** Simplified list of RNP constituents. **b** Different types of molecule-to-molecule interaction mode. **c** Multivalent interactions that may occur in different RNP context. **d** Defective molecular assembly that may depict liquid-to-solid transition of RNP components in pathogenic conditions

differences have been highlighted in the composition, assembly, and dynamics of SGs and SG-like cytoplasmic foci that are also induced by stress. For example, while sorbitol stress recruits TDP-43 to canonical SGs, sodium arsenite stress does not [12, 13].

Similar to mammalian SGs, *C. elegans* P granules are also composed of RNA and certain conserved classes of proteins, many of which are disordered [7]. These include DEAD-box helicases LAF-1, GLH-1 through 4; RNA-binding proteins MEG-3, PGL-1 through 3. Overall, the conserved presence of intrinsically disordered and RNA-binding proteins in both SGs and P granules underscores the importance of RNA remodeling activities in RNP granules.

3.1.2 Mechanisms of RNP Granule Formation: Protein–Protein and Protein–RNA Interactions Are Both Driving Forces

In the recent past, molecular mechanisms underlying RNP granule assembly have been intensely investigated, from both the biophysics and material science perspective. In in vitro studies, granule/pathological inclusion forming proteins, such as FUS, have been shown to access different material states [11, 14]. Initially, FUS phase separates into metastable droplets that behave as liquids, then with time and increased local protein concentration, this liquid phase quickly matures to an intermediate hydrogel-like state, finally converting to stable solid-like fibers (structurally resembling amyloid fibers in neurodegeneration), and this conversion is accelerated by ALS-implicated protein mutations in FUS. In vivo, however, in the healthy state, RNP granules are thought to maintain liquid-like state; for example, liquid properties of P granules have been established, based on fusion, dripping, and rapid diffusion rate of components between inside and outside of these granules [15]. In vivo SGs, on the other hand, have been envisioned as a coexistence of densely packed state and liquid states, evidenced by the presence of a stable core surrounded by a dynamic shell [9].

Mostly derived from cell-free in vitro studies, at present, the converging understanding is that IDR protein-driven LLPS is the key behind liquid–liquid demixing and formation of reversible RNP granules that coexist with its surrounding components [6]. The critical requirement here is multivalent weak and transient interactions (provided by the labile interactions between IDR proteins), that are strong enough to hold the RNP assemblies together, but are not so strong as to arrest dynamics and reversibility of these structures. Because these RNP granules contain many different proteins, protein–protein interaction between non-IDR proteins also contributes toward building and maintenance of RNP granules. Heat shock proteins and chaperones that constitute SG of yeast and mammalian cells [9] are thought to counteract the aggregation-prone tendency of IDR-containing proteins by resolving misfolded proteins and dispersing granule components [16]. A recent study shows that ATP can also act as a hydrotrope to solubilize the molecularly crowded and compacted state of cellular granules [17].

Increasingly critical role of RNA in RNP granule nucleation and dynamics: Many RBPs present in RNP granules, apart from possessing RRMs, also possess IDRs, for example, LAF-1, FUS, TDP-43, MEG-3, and Whi3. A central question here is: Does RNA play an active role in nucleation and assembly of RNP granules, or is it just an inevitable consequence of RNA tagging along the RBP? Recent reports allude to both the active and regulatory roles for RNA in RNP assembly, albeit through multi-faceted molecular mechanisms, as highlighted below: (i) *RNA impacts assembly of LLPS droplets*. RNA seeds FUS higher-order assemblies (visible as ropey structures in transmission electron microscopy images) even at low protein concentrations, suggesting that RNA promotes the phase separation of FUS [18]. In vivo, in *Drosophila* fly model and in mammalian neuronal cells, RNA-binding ability of FUS is essential for ALS mutation containing FUS to show neurodegenerative phenotype or to localize to cytoplasmic SGs, respectively [19]. Similarly, disrupting RNA-binding ability of TDP-43 mutants (ALS-linked) rescued TDP-43 mediated cellular toxicity [20]. Whi3 is a fungal RNA-binding IDR protein (responsible for asynchronous nuclear division via spatial patterning of RNA transcripts), whose phase separation is driven by its cellular mRNA target sequences [21]. Furthermore, mRNA structure is critical for assembling of distinct Whi3 droplets and protein-driven RNA conformational changes for maintaining such identity [22]. (ii) Strikingly, a recent pioneering study reports on how *RNA by itself can phase separate*. This particular RNA arises from repeat expansion at C9 or f72 (chromosome 9 open reading frame 72) which is responsible for causing high percentage of both familial and sporadic ALS [23, 24]. The expanded form of RNA was shown to phase transition into nuclear foci with strength directly proportional to the length and secondary structure of the repeat RNA [25]. (iii) *RNA also tunes dynamics of the RNP granules post-nucleation*. RNA fluidizes LAF-1 liquid droplets in vitro, with a concurrent increase in the protein–RNA dynamics probed by single-molecule assay [26]. In contrast, Whi3 droplet viscosity was increased and dynamics decreased in the presence of a specific mRNA [21]. Therefore, RNA can up or down regulate the granule fluidity depending on the molecular context. (iv) *RNA regulates phase separation of IDR proteins in cells*. Maharana et al. [27] showed that high RNA:protein ratio keeps these proteins soluble in nucleus, while low ratio promotes LLPS in the cytoplasm, a hallmark in ALS patients.

The overall current understanding is that at the granule assembly stage, RNA provides a platform which leads to recruitment of multiple RBPs, thus enabling multivalent interactions among RNPs (Fig. 3.1). These multivalent interactions, in turn, increase the local protein concentration, which increases the propensity of interaction between the IDRs of these RBPs, allowing for clustering into a stable nucleation core. Once established, such core can incorporate other proteins and phase separate into cellular granules [28]. Post-assembly, RNA may continue to have a critical role in modulating granule fluidity, plausibly via tuning dynamics of RNA–RBP interaction. Thus, a synergistic role of RNA and protein in regulating RNP granule assembly, properties, and maintenance is increasingly been acknowledged.

3.1.3 Stages of RNP Granule Life and Implications in Disease

As mentioned in the previous section, evidence suggests that cellular RNP granules such as SGs and P granules are liquid-like reversible structures. In vitro however, the disordered constituent proteins of SGs, such as FUS and TDP-43, can quickly convert from liquid to hydrogel to fibers. In agreement with the finding that SGs absorb pathogenic inclusions [29], several SG proteins, including FUS and TDP-43, are present in aggregates/inclusions that are present in the motor neurons of ALS and FTD patients.

In fact, pathological inclusions are believed to originate from misregulated SGs [10]. Aberrations in interactions between IDR-containing RBPs and other SG proteins potentially can convert the weak transient interactions (that are responsible for LLPS and normal liquid-like SG dynamics) into more ordered solid-like interactions that cause loss/misregulation of SG fluidity and/or disassembly. Although there is lack of direct evidence for existence of the gel and fiber states in vivo for FUS and TDP-43, it is generally perceived that while the liquid-like state of granules represents the normal/default situation in cells, the fiber-like states resemble the beta-sheet structure of amyloid fibers that are found in the aggregates that occur in diseased individuals. Many ALS and FTD patient mutations in FUS and TDP-43 map to their RRM and IDR domains [30], strongly suggesting the deleterious effect caused by RNA-binding defect and IDR-driven aggregation in diseased state. On the other hand, the ALS-associated long multiple repeat RNAs, C9 or f72, promote multivalent intermolecular interactions responsible for LLPS of the RNAs into liquid droplets, and then into gels that manifest as RNA foci, a hallmark of C9 or f72 associated ALS [25].

While the quick coalescence of RNA and RBPs into SGs is critical for preserving cellular processes during stress, it comes with the high risk of aggregating these proteins into pathological inclusions. *How does the cell manage this risk and prevent the abnormal aggregation?* Two proposed avenues by which cells do this are: (i) *Balanced cross-talk between RNA and RBP quality control* [31]. Spinocerebellar ataxia type 31 (SCA31) is characterized by toxicity arising from RNA foci formed by expanded repeats of UGGAA and the aberrant proteins produced from non-canonical RAN translation (non-AUG translation) from the expanded RNAs. Ishiguro et al. showed that FUS and TDP-43 act as RNA chaperones by directly binding to the expanded UGGAA RNAs, resolving the folded structures, leading to reduced RNA foci and suppressed neurotoxicity [32]. ATP-dependent RNA helicases such as DEAD-box proteins were also found to unwind expanded RNA repeats and rescue toxicity. This group proposed a model in which not only RBPs can mitigate RNA toxicity, but also non-expanded RNAs can rescue mutant RBP-mediated toxicity. Thus, mutation in either the RNAs or the RBPs can perturb the balance in protein–RNA homeostasis, causing aggregation and toxicity. In addition, the protein quality control pathways such as molecular chaperones, protein degradation pathways, and prevention of mistranslation at the levels of aminoacyl-tRNA synthetases and ribosomes also contribute toward RNP homeostasis [33]. (ii) *Small molecules in cells, such as ATP*. Early study by Brangwynne et al. demonstrated that ATP removal from *C. elegans* induced loss of liquid-like property in P granules, suggesting the role of ATP or ATP-mediated processes in fluidizing RNP granules [15]. ATP depletion experiments in mammalian cells showed that ATP is required during SG assembly and also in maintaining granule fluidity [9]. A recent in vitro analyses shed insight into how cellular ATP in high concentrations (similar to the physiological concentration of 5–10 mM) may act as a biological hydrotrope whereby the amphipathic property of ATP induces solubilization of aggregation-prone cellular proteins such as FUS [17].

3.2 Current Methods in Probing RNP Granules: Strengths and Limitations

Early genetic and cellular studies on stress response provided clear evidence of the existence of SGs in cells. Evidence indicates that SG formation is vital to cellular survival under stressed condition [34]. Genetic studies also identified key players and mutations in the RNP components that can cause neurodegenerative diseases. Recent research effort has focused on understanding the material properties of in vitro droplets and cellular granules, using a combination of simple and sophisticated methods, as outlined here (Fig. 3.2): (i) Bright-field imaging of granule-forming IDR protein droplets, turbidity measurements by optical density (300–600 nm) (Fig. 3.2a, b) under different conditions such as temperature, salt, and protein concentration have provided valuable information about the propensity and size of protein droplets, enabling construction of a phase diagram which defines conditions that promote LLPS-driven droplet formation [26, 35]. Droplet fusion events have also been studied more precisely using FUS protein and optical tweezers [11]. (ii) Microrheology experiments have been developed to measure the viscosity and elasticity of droplets (Fig. 3.2d), shedding light on fluidity of the droplets under varying conditions such as the presence or absence of RNA [21]. (iii) Fluorescence recovery after photobleaching (FRAP) of both in vitro droplets and cellular granules (Fig. 3.2c) has also yielded information about the diffusion kinetics [11, 15, 21, 26]. (iv) Conventional biochemical methods including SDS-PAGE and Western blotting have been employed to distinguish the hydrogel-like state from liquid droplet and solid fiber [14]. In combination with electron microscopy and X-ray diffraction, these hydrogels were deduced to contain homotypic polymerized fibers that were dynamic, unlike disease-featuring amyloid fibers. Also, oligomerization stoichiometry of IDR proteins has been probed by electrophoretic mobility shift assay (EMSA) (Fig. 3.2f) [36]. (v) Recently, Roy Parker's lab has devised a SG isolation method [37]. While mass-spectrometric analyses (Fig. 3.2e) of isolated granules have revealed the diverse proteome of mammalian SGs [9], time-course fluorescence microscopy analyses have revealed that SG assembly is a multi-step process in which the stable core forms first, followed by the dynamic shell [28].

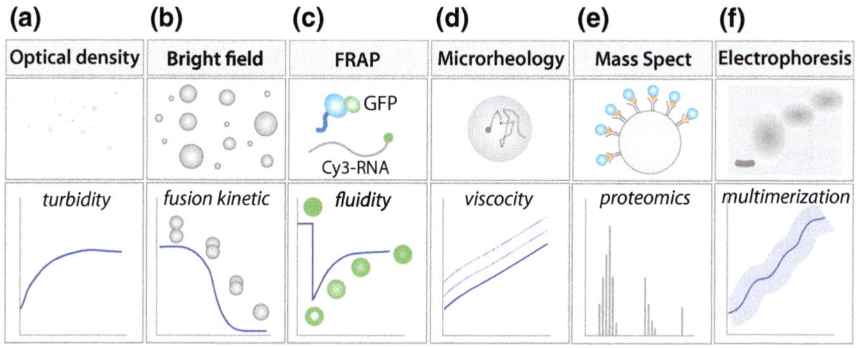

Fig. 3.2 Biophysical ensemble measurement. **a** Formation of LLPS can be measured by optical density at 350–600 nm. **b** Bright-field imaging can be used to track the growth and fusion kinetic of in vitro droplets. **c** FRAP analysis indicates fluidity of droplet content. **d** Microrheology measures viscoelasticity of droplets. **e** Mass spectrometry reveals constituents of RNP granules. **f** Electrophoresis probes' stages of protein/RNA multimerization

The above-mentioned methods have undoubtedly provided valuable mesoscale information such as droplet formation conditions, size, fusion kinetics, viscosity, and diffusion parameters, some molecular mechanisms underlying RNP nucleation and changes in granule properties post-nucleation. However, most of our current knowledge derived from these methods lack molecular details of RNA–protein interaction involved in the early stages of RNP granule nucleation and assembly. In the next part of this chapter, we introduce a few such biophysical methods, including single-molecule fluorescence detection, which are ideally suited to address the molecular basis of RNP formation [26, 36].

3.3 Methods to Probe Initial Phases of RNP Assembly

3.3.1 Measuring RNA–Protein Interaction Across the Phase Boundary: Single-Molecule FRET and EMSA

One of the very early stages in RNP granule assembly likely involves discrete steps of RNA–protein and protein–protein interaction in the soluble phase which transitions to the liquid-like phase separation, which ages to more mature forms of hydrogel-like and solid state such as fibers. In vitro, the conditions that lead to the onset of phase separations such as temperature, salt, and protein concentrations can be tuned for each protein or protein/RNA system to generate a phase diagram which can display the clear partition between the soluble and the LLPS space. Such analyses have been done for proteins including FUS, LAF1, and Whi3 [21, 26, 35].

Understanding RNA–protein interaction at the onset of granule assembly necessitates first understanding how the protein interacts with RNA in its soluble phase. To probe the interaction between single RNA and single protein, we used single-molecule FRET (smFRET) assay based on total internal reflection fluorescence (TIRF) microscopy (Fig. 3.3) [38]. In addition, we applied EMSA (Fig. 3.2f) [39] to determine the stoichiometry of protein–RNA complex. Combination of these two approaches allows one to probe single RNA–protein interaction as a function of granule promoting parameters (protein and salt concentration) and map it to corresponding phase space as demonstrated in our previous work [26, 36].

In our previous study, we employed LAF-1, an IDR-containing DEAD-box RNA helicase present in P granules of *C. elegans* as a model protein. Purified LAF-1 phase separates in vitro, driven by its intrinsically disordered *N*-terminal RGG-rich domain that is also an RNA-binding domain. EMSA experiments revealed that LAF-1 binds specifically to single-stranded (ss)RNA. So, our model RNA substrates consisted of ssRNA overhang of 15–50 nucleotides (poly U sequence) in a format of partially duplexed RNA (Fig. 3.3a). We refer to these substrates as U15, U30, U40, U50, depending on the length of the poly U overhang. One of the RNA strands is biotinylated so that the RNA substrate can be surface immobilized on the PEG-passivated quartz slide to be used on the TIR microscope set up. Each RNA substrate is dual

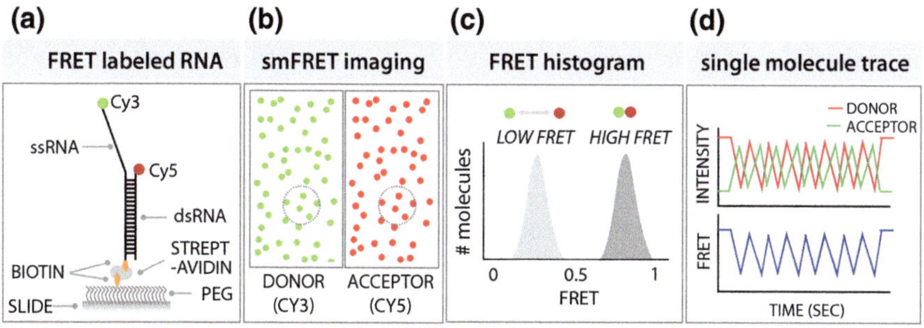

Fig. 3.3 smFRET detection. **a** FRET-RNA or DNA substrate immobilized to PEG surface. **b** Cy3 and Cy5 signal from same set of molecules (circle). **c** FRET values collected from thousands of molecules are built into FRET histogram. **d** Individual smFRET traces report on time-dependent change in FRET

labeled with a pair of FRET-suitable fluorophores which are arranged such that FRET reports on how LAF-1-binding impacts the conformation of ssRNA (Fig. 3.3b–d).

Varying LAF-1 concentration from low to high (corresponding to the transition from soluble phase to LLPS regime based on the LAF-1 phase diagram) was applied to fluorophore-labeled RNA substrates for EMSA analysis (Fig. 3.4a, b). Only one shifted band (relative to the unbound RNA only band) was observed for U30 across the protein concentrations, indicating a monomer protein binding to RNA. For U40 and U50, in addition to this band, a super-shifted band was observed in high protein

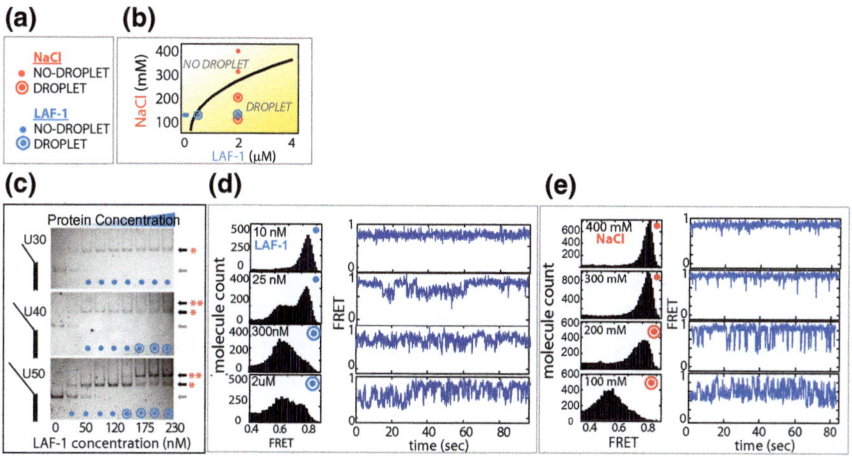

Fig. 3.4 LAF-1 induces dynamics on ssRNA in droplet-forming condition. **a** Color and shape key for droplet versus non-droplet-forming conditions. **b** Experimental conditions cutting across phase boundary in [LAF-1] and [NACl]. **c** Droplet-forming conditions coincide with dimerization of LAF-1 denoted by double red asterix. **d**, **e** In droplet-forming condition of high [LAF-1], LAF-1 induces dynamic mobility on ssRNA, evidenced by FRET fluctuation

concentrations, indicating multimer protein binding to RNA (Fig. 3.4c). In light of the applied LAF-1 concentrations required for soluble to LLPS transition, this set of data suggests that U30 may not be long enough to accommodate more than one protein, yet U40 and U50 have sufficient length to recruit multimers of proteins which can promote the droplet assembly.

Is there any change in these protein–RNA interactions going from soluble to phase separation? Here is where the smFRET measurements offer unique advantage. We applied different protein concentrations to surface immobilized FRET-labeled RNA. The intensities of donor (Cy3) and acceptor (Cy5) are collected from approximately 300–400 RNA molecules per field of view, i.e., in one movie. Such data can be analyzed in two ways: (i) FRET efficiencies collected from thousands of U50 molecules (from 10 to 20 short movies) are built into a histogram which displays the overall FRET distribution; (ii) individual smFRET traces taken for 2–3 min displays how FRET (calculated from intensity of the donor and acceptor dyes) changes over time, which is interpreted as the conformational changes within individual RNA molecules as the proteins act upon them. These two evaluations together give us a clear picture regarding not only the binding mode of the protein to the RNA, but also the rare glimpses of molecular details of this interaction intractable by ensemble methods. The U50 RNA alone yields a low FRET peak due to the distance between the two dyes separated by 50 ribonucleotides. Application of low LAF-1 or high salt concentrations (that represent soluble phase) to U50 RNA shifts the FRET histogram peak from low FRET (unbound U50 RNA) to high FRET (representing LAF-1-bound U50 RNA) (Fig. 3.4d, e). The time traces of individual U50 molecules show a shift from low to high FRET immediately after the protein addition and the signal remaining stable over time. The EMSA data taken in the same condition shows a single band shift, representing monomer-bound U50 fraction. This indicates that a monomer LAF-1 binding induces tight compaction of the RNA (bringing the two dyes into a close proximity) that is stable over time. As LAF-1 protein concentration increases or salt concentration decreases, approaching and crossing the phase boundary, a broad mid-FRET peak appears in addition to the high FRET peak. The single-molecule traces exhibit dynamic FRET fluctuation interspersed with a static high FRET state (Fig. 3.4d, e). In this condition, EMSA analysis reveals a mixture of a monomer-bound (single shift) and multimer-bound (double shift) stoichiometric states (Fig. 3.4c), reflecting the coexistence of a monomer and multimer-bound states generating static high FRET and dynamic fluctuating FRET, respectively. When the protein concentrations correspond to the inside of phase boundary, EMSA showed primarily double shift and smFRET traces displayed majority of molecules exhibiting FRET fluctuations.

Thus, an ensemble biochemical assay such as EMSA, biophysical measurements including microrheology, viscoelasticity, and the smFRET assay can be combined to extract unique material properties of RNP droplet and the underlying molecular details involved in the formation of LLPS, especially at the early stages of RNP assembly. In case of LAF-1, such a strategy helped us understand that as conditions (concentrations of protein, salt, etc.) transition from soluble to phase-separated LAF-1 RNP droplet, monomer-bound and tightly wrapped RNA evolves to multimer protein which dynamically interacts with RNA, likely representing a state that is ready to assemble into RNP droplets. The dynamicity may also contribute to the droplet fluidizing effect that RNA has been observed to have on LAF-1 RNP droplets. We envision this strategy to be effective with other IDR granule-forming proteins as well, such as FUS and MEG-3 (unpublished data).

3.3.2 RNA Annealing Assay as a Proxy for RNP–RNP Interaction in RNP Granules

In the context of RNP granules, RNP–RNP interaction is undoubtedly a key factor in regulating all stages of granule life. We devised an assay that can potentially test for this level of interaction [36]. We posited that for two complementary strands of RNA to hybridize, two sets of RNA–protein complexes need to come together, i.e., requiring RNP–RNP contact. We established an annealing assay in which we immobilized a partially duplexed RNA (with ss overhang of mixed sequence) that is FRET dye (Cy3, Cy5) labeled, exhibiting high FRET. We apply pre-incubated mixture of complementary ssRNA and LAF-1 (Fig. 3.5a). In the pre-incubated mix, while some LAF-1 is expected to be in complex with the ssRNA, some protein is expected to be free to interact with the immobilized RNA on surface. The annealing between the two complementary RNA is expected to result in a decrease in FRET since the dyes in the annealed substrate will now be separated by duplexed RNA (Fig. 3.5b). We subjected various conditions of LAF-1 including its *N*- and *C*-terminal truncation

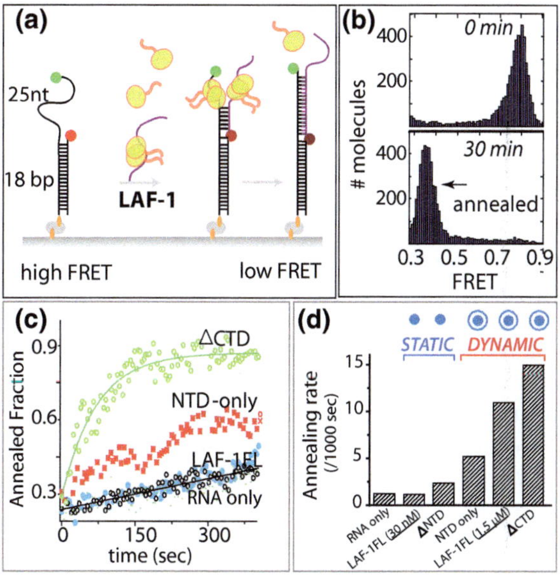

Fig. 3.5 LAF-1-RNA dynamics promote RNA annealing. **a** High FRET converts to low FRET upon RNA annealing. **b** FRET histogram before (top) and after (bottom) annealing. **c** Kinetic analysis of RNA annealing reaction. **d** Annealing rate for various mutants that represent static versus dynamic LAF-1-RNA interaction

mutants and found that RNA annealing was greatly enhanced in the conditions that promote droplet formation and dynamic RNA–protein interactions, but substantially diminished for out-of-LLPS conditions that induce tight RNA compaction by the protein (Fig. 3.5c, d). Thus, dynamic RNA–protein interactions (when approaching granule-forming conditions) promote RNP–RNP interactions between LAF-1-RNA complexes. The correlation between (i) monomer to multimer stoichiometric transition, (ii) static to dynamic change in RNA–protein interaction, (iii) defective to efficient RNA annealing, and the (iv) soluble to LLPS reflect that this set of measurement could serve as reporter assays that define the underlying molecular transactions that contribute to the RNP droplet assembly.

3.3.3 Measuring Size of In Vitro Droplets and Cellular Granules at Nucleation: Dynamic Light Scattering and Single-Molecule Pull-Down Assay

In vitro droplets from purified IDR, proteins have been evaluated by DIC imaging, their fluidity measured by FRAP and microrheology analysis as discussed above. These methods, however, do not offer insight into the size of the assemblies at the very early stages at the onset of nucleation. Dynamic light scattering (DLS) is sensitive enough to detect monomer to multimer transition of protein condensation at the onset of droplet assembly, but becomes unsuitable once stable droplets (>1 μm in radius) have formed. DLS has been used to probe oligomer size growth and kinetics of assembly for poly A binding protein (Pab1 in yeast) which form into granules [40]. For such purposes, DLS can yield two parameters: (i) hydrodynamic radius (R_h) for the IDR protein in the soluble phase, by batch mode DLS which estimates size and size distribution of oligomers (in the range of 0.5–1000 nm radius). The highly sensitivity capturing of light scattering pattern that changes according to the size of protein particles makes DLS an apt method to track size and growth of granules at the very early stages of droplet nucleation, far beyond the detection limit of DIC imaging (Fig. 3.6a, red arrow indicating increasing protein concentration, moving to LLPS favorable condition). (ii) Continuous thermal scans of IDR proteins can reveal the temperature of aggregation onset, defining the lower critical solution temperature (LCST) above which the protein will phase separate. This measurement mode is useful for comparing disease mutants of granule proteins, and also for assessing how the presence of RNA may impact the LCST and hence the onset of aggregation. However, as mentioned before, beyond a certain radius (1000 nm), detection of aggregates by DLS is not reliable, thus making it unsuitable to probe droplets that have already assembled. Furthermore, the resolution limit of batch DLS is a factor of 2–5 in size, making it difficult to assign precise stoichiometric state, which can be assessed better by the method introduced below.

Single-molecule pull-down (SiMPull) assay [41, 42] is a unique and powerful single-molecule technique that combines traditional pull-down assay principles with

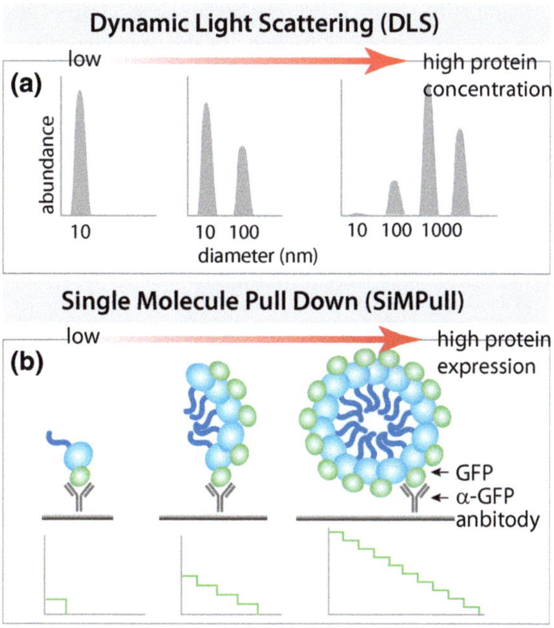

Fig. 3.6 Probing molecular assembly of granules. **a** Dynamic light scattering is useful in measuring early phase of molecular assembly in vitro droplets of purified protein. **b** Single-molecule pull-down assay can reveal the multimeric state of target proteins in cellular granules by photobleaching

single-molecule fluorescence microscopy, permitting direct visualization of individual protein complexes directly pulled down from cell lysate, thus constituting a method which is non-perturbing and preserving native cellular context. Upon expressing a target protein fused to a fluorescence marker protein such as GFP or RFP, SiMPull analysis can reveal how many of the labeled proteins are present in cellular protein complexes. The key in this assay is the selective capture of the protein of interest from a cell lysate via an antibody. We discuss an experimental design here to illustrate how SiMPull can be applied to reveal protein oligomeric state in cellular granules, using SGs as an example. Mammalian cells typically used in SG studies such as HEK293, U2OS, or HeLa cells may also be used for this assay. A fluorescently tagged version (such a GFP, YFP, RFP) of the known SG protein of interest can be expressed in the mammalian cells of choice. The cells can be subjected to stressed or non-stressed conditions and SG formation (bright fluorescence puncta) can be checked by fluorescence microscopy. The cell lysate can then be applied onto the single-molecule imaging surface (composed of flow chambers constructed on a sandwich of PEG-passivated slide and coverslip). The imaging surface can be coated with the specific antibody against the protein of interest (anti-GFP antibody, for example) and the cell lysate can be applied. The target protein and protein-containing complexes will be captured by the antibody. Because the target protein is tagged with GFP at either *N*- or *C*-terminus, counting the number of photobleaching steps in each spot can yield stoichiometric information about the protein-containing complex unit. In the absence of stress, when there is no granule formation, the protein is expected to be in soluble phase (Fig. 3.6b). The SiMPull image is likely to be occupied by low-intensity spots, depicting monomeric state of the protein (deduced from analyses that are described below). In the presence of stress, we expect to also capture bright fluorescent spots representing clusters of proteins on route to assemble into granules. The total intensity of each of the high-intensity spots will be proportional to the number of fluorescent protein units present in that granule. Therefore, intensities of individual spots alone can reflect the oligomeric status of the proteins. In addition, tracking the number of photobleaching steps for conditions with and without stress can lead to more accurate analysis to distinguish mono- di-, trimeric, and higher oligomeric cellular granules that may represent clusters that form in the early stages of granule assembly. In addition, the SiMPull assay may also be expanded to probe the granule assembly of multiple granule-forming proteins by co-expressing with different fluorescent proteins in cells.

3.3.4 Super-Resolution Imaging to Reveal Granule Structure

Imaging cellular granules by expressing fluorescently tagged protein or in vitro imaging of LLPS droplets formed by purified protein have been a simple but powerful tool for initial studies in the field. The length scale of RNP granules ranges from 100 nm to several micrometers, which is roughly around the limit of conventional diffraction limited microscopy (~300 nm). Recently, new findings suggest that the RNP granules may have substructures: within the nucleolus, and two proteins' phase separate into two layers of LLPS, driven by different surface tension [43]; a proteomic analysis also revealed that SGs have substructures, consisting of a stable core and a dynamic shell [9]. These studies clearly indicate a more complex level of RNP granule architecture that requires further investigation. However, the substructures are often too small to be observed clearly with a conventional microscope.

Super-resolution imaging techniques developed in the last two decades provides an ideal tool to resolve the substructures within RNP granules which are inaccessible with a conventional microscope. Two important branches of super-resolution imaging technique are stimulated emission depletion (STED) microscopy and localization microscopy. Localization microscopy includes photoactivated localization microscopy (PALM) [44] and stochastic optical reconstruction microscopy (STORM) [45]. STED microscopy was developed by Hell, S. group [46]. Here, we will briefly discuss the applicability and the potential of STED microscopy to reveal granule substructures.

Briefly, STED microscopy uses two lasers instead of one to achieve super-resolution above the optical diffraction limit. The excitation laser is used to excite fluorophores in the same manner as the conventional confocal imaging, and the other STED laser is used to deplete the fluorophore in the shape of a donut, making the fluorophores emit photons only at the center of the donut, hence achieving super-resolution. For STED imaging, the GFP tagged protein is less ideal than the immunolabeling using antibodies conjugated with organic fluorophores. This method is typically better for higher signal-to-noise ratio due to the brightness of the organic fluorophores. The choice of fluorophores is subject to the specific experimental design and scheme. STED microscopy can be beneficial in capturing structural details of granules because the typical size of membraneless granules in cells is sub-micron.

3.4 Concluding Thoughts

In light of the tremendous diversity that has been observed in methods to study granules, we reiterate two important points: Firstly, methods to study granules must be carefully selected according to the stage of the granule's life that is being targeted. Secondly, a strategic combination of methods is often more powerful to extract the maximum and most accurate information about a particular stage of granule life. While bulk and mesoscale methods will continue to hold an important place in the granule field, because they report on the material property of granules by relatively simple means, methods accessing finer molecular details of interactions will expand and grow, shedding insight about the granule assembly and dynamics. Although we did not discuss here, atomic level probing for the IDR proteins using NMR and hydrogen exchange-coupled mass spectrometry has also been employed to define molecular coordinate of protein conformations and dynamics. However, we believe single-molecule methods, including smFRET and SiMPull, which have found novel applications in granule studies, have the potential to access unprecedented fine molecular details that, until recently, have remained inaccessible by other methods.

Acknowledgements This project was supported by Catalyst Research Award from Johns Hopkins University, R01 GM115631 and R01 CA 207342 for all members at Johns Hopkins University.

References

1. Guo, L., & Shorter, J. (2015). It's raining liquids: RNA tunes viscoelasticity and dynamics of membraneless organelles. *Molecular Cell, 60,* 189–192.
2. Larson, A. G., Elnatan, D., Keenen, M. M., Trnka, M. J., Johnston, J. B., Burlingame, A. L., et al. (2017). Liquid droplet formation by HP1α suggests a role for phase separation in heterochromatin. *Nature, 547,* 236–240.
3. Strom, A. R., Emelyanov, A. V., Mir, M., Fyodorov, D. V., Darzacq, X., & Karpen, G. H. (2017). Phase separation drives heterochromatin domain formation. *Nature, 547,* 241–245.
4. Woodruff, J. B., Ferreira Gomes, B., Widlund, P. O., Mahamid, J., Honigmann, A., & Hyman, P. O. (2017). The centrosome is a selective condensate that nucleates microtubules by concentrating tubulin. *Cell, 169*, 1066–1077 e1010.
5. Spector, D. L. (2006). SnapShot: Cellular bodies. *Cell, 127,* 1071.
6. Protter, D. S., & Parker, R. (2016). Principles and properties of stress granules. *Trends in Cell Biology, 26,* 668–679.
7. Updike, D., & Strome, S. (2010). P granule assembly and function in *Caenorhabditis elegans* germ cells. *Journal of Andrology, 31,* 53–60.
8. Anderson, P., & Kedersha, N. (2006). RNA granules. *The Journal of Cell Biology, 172,* 803–808.

9. Jain, S., Wheeler, J. R., Walters, R. W., Agrawal, A., Barsic, A., & Parker, R. (2016). ATPase-modulated stress granules contain a diverse proteome and substructure. *Cell, 164,* 487–498.
10. Aulas, A., & Vande Velde, C. (2015). Alterations in stress granule dynamics driven by TDP-43 and FUS: A link to pathological inclusions in ALS? *Frontiers in Cellular Neuroscience, 9,* 423.
11. Patel, A., Lee, H. O., Jawerth, L., Maharana, S., Jahnel, M., Hein, M. Y., et al. (2015). A liquid-to-solid phase transition of the ALS protein FUS accelerated by disease mutation. *Cell, 162,* 1066–1077.
12. Aulas, A., Fay, M. M., Lyons, S. M., Achorn, C. A., Kedersha, N., Anderson, P., et al. (2017). Stress-specific differences in assembly and composition of stress granules and related foci. *Journal of Cell Science, 130,* 927–937.
13. Dewey, C. M., Cenik, B., Sephton, C. F., Dries, D. R., Mayer, P., 3rd, Good, S. K., et al. (2011). TDP-43 is directed to stress granules by sorbitol, a novel physiological osmotic and oxidative stressor. *Molecular and Cellular Biology, 31,* 1098–1108.
14. Kato, M., Han, T. W., Xie, S., Shi, K., Du, X., Wu, L. C., et al. (2012). Cell-free formation of RNA granules: Low complexity sequence domains form dynamic fibers within hydrogels. *Cell, 149,* 753–767.
15. Brangwynne, C. P., Eckmann, C. R., Courson, D. S., Rybarska, A., Hoege, C., Gharakhani, J., et al. (2009). Germline P granules are liquid droplets that localize by controlled dissolution/condensation. *Science, 324,* 1729–1732.
16. Kroschwald, S., Maharana, S., Mateju, D., Malinovska, L., Nuske, E., Poser, I., et al. (2015). Promiscuous interactions and protein disaggregases determine the material state of stress-inducible RNP granules. *eLife, 4,* e06807.
17. Patel, A., Malinovska, L., Saha, S., Wang, J., Alberti, S., Krishnan, Y., et al. (2017). ATP as a biological hydrotrope. *Science, 356,* 753–756.
18. Schwartz, J. C., Wang, X., Podell, E. R., & Cech, T. R. (2013). RNA seeds higher-order assembly of FUS protein. *Cell Reports, 5,* 918–925.
19. Daigle, J. G., Lanson, N. A., Jr., Smith, R. B., Casci, I., Maltare, A., Monaghan, J., et al. (2013). RNA-binding ability of FUS regulates neurodegeneration, cytoplasmic mislocalization and incorporation into stress granules associated with FUS carrying ALS-linked mutations. *Human Molecular Genetics, 22,* 1193–1205.
20. Voigt, A., Herholz, D., Fiesel, F. C., Kaur, K., Muller, D., Karsten, P., et al. (2010). TDP-43-mediated neuron loss in vivo requires RNA-binding activity. *PloS One, 5,* e12247.
21. Zhang, H., Elbaum-Garfinkle, S., Langdon, E. M., Taylor, N., Occhipinti, P., Bridges, A. A., et al. (2015). RNA controls PolyQ protein phase transitions. *Molecular Cell, 60,* 220–230.
22. Langdon, E. M., Qiu, Y., Ghanbari Niaki, A., McLaughlin, G. A., Weidmann, C., Gerbich, T. M., et al. (2018). mRNA structure determines specificity of a polyQ-driven phase separation. *Science.*
23. DeJesus-Hernandez, M., Mackenzie, I. R., Boeve, B. F., Boxer, A. L., Baker, M., Rutherford, N. J., et al. (2011). Expanded GGGGCC hexanucleotide repeat in noncoding region of C9ORF72 causes chromosome 9p-linked FTD and ALS. *Neuron, 72,* 245–256.
24. Renton, A. E., Majounie, E., Waite, A., Simon-Sanchez, J., Rollinson, S., Gibbs, J. R., et al. (2011). A hexanucleotide repeat expansion in C9ORF72 is the cause of chromosome 9p21-linked ALS-FTD. *Neuron, 72,* 257–268.
25. Jain, A., & Vale, R. D. (2017). RNA phase transitions in repeat expansion disorders. *Nature, 546,* 243–247.
26. Elbaum-Garfinkle, S., Kim, Y., Szczepaniak, K., Chen, C. C., Eckmann, C. R., Myong, S., et al. (2015). The disordered P granule protein LAF-1 drives phase separation into droplets with tunable viscosity and dynamics. *Proceedings of the National Academy of Sciences of the United States of America, 112,* 7189–7194.
27. Maharana, S., Wang, J., Papadopoulos, D. K., Richter, D., Pozniakovsky, A., Poser, I., et al. (2018). RNA buffers the phase separation behavior of prion-like RNA binding proteins. *Science.*
28. Wheeler, J. R., Matheny, T., Jain, S., Abrisch, R., & Parker, R. (2016). Distinct stages in stress granule assembly and disassembly. *eLife, 5.*

29. Li, Y. R., King, O. D., Shorter, J., & Gitler, A. D. (2013). Stress granules as crucibles of ALS pathogenesis. *The Journal of Cell Biology, 201,* 361–372.
30. Lagier-Tourenne, C., Polymenidou, M., & Cleveland, D. W. (2010). TDP-43 and FUS/TLS: emerging roles in RNA processing and neurodegeneration. *Human Molecular Genetics, 19,* R46–R64.
31. Ishiguro, T., Sato, N., Ueyama, M., Fujikake, N., Sellier, C., Kanegami, A., et al. (2017). Regulatory role of RNA chaperone TDP-43 for RNA misfolding and repeat-associated translation in SCA31. *Neuron, 94,* 108–124 e107.
32. Mateju, D., Franzmann, T. M., Patel, A., Kopach, A., Boczek, E. E., Maharana, S., et al. (2017). An aberrant phase transition of stress granules triggered by misfolded protein and prevented by chaperone function. *EMBO Journal, 36,* 1669–1687.
33. Alberti, S., Mateju, D., Mediani, L., & Carra, S. (2017). Granulostasis: Protein quality control of RNP granules. *Frontiers in Molecular Neuroscience, 10,* 84.
34. Lavut, A., & Raveh, D. (2012). Sequestration of highly expressed mRNAs in cytoplasmic granules, P-bodies, and stress granules enhances cell viability. *PLoS Genetics, 8,* e1002527.
35. Burke, K. A., Janke, A. M., Rhine, C. L., & Fawzi, N. L. (2015). Residue-by-residue view of in vitro FUS granules that bind the C-terminal domain of RNA polymerase II. *Molecular Cell, 60,* 231–241.
36. Kim, Y., & Myong, S. (2016). RNA remodeling activity of DEAD box proteins tuned by protein concentration, RNA length, and ATP. *Molecular Cell, 63,* 865–876.
37. Wheeler, J. R., Jain, S., Khong, A, & Parker, R. (2017). Isolation of yeast and mammalian stress granule cores. *Methods.*
38. Roy, R., Hohng, S., & Ha, T. (2008). A practical guide to single-molecule FRET. *Nature Methods, 5,* 507–516.
39. Hellman, L. M., & Fried, M. G. (2007). Electrophoretic mobility shift assay (EMSA) for detecting protein-nucleic acid interactions. *Nature Protocols, 2,* 1849–1861.
40. Riback, J. A., Katanski, C. D., Kear-Scott, J. L., Pilipenko, E. V., Rojek, A. E., Sosnick, T. R., & Drummond, D. A. (2017). Stress-triggered phase separation is an adaptive, evolutionarily tuned response. *Cell, 168,* 1028–1040 e1019.
41. Jain, A., Liu, R., Ramani, B., Arauz, E., Ishitsuka, Y., Ragunathan, K., et al. (2011). Probing cellular protein complexes using single-molecule pull-down. *Nature, 473,* 484–488.
42. Jain, A., Liu, R., Xiang, Y. K., & Ha, T. (2012). Single-molecule pull-down for studying protein interactions. *Nature Protocols, 7,* 445–452.
43. Feric, M., Vaidya, N., Harmon, T. S., Mitrea, D. M., Zhu, L., Richardson, T. M., et al. (2016). Coexisting liquid phases underlie nucleolar subcompartments. *Cell, 165,* 1686–1697.
44. Betzig, E., Patterson, G. H., Sougrat, R., Lindwasser, O. W., Olenych, S., Bonifacino, J. S., et al. (2006). Imaging intracellular fluorescent proteins at nanometer resolution. *Science, 313,* 1642–1645.
45. Rust, M. J., Bates, M., & Zhuang, X. (2006). Sub-diffraction-limit imaging by stochastic optical reconstruction microscopy (STORM). *Nature Methods, 3,* 793–795.
46. Klar, T. A., & Hell, S. W. (1999). Subdiffraction resolution in far-field fluorescence microscopy. *Optics Letters, 24,* 954–956.

Chapter 4
Single-Molecule Studies of Exonucleases: Following Cleavage Actions One Step at a Time

Gwangrog Lee

4.1 Introduction

Several metabolic cascades start with the synthesis of the relevant molecules, such as transcription factors and signaling molecules, and end with their degradation. In this perspective, synthesis and degradation are thus ubiquitous processes orchestrating cellular homeostasis to maintain life. Deconstruction is necessary for the sake of construction in terms of recycling limited cellular components. Especially, degradation processes are carried out by different kinds of degrading enzymes. For instance, nucleic acids are hydrolyzed by a group of nucleases that cleave phosphodiester bonds between two adjacent nucleotides in a nucleic acid strand. The deconstructive cleavage must be tightly regulated by proper signals and checkpoints [1]. Otherwise, they may lead to unwanted and uncontrolled degradation.

Nucleic acids, such as DNA and RNA, are epigenetically modified and spontaneously damaged. Nuclease activities are necessary to restore modified DNA to its unmodified state or to degrade modified RNA for a regulatory purpose. Owing to their significance, nucleases have been extensively studied for several decades [2–9]. However, mechanistic details of these enzymes have not been unraveled yet, owing to the difficulty in acquiring intramolecular dynamics information. Dynamics information is critical for understanding enzyme function [10, 11] because domain movements not only dictate each step of the enzyme, but also determine the overall timescale of the catalytic reaction that should harmonize with other cellular processes. Since most enzymatic activities consist of complicated dynamics and coordination of internal motions [12, 13] between their domains, monitoring molecular actions is challenging via traditional methods. Individual molecules in an ensemble are subjected to different microscopic states of time-dependent motions resulting from the interaction with random thermal fluctuations of the system [14, 15] and chemical and

G. Lee (✉)
Life Sciences, Gwangju Institute of Science and Technology, Gwangju 500-712, Korea
e-mail: glee@gist.ac.kr

C. Joo and D. Rueda (eds.), *Biophysics of RNA-Protein Interactions*,
Biological and Medical Physics, Biomedical Engineering,
https://doi.org/10.1007/978-1-4939-9726-8_4

physical heterogeneity [16–19] during the enzymatic reaction. However, the average of multimodal behavior over a whole population smears out each characteristic dynamic state, which contributes to the overall rate of the enzymatic reaction. Thus, such an ensemble averaging misrepresents the true nature of dynamic information of the system. A possible method to overcome the averaging problem is to observe individual molecules and measure one molecule at a time. The absence of a need for synchronization of all molecules reveals detailed dynamic information without additional processes for de-phasing. In this way, single-molecule data have provided a wealth of information about kinetics [20–23] as well as thermodynamics of systems [24–26].

In this review, I discuss recent progress and new findings from single-molecule studies that have increased our understanding of exonucleases. Simply detecting one molecule at a time provides unprecedented accuracy and clarity. This straightforward method has been used in many biological systems such as ion channels [27, 28], vesicle fusion [29–31], DNA replication [32–37], recombination [38–43], repair [44–46], RNA transcription [14, 47–52], ribosome dynamics [12, 53, 54], RNA degradation [55, 56], protein degradation [57–60], muscle contraction [61–63], and ATP synthesis [64, 65]. These studies have discovered important dynamics of biological macromolecules such as enzymatic dynamics disorders [66], heterogeneity [16–19], transient intermediates [67–69], and hidden complexity [70]. The history of single-molecule studies has changed our approach toward biological processes. For example, canonical biology describes enzymatic reactions as catalytic roles where enzymes lower the activation energy by stabilizing the enzyme–substrate complex, thus accelerating chemical reactions as described in biochemistry books. However, high-resolution data obtained by single-molecule studies make us rethink the meaning of dynamics data, leading us to consider not only kinetics but also other parameters such as distance changes, amount of force exerted, and the overall coordination mechanics. This novel approach has given rise to a new field called mechanochemical biology, which focuses on global system-level explanations of a mechanistic chemo-mechanical coupling [56, 71] between enzymatic action and catalytic hydrolysis.

4.2 Single-Molecule Methods to Study Nuclease Mechanisms

The ensemble techniques most commonly used for studying cleavage reactions by nucleases are gel electrophoresis-based [8, 72, 73] and fluorescent-based methods [74, 75]. These techniques measure the accumulation of the final products or the relaxation of the initial substrate amount over time. Then, steady-state kinetic parameters can be extracted by fitting the conversion of the initial substrates to the final products. In contrast, single-molecule techniques directly detect a structural change in the substrate induced by the activity of enzymes or a conformational change within the enzyme complex. The kinetic values obtained from individual enzymes are heterogeneous, but the similar parameters in an ensemble can be derived by averaging all the values. However, ensemble measurement often underestimates the true kinetic parameter value compared to single-molecule techniques because the protein sample contains some degree of inactivated proteins that do not contribute to the accumulation of the final products.

The single-molecule methods being used are force-based measurements (e.g., optical tweezers and flow stretch) and force-free (e.g., single-molecule FRET and single fluorescence particle tracking). For the former class, the force can be directly applied onto motor proteins or their substrates. This allows researchers to precisely manipulate, on demand, either protein movement itself in an enzymatic cycle or the thermodynamic stability of individual substrates. The force application onto the proteins slows down the reaction and thus enables the observation of intermediate states present along the reaction coordination. Even through force-based methods are laborious because the experiments are typically performed one measurement at a time using one force setup, the analysis is straightforward since stretching by force restricts and aligns all possible motions onto the stretching-favor direction. For this reason, these methods have yielded many fruitful discoveries over the last two decades, including the mechanisms of AAA+ ATPases in DNA packing machinery [76–78], DNA unwinding [79–81], ribosomal dynamics [54, 82, 83], and protein unfolding for degradation [57, 58]. This recent development in single-molecule techniques thus provides a new platform for accurate measurements, allowing the examination of complicated exonuclease activity which would not have been possible without the advance of single-molecule techniques.

4.2.1 Optical Tweezers

Optical tweezers possess excellent sensitivity to probe length and force changes induced by enzymatic activity. An optical trap can be achieved by a focused beam on a dielectric bead, made of a material such as polystyrene. A dielectric particle in the proximity of the focused beam is subjected to a three-dimensional restoring force with the tendency of reverting toward the center of the focused beam, and thus, the stiffness of the optical trap serves as a force sensor [84]. Two different trapping geometries (e.g., a dual-trap and a single trap) have been used to monitor real-time trajectories of individual nucleases. In the case of a single trap, one bead tethered to one end of a DNA handle (green) is held in a focused laser beam whereas the second end of the handle is conjugated to the enzymes (orange) of interest via a ligand–receptor interaction on a microscope slide surface (Fig. 4.1a). In the case of the dual trap, the substrates (green) and hydrolases (orange) are tethered on different polystyrene beads, each via biotin–avidin and digoxigenin–anti-digoxigenin interactions (Fig. 4.1b). The dual-trap experiences much lower drift and vibration compared to the single trap because the experimental system is completely uncoupled from the stage, and both the beads suspended are subjected to the same Brownian motion in liquid.

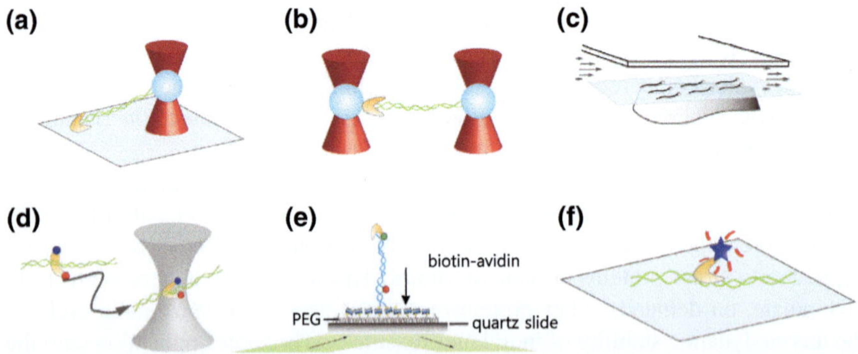

Fig. 4.1 Graphical representation of single-molecule techniques used to monitor activities of various nucleases (not to scale). **a** Schematic representation of a surface-tethered single trap: a nuclease, immobilized onto the surface, is engaged by DNA through its active site, and the other free end of DNA is tethered to the bead, trapped in a focused laser beam. The coupling to the surface compromises its spatial resolution, but this approach is often used owing to its easy instrumentation and simple experimental scheme as compared to the dual trap. **b** Geometry of dual-trap optical tweezers: one end of the DNA is attached to a bead held in the first focused laser beam, whereas the other free end is directly attached to the bead kept in the second optical trap. In this assay, the nuclease (orange), immobilized on the surface of the first trapped bead, pulls the other trapped bead through DNA tethering during the reaction. To maintain the trap at constant force, the position of the second trapped bead is adjusted while the first bead is kept at the same position. Thus, the change in distance between the two beads reflects the activity of the enzyme. **c** Flow stretch methods: this technique has been used to study enzymes such as helicases and nucleases that convert dsDNA to ssDNA. The conformational properties of dsDNA and ssDNA are significantly different in solution, and thus the conversion of dsDNA to ssDNA shortens the distance between the two ends of DNA. The technique has the high-throughput capability, but has a low spatial resolution. **d** Non-tethered diffusion single-molecule FRET: confocal microscopy is used to detect individual diffusing molecules labeled with fluorescent dyes. This technique is superior for quantifying heterogeneous subpopulations within a sample. **e** Surface-tethered single-molecule FRET: observing single proteins labeled with a single fluorescent dye requires substantial suppression of background fluorescence. Total internal reflection fluorescence (TIRF) microscopy is used to achieve a high signal-to-noise ratio. Substrates are attached to a polymer-coated surface via biotin–avidin interaction, and fluorescent signal changes from either substrates or proteins are monitored in real time. This technique allows us to study the activity of enzymes in a force-free condition. **f** Single-molecule fluorescent particle tracking: DNA molecules are directly attached at one end to the surface and extended by laminar flow within an evanescent field, resulting in a DNA arch-like stretch. Proteins labeled with a fluorescent dye are added to monitor their movement along their substrates. The technique allows the study of different types of diffusion and the observation of hundreds of aligned DNA molecules in real time within a single imaging area. In addition, quantum dot labeling allows the observation of diffusing molecules for a long period without photobleaching and reductions in intensity

There are two operation modes: passive and active force clamp modes. Under passive force clamp mode, the position of a bead was continuously adjusted to maintain a target tension in response to a change in substrate length, caused by the movement of a nuclease during degradation. In contrast, under an active force clamp mode, the distance between the two beads increases with a constant speed, allowing changes in force in response to the structural changes of the substrate. For all the cases, the accuracy of optical tweezers is of the order of nanometers, sub-milliseconds, and pico-newtons [85], serving as a perfect tool to follow enzymatic activities during biological processes. In addition, the capability to manipulate enzyme activities and substrate stabilities by force allows us to deeply investigate a change in the free energy landscape of a complex biological system along the reaction coordination. The platform of force measurements has revolutionized our mechanistic understanding of many motor systems.

4.2.2 Flow Stretching

Flow stretching has been applied to study exonuclease mechanisms. In the flow-stretching technique, one end of DNA is immobilized to a slide surface and the other end is tethered to a micron bead (Fig. 4.1c). A laminar flow is used to stretch DNA substrates by exerting a hydrodynamic drag force less than 6 pN on the micron bead [86]. To minimize the nonspecific binding of beads onto the surface, a paramagnetic microsphere is slightly suspended over the surface via a magnetic field applied above the flow chamber. The micrometer size of the bead is large enough to permit accurate determination of the bead position over time. Its position is then monitored in real time to assess the extent of DNA degradation performed by the exonuclease. Upon degradation, a rigid double-stranded (ds) DNA is converted into a flexible single-stranded (ss) DNA, shortening the end-to-end distance of DNA. Based on the nature of the stretching curve, in which the distance between the two ends of ssDNA is shorter than that of dsDNA under forces below ~6 pN, the shortening in length gives an estimation of DNA degradation of the order of ~10 [86] to 400 base pairs [66] of spatial resolution. The advantages of flow stretching include its high-throughput capability (several hundred molecules measured in parallel) compared to optical tweezers, and its ability to detect a long distance is an advantage over single-molecule FRET. This method was recently combined with fluorescent particle tracking such that the DNA degradation and the intramolecular change in a fluorescently labeled enzyme complex are detected simultaneously.

4.2.3 Single-Molecule Fluorescence Resonance Energy Transfer (smFRET)

smFRET is performed by either a non-tethered diffusion [87, 88] or a surface-tethered assay [89, 90]. For the former assay, the observation is limited by the diffusion time passing through the detection space, but it is suited for a subpopulation analysis that simultaneously displays heterogeneity and intermediates of individual molecules in the ensemble (Fig. 4.1d). In contrast, the latter is capable of obtaining time trajectories of biological reactions, but the issues of surface effects and nonspecific binding arise, since the reaction is performed near the surface (Fig. 4.1e). Hence, surface passivation with polyethylene glycol (PEG) has been adopted to minimize nonspecific protein-surface binding, and numerous tests have been intensively conducted to make sure that the results obtained from surface-tethered assays were consistent with those obtained from other surface-free assays. This turns out to be right for most cases of DNA–protein interactions. Additionally, a low fluorescence background is critical to distinguish single molecules labeled with single fluorophores, and thus, the background is typically reduced by confocal or total internal reflection (TIR) excitation techniques. So far, the most popular method used for smFRET studies on nucleases is the surface tethering method combined with TIR excitation and PEG passivation.

Specific tethering is usually achieved by the interaction between biotin on a substrate and avidin on the PEG polymer surface. The fluorescent molecules to be immobilized are diluted to attain desired surface density where individual molecules are well separated and distinguished within a diffraction limit. Green and red fluorescent dyes are typically used as a donor and acceptor pair for FRET experiments. FRET occurs when the two dyes are within 20–80 Å of each other, and the spectra of the donor emission and the acceptor absorption overlap. For this reason, FRET is a strong function of the distance between the two dyes. The efficiency of FRET is defined by a ratio, $E = 1/[1 + (R/R_o)^6]$, where R is the distance between the donor and acceptor, and R_o is the value when the efficiency becomes 50%. The FRET-sensitive region is adjustable by choosing different combinations of donors and acceptors. This method is thus simple and reliable for studying molecular systems driven by linear motion (e.g., 1D diffusion, degradation, and translocation), but its limitation is that it can sense 3D dynamic systems with only two fluorescent probes. Hence, three-color [91] and four-color [92] FRET has been developed to examine complicated biological motions by observing several degrees of freedom or relative motions of different positions at once. However, the analysis of these methods remains challenging. The advantages and challenges of multi-color FRET techniques are discussed in a recent review article [93].

4.2.4 Single-Molecule Fluorescent Particle Tracking (smFPT)

smFPT has been used to study nuclease activities. This technique is an excellent tool for studying the long range of a directed movement by motor proteins along a cellular track or of a random diffusion by ATP-independent enzymes in 3D space (Fig. 4.1f). The resolution of the technique strongly relies on the accuracy of particle localization, and drift and vibration of the measurement stage. In particular, the localization accuracy for fluorescence particles is determined by the total number of collected photons. If given ~20,000 photons from individual fluorescent particles, the resolving power reaches ~1 nm via Gaussian fitting of one point spread function (PSF), the method so-called FIONA (fluorescence imaging with 1 nm accuracy), demonstrated by Selvin et al. [62]. However, maintaining a high spatial resolution compromises temporal resolution and observation time because of the limited number of fluorescence emission cycles. For example, Cy3 as a typical organic dye emits photons only for four minutes at a rate of 20,000 photons/s [94]. smFPT has often been combined with substrate alignment techniques, e.g., DNA [95] and Actin curtain [96], which exploit nanofabrication, surface chemistry, and microfluidics to observe hundreds of molecules in real time. In short, this combination with the alignment approach not only simplifies 3D to 1D analysis but also enables high-throughput data acquisition.

4.3 Molecular Bases of Nucleic Acid Degradation by Nucleases

4.3.1 Classes of Nucleases

Nucleases are enzymes that specifically catalyze the degradation of a phosphodiester bond in a certain nucleic acid substrate. They generally promote the hydrolysis of a bond by activating a nucleophile and stabilizing the catalytically competent intermediate by forming a transition state [97–100]. Hydrolysis reactions typically require divalent cations to carry out a metallic coordination for the nucleophile attack. This activity is ubiquitous in many cellular processes. Exemplary enzymes involved in those cleavage reactions include lipases, phosphatases, nucleases, glycosidases, peptidases. I have limited the scope of this review to nucleases, since single-molecule techniques have been successfully applied to these enzymes.

Nucleases have been grouped into three categories based on their characteristics of [1] metal ion dependence; [2] substrate sequence specificity; and [3] types of hydrolysis, e.g., endo- and exonucleases. Metal ion-dependent nucleases generally create 3′ OH and 5′ phosphates upon cleavage reactions, while metal ion-independent enzymes such as ribozymes usually generate 2′, 3′-cyclic phosphate products. Regarding the manner in which cutting is performed, endonucleases cut their substrates in the middle of the single-stranded DNA or RNA. For this reason, it has been difficult to dissect the cutting activity of these endo-hydrolases. In contrast, exonucleases continuously cut a bond between two neighboring residues from one of the two ends and move by a distinct length (i.e., the unit of the polymer chain). Thus, the cutting site is located each time at the catalytic site of the enzyme. In this sense, exonucleases can be considered as motor-like proteins, in that they translocate on a cellular substrate, powered by the hydrolysis of their own track. The enzymatic movement along the chain of the substrate is presumably driven by intramolecular physical motion, coupled with the chemical energy released from the hydrolytic reaction. This motion may exert a force of the order of several tens of pico-newtons (pN), which sometimes induces the mechanical denaturation of the substrate.

4.3.2 One- and Two-Metal Ion Chemistry for Cleavage Reactions

A key catalytic feature of nucleases is the incorporation of divalent metal ions. Because of the abundance of Mg^{2+} ions in cell nuclei, Mg^{2+} ions are typically accommodated as a metal ion cofactor at the catalytic core. Nucleases perform the phosphoryl transfer reaction via the octahedral geometry of Mg^{2+}, in which one- or two-metal ions are typically coordinated by six ligands in total, but typically by three different kinds of groups: (1) conserved carboxylate residues (Asp and Glu) at nuclease active sites, (2) scissile phosphates of nucleic acids, and (3) water molecules. One- or two-metal ion catalysis reactions are found in nucleic acid nucleases. In the two-metal ion catalysis, metal ion A deprotonates a water molecule to initiate a nucleophile attack via a hydroxide ion, whereas metal ion B stabilizes the pentavalent phosphate intermediate by transiently bridging the nucleophile and the phosphate, further leading to the cleavage reaction (Fig. 4.2a). Afterward, metal ion B dissociates and destabilizes the substrate-enzyme complex. Thus, the role of metal ion B balances the needs of the chemical transition state and allows timely product release. In contrast, one-metal catalysis contains metal ion B, but not metal ion A (Fig. 4.2b). The system often utilizes a histidine residue of proteins to achieve a nucleophilic attack, replacing the role of metal ion A, whereas one-metal ion B exists and plays the same role in stabilizing the pentacovalent intermediate.

Despite many mechanistic studies, key questions remain unanswered. What are the functional roles of each metal ion during endo or exonuclease activity? How do the two-metal ions coordinate catalysis? Is metal ion A sufficient to yield nucleophile formation? Then why do enzymes function when the coordinating residues of the metal ion B are mutated, but not the residues that chelate metal ion B? This suggests an alternative mechanism for nucleophile generation without metal ion A, and metal ion B alone is enough to carry out the phosphoryl transfer reaction by stabilizing the transition state. How do metal ion dynamics influence overall enzymatic activity? It is

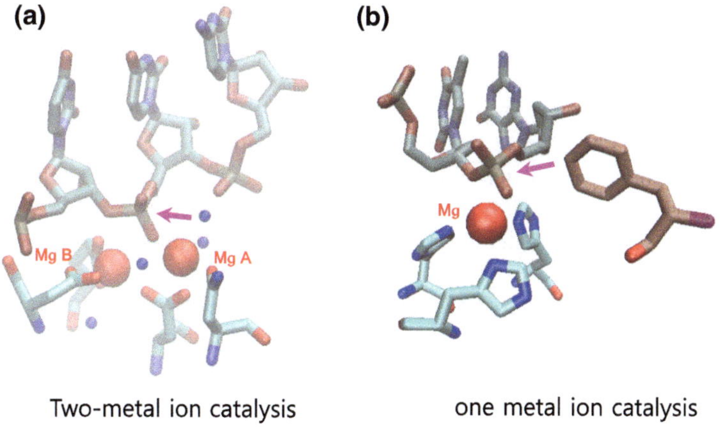

Fig. 4.2 Mechanism of metal ion-dependent catalysis. Metal ions incorporated by nucleases play key roles in catalyzing phosphodiester bond breakage. Mg^{2+} prefers to form the octahedral coordination by water molecules but exchanged by conserved carboxylate acidic residues (Asp and Glu). **a** Two-metal ion catalysis mechanism (RNase H: PDB ID, 1ZBL). The metal ion A deprotonates a water molecule (red arrow) to initiate a nucleophile attack by generating a hydroxide ion, whereas the metal ion B stabilizes the transient pentavalent phosphate upon nucleophile attack onto the scissile phosphate; thus, both ions facilitate the cleavage reaction. **b** One-metal catalysis mechanism. Metal ion A is replaced with a His residue as a general base to deprotonate and activate nucleophilic water, but the metal ion B is present at the equivalent position as that in two-metal ion catalysis, playing the same role in stabilizing the pentacovalent intermediate

well known that the incorporation of metal ions into the active site enhances enzyme binding affinity. This suggests that a subtle structural change during the formation of ligands around metal ions triggers a transition to form a catalytically competent complex. Binding affinity to the substrate defines the processivity of the enzyme.

4.3.3 Endo- Versus Exonucleases

Nucleases play an important role in the metabolism of nucleic acids in various substrates, e.g., sequence-specific endonuclease, topology-specific topoisomerase, damage-sensitive nuclease, and structure-specific flap or apurinic/apyrimidinic (AP) endonucleases.

Nucleases are classified as either endo or exonucleases. Exonucleases begin acting from a free terminus of DNA or RNA, producing mono- and/or oligonucleotides of regular sizes, whereas endonucleases act anywhere within DNA or RNA, producing oligonucleotides. For this reason, exonucleases can act on a linear nucleotide chain but not on a closed circular DNA or RNA, whereas endonucleases can act on both closed and linear molecules. Exonucleases repeatedly digest their substrates from either the 3′ or 5′ end in a stepwise manner, whereas endonucleases work randomly on internal phosphodiester bonds within a chain, producing nucleic acid fragments with different lengths.

Some nucleases possess nucleolytic function with single or separated catalytic domains. Examples with both endo- and exofunctions include exonuclease III, exonuclease V, RNase H, and Rrp44, but the mechanistic basis for their endo- and exonuclease activities remains to be determined. A property for performing as an exonuclease might be exophilic so that it possesses strong affinity toward open-free termini that contain either a 5′ phosphate or a 3′ hydroxyl group. This may allow exonucleases to translocate in a processive or distributive manner.

Exonucleases may start to digest a DNA chain either from the 3′ end or from the 5′ end. They may further be classified based on a preference for single- or double-stranded structure and further categorized into sequence-specific and structure-specific nucleases. Exonucleases possess the ability to continuously digest nucleic acids from the 5′-end or from the 3′-end based on their binding preference of substrate terminus. The directionality or mode of cleavage action is also correlated with the ability of the enzyme to translocate along its substrate.

4.3.4 Structural Basis of Processive Versus Distributive Degradation

Exonucleases carry out either distributive or processive activity. Processive activity is defined by the ability to completely degrade a nucleic acid molecule before acting on a new nucleic acid chain. Enzymes unable to complete the degradation of their substrate molecules are thought to be distributive.

To perform a processive activity, a topological linkage or stable affinity between an enzyme and its substrate is a key feature. Ring-shaped oligomerization is a common strategy involved in processive reactions of nucleic acid substrates, such as λ and RecE exonucleases in DNA degradation, and RNA exosome and polynucleotide phosphorylase in RNA degradation [101]. As observed for monomeric nucleases, the similar topological coupling can be accomplished by plugging a single-stranded tail of nucleic acids through the narrow entrance to the active site (see ssRNA inserted into the proteins in Fig. 4.3a, b). This physical threading prevents dissociation of the enzyme from the substrate and aids a series of nucleic acid degradation. This phenomenon thus provides the structural basis of a stable interaction of the enzyme-substrate complex, allowing processive degradation. Two processive ribonucleases are exemplified as shown in Fig. 4.3a, b (Rrp44 and XRN1, respectively), whereas two examples of distributive enzymes are displayed in Fig. 4.3c, d (RNase H and exonuclease III, respectively). The main difference is the degree of coupling between the enzymes and their substrates. However, both processive and distributive activities are carefully optimized to coordinate timely degradation and accurate processing during nucleic acid metabolism. Functional cooperation in modulating enzyme activity is especially important for those enzymes carrying multi-catalytic domains.

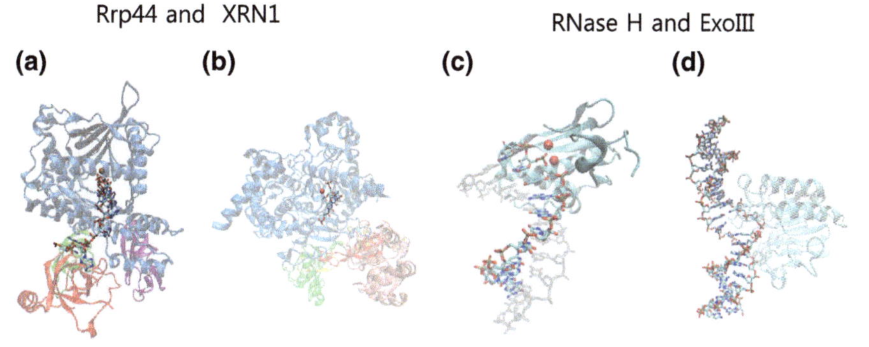

Fig. 4.3 Structural difference between processive and distributive exonucleases. The degree of topological coupling between an enzyme and its substrate determines the appropriate binding affinity for either processive or distributive enzyme. **a–b** Two processive exoribonucleases (Rrp44 (**a**) and XRN1(**b**), respectively). Plugging of ssRNA into the active site of enzymes prevents dissociation of the enzymes from their substrates, enabling processive degradation. **c–d** Two distributive exoribonucleases (RNase H (**c**) and exonuclease III (**d**), respectively). Both nucleases perform distributive degradation due to a lack of tight coupling between the enzyme and its substrate

4.3.5 Multiple Phases of the Exonucleolytic Cycle

Enzymes are continuously recycled as they convert substrates to products. During exonuclease activity, the enzymes perform a series of chemo-mechanical actions, such as binding, fraying, cutting, translocation, and dissociation. For this reason, enzyme actions are dissected into characteristic multi-phases that are temporally distinctly separated, such as initiation, elongation, and termination. In particular, exonuclease-digesting dsDNA or dsRNA possesses the ability to unpair the terminal junction and translocate along the substrate in a directional manner, similar to helicase activity. Helicases unwind double-stranded DNA or RNA using chemical energy released from ATP hydrolysis, while exonucleases accomplish this by exonuclease hydrolysis-coupled translocation. Thus, these enzymes are powered by digesting their own substrates. The chemo-mechanical conversion between hydrolysis and translocation for next cleavage might be dictated by conformational changes.

In general, the degradation rate should be determined by the time taken for hydrolytic steps plus the time taken for translocation steps. In the case of duplex degradation, melting might be a rate-limiting step and may precede translation [66]. Highly coordinated conformational transition may execute melting to overcome the energy barrier between melted and annealed states. Key protein residues serving as a wedge destabilize the duplex junction and accelerate the separation of duplex structures. In the coordinated reaction, all steps should be characteristically unique and optimized for efficient catalysis. All steps have been evolutionally tuned for timely coordination with other reactions. Identifying a rate-limiting step in the reaction cycle provides a molecular mechanism by which a nuclease catalyzes multiple cleavage and melting reactions by the transition conformation.

A specific substrate structure would provide a nuclease its specificity for recognition and then subsequent enzymatic activation, known as a catalytically competent complex formation. This structural preference is the key molecular basis to prevent a catastrophic destructive rampage and exists to regulate nuclease activity because uncontrolled activity leads to genomic instability.

4.3.6 Understanding of Processivity and Multiple Phases by Lambda Exonuclease

Even though λ exonuclease (λ-exo) is not an exoribonuclease but rather an exodeoxyribonuclease, I will discuss it due to the catalytic similarity to provide many structural and functional insights into processivity and its multi-step reaction involved in nuclease activity. λ-exo rapidly digests one strand of duplex DNA in the 5′-3′ direction, producing a DNA intermediate with a 3′ single-stranded (ss) overhang to initiate homologous recombination (Fig. 4.4a) [102]. The hydrolytic reaction requires Mg^{2+} as a cofactor, but the translocation of λ-exo along the DNA does not require ATP, unlike other ATP-dependent motors. In fact, translocation is powered by the energy released from the cleavage reaction.

λ-exo is known to be a highly processive enzyme. The homo-trimeric ring of λ-exo is believed to be the origin of high processivity through the encirclement of the ring around the ssDNA generated [7]. The processivity of λ-exo is more than 3000 nucleotides per attempt [103]. Thus, numerous studies have focused on λ-exo as a model to uncover the mechanism of processive degradation. Several single-molecule techniques such as flow stretching [66], optical tweezers [104], and single-molecule FRET [101] have been successfully applied to study the processivity and the enzymatic kinetics of λ-exo. These studies considered the fact that the activity of λ-exo converts dsDNA to ssDNA, which results in a decrease in the time-averaged distance between the two ends of DNA. Thus, the hydrolysis rate of a single λ-exo can be monitored in real time by measuring the conversion from dsDNA to ssDNA.

The kinetics of single λ-exo was first studied by a flow-stretching method [66] to examine the enzymatic reaction in real time at the single-molecule level (Fig. 4.4b). The study revealed two findings that had never been observed previously in ensemble experiments: (1) The hydrolytic rate is strongly dependent on the local sequence composition of the substrate DNA, suggesting that base melting in the catalytic cycle is a rate-limiting step. (2) The catalytic rates of the individual λ-exo molecules displayed large fluctuations during the processive reaction, known as a dynamic disorder (bottom right in Fig. 4.4b).

Unlike ensemble measurements, single-molecule techniques actually allow the investigation of different individual enzymatic activities. In an effort to identify the fluctuation of enzymatic activity, researchers use different explanations such as heterogeneity, dynamic disorder, and static disorder. Heterogeneity in enzymology is a general term that describes fluctuations in molecular activities. Static disorder

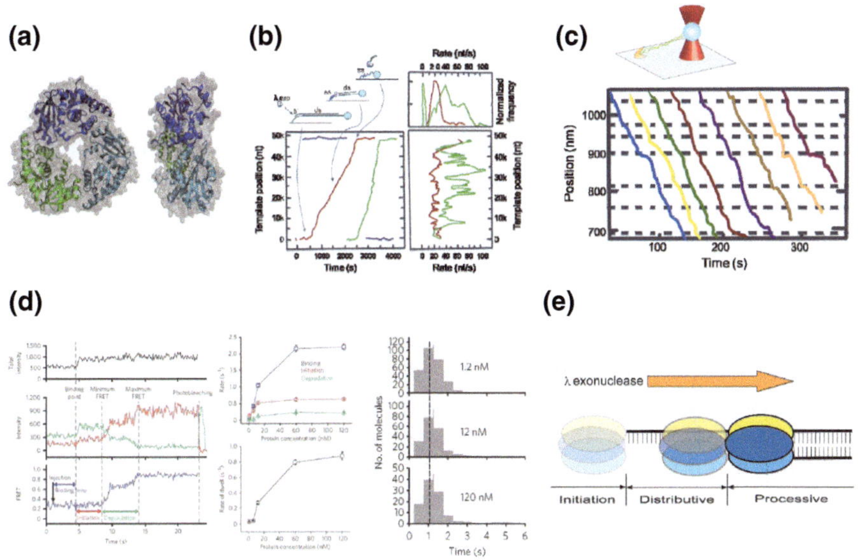

Fig. 4.4 Single-molecule observation of DNA degradation by λ exonuclease. **a** Top-to-bottom view (left) and side view (right) of the crystal structure of λ exonuclease. **b** Experimental scheme and data obtained by a flow-stretching method. The change in the bead position is monitored in real time to estimate the degree of DNA degradation (top left). Two representative traces (red and blue), displaying DNA degradation with ~400-nucleotide resolution (bottom left). Time derivatives of the two traces (bottom right). Histograms of the degradation rates derived from the two traces shown in the bottom right (top right). The black curve is the degradation rate calculated from experimental uncertainty. **c** Sequence-specific pauses measured from optical tweezers during DNA degradation. **d** λ-exonuclease performs concentration-dependent initiation and degradation before complete engagement to DNA. Single-molecule FRET-time trajectory illustrating how binding, initiation, and degradation times are determined (left). Inversed characteristic times of binding, initiation, and degradation as a function of protein concentration (middle top). Pause duration versus protein concentration (middle bottom). The protein concentration-dependent degradation suggests that the protein dissociates from the substrate. The histograms of degradation time show concentration independence of processive degradation after it is stably engaged with the DNA substrate (right). **e** Model with three phases of DNA degradation by λ exonuclease: initiation, distributive degradation, and processive degradation. **a**, **d**, **e** Reprinted with permission from Ref. [101]. **b** reprinted with permission from Ref., [66]. **c** reprinted with permission from Ref. [104]

[105] and dynamic disorder [66] are origins of this heterogeneity. Static disorder can be explained by the fact that the structures of individual proteins are slightly different, e.g., many long-lived conformers in a rugged protein folding landscape (called a structural memory effect). In contrast, the dynamic disorder is attributed to fluctuations in dynamic conformers due to differences in enzyme structure by physical interactions (e.g., physical binding geometry) and chemical interactions (e.g., electrostatic, dipole-dipole, and van der Waals). Molecular heterogeneity might be important to understand the intrinsic properties of nucleases.

The enzymatic cycle of λ-exo consists of hydrolytic scission, base pair melting, nucleic release, and 5′ to 3′ translocation along the DNA. However, the order of this reaction cycle has remained unknown. The rate-limiting step during degradation is either the pre- or post-cleavage melting of the terminal base pair. A recent structural study [106] showed that melting occurs before cleavage. Together with all the data available, the order of the whole reaction cycle might be determined as follows: base pair melting → scission → nucleotide release → 5′ to 3′ translocation, where the base pair melting is chemo-mechanically coupled with the 5′ to 3′ translocation.

The processive reaction of λ-exo was also studied by the optical tweezers technique [104]. One study found that the degradation rate of λ-exo over a long DNA molecule was not gradual, but rather showed frequent pauses (Fig. 4.4c). The pauses were rescued after various time delays. Close examination of individual traces revealed that the pauses appeared in a specific sequence-dependent manner [104]. GGCGA was identified as a common sequence motif resulting in the pauses, which is also found in the left cohesive end of the phage λ gene (cosNL: GGGC**GGCGA**CCTC). Thus, it was proposed that this sequence may serve as a possible inhibitory regulation site for lambda recombination.

The follow-up smFRET study further examined the pause and rescue mechanism [101] by which the enzyme is trapped and later rescued from the pause state. The smFRET study found that λ-exo often moved backward before it passed the pause site, reminiscent of RNA polymerase backtracking. The backtracking of λ-exo may be driven by diffusion because DNA hydrolysis could not be involved in the backward movement. The pause rescue process was rate-limiting with a time constant of ~24.2 s [101]. The propensity of λ-exo to pause at specific sequences might be due to the fact that the enzyme must register each nucleotide to digest DNA, one at a time. The precise visit of each nucleotide may cause λ-exo to strongly bind the particular DNA sequence so that the enzyme pauses in a sequence-dependent manner. A previous study proposed that a residue near the protein active site may intercalate between two adjacent guanosine bases so that a tight ring-stacking interaction forms along the pause sequence [104].

The studies using flow stretching and optical tweezers only focused on the processive degradation phase of the enzyme. However, many enzymatic reactions consist of multiple phases such as initiation, elongation (processive phase), and termination [107, 108]. The activities of nucleases have not been dissected, and nucleases were only classified as either distributive or processive enzymes. The smFRET study [101] dissected the activity of λ-exo, and binding, initiation, and degradation times were temporally assigned (left panel in Fig. 4.4d). Their inversed characteristic times, before λ-exo was completely engaged with its DNA substrate, showed protein concentration dependence (middle panel in Fig. 4.4d). In contrast, the degradation, after λ-exo was completely engaged with the substrate, was concentration-independent, indicating processive degradation (right panel in Fig. 4.4d). Taken together, the results of the study revealed that λ-exo performs three distinct phases: initiation (forming a functional complex), distributive (displaying a high tendency to dissociate), and processive degradation phases (Fig. 4.4e). Before the complete threading of the 3′ non-hydrolyzed strand through the trimeric ring, the DNA substrate is degraded in a distributive manner, and processive degradation then begins upon complete threading.

The smFRET study also found that the degradation rate at a mismatched base pair is ~fivefold slower than that at an intact base pair, suggesting that base-pairing and stacking interactions play a stimulatory role during the DNA degradation by λ-exo, possibly by aligning the 5′ end of the degradation strand toward the active site of the protein. This result supports the idea that B-form helicity serves as a guidance, directing the 5′ end to the active site. It is unknown how the three active sites of the enzyme are coordinated during degradation. It is possible that the three active sites are sequentially rotated along the B-form helicity, similar to the movement of a screw threading. In this way, the activity of the enzyme would be enhanced compared to the activity of random hydrolysis by the three active sites.

In summary, the single-molecule studies on λ-exo have provided a new molecular-level insight into the mechanism of exonuclease underpinning nuclease degradation process and revealed a new method of conceptualizing the molecular bases for processive enzymes related to the time it takes for the protein to form a competent complex on DNA that efficiently carries out catalysis. In particular, the smFRET study provided a whole picture of the degradation reaction beyond mere processive degradation.

4.3.7 ssRNA Degradation and Processivity by Archaeal Exosome

RNA exosomes function in the biogenesis, turnover, and processing of various RNA species in eukaryotes and archaea. In eukaryotes, the exosome comprises a 9-subunit core and is equipped with either Rrp44, Rrp6, or both subunits, making a ten- and eleven-subunit complex [5, 109]. The nine-subunit core is formed into a double ring-like structure [110], functioning as a multi-functional scaffold while Rrp44 and Rrp6 serve as $3' \rightarrow 5'$ catalytically active RNases. Exosome threads an unfolded RNA substrate into its internal channel of the nine-subunit core for degradation. To do so, the eukaryotic exosome interacts with two ATP-dependent RNA unfoldases: the cytoplasmic SKI complex and the nuclear TRAMP complex. The RNA unfoldases remove secondary structures of RNA and feed the unstructured RNA chain through the channel of the exosome. Thus, the system of the RNA exosome is quite similar to that of the proteasome [111] in a sense that the unfoldase eliminates secondary structures of substrates, whereas the RNA exosome degrades unstructured linear chains via hydrolysis afterward. Without the aid of ATP-dependent unfoldases, the exosome is also able to remove various secondary structures by pulling the 5′ strand at the constricted entrance pore of the channel using processive degradation activity. Tension would be generated by a series of translocations, one nucleotide at a time, and accumulated via geometric occlusion of the protein–substrate complex. The complex can be deformed accordingly until RNA structures break by bursting. Upon unfolding of structured RNA, the tension is released, based on a spring-loaded mechanism, as previously proposed for several helicases.

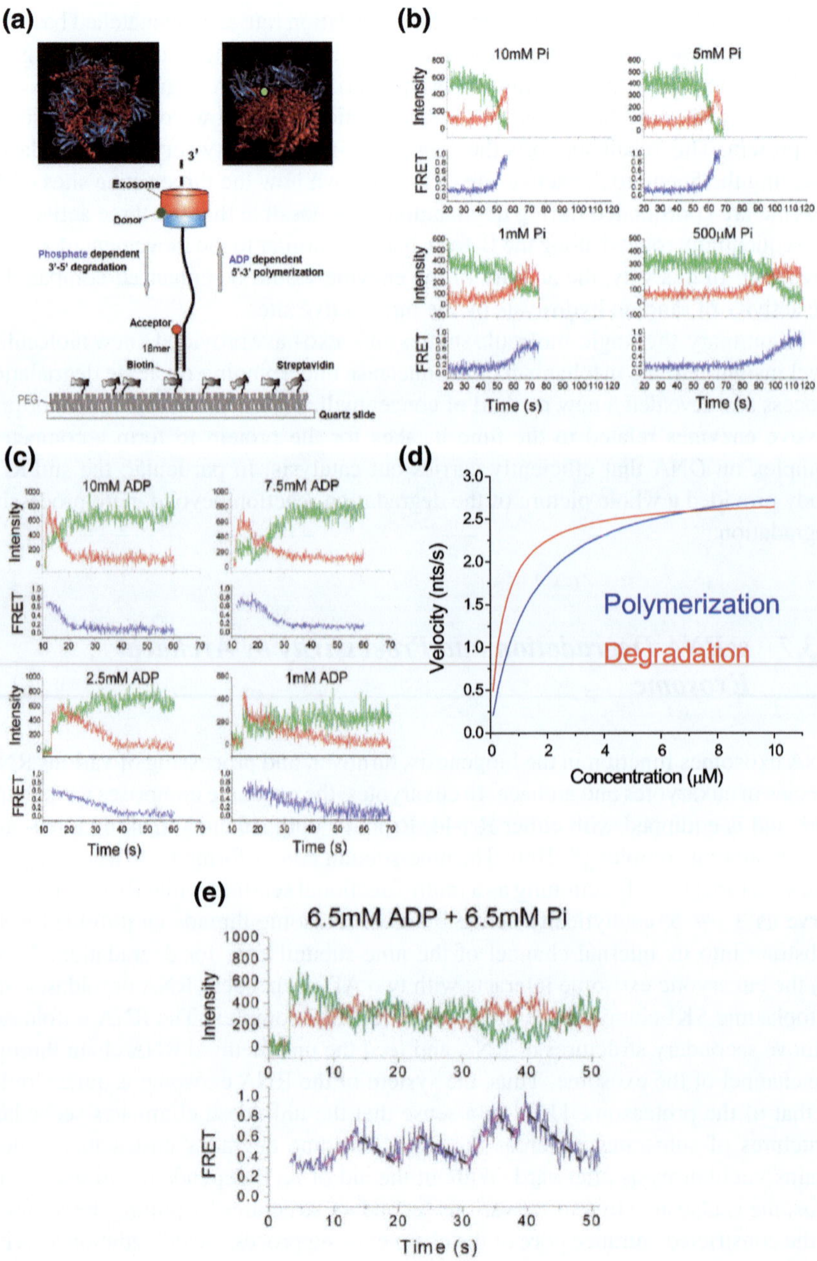
(a)
3'
Exosome
Donor
Phosphate dependent 3'-5' degradation
ADP dependent 5'-3' polymerization
Acceptor
18mer
Biotin
Streptavidin
PEG
Quartz slide
(b)
10mM Pi
5mM Pi
1mM Pi
500μM Pi
Intensity
FRET
Time (s)
(c)
10mM ADP
7.5mM ADP
2.5mM ADP
1mM ADP
Intensity
FRET
Time (s)
(d)
Velocity (nts/s)
Polymerization
Degradation
Concentration (μM)
(e)
6.5mM ADP + 6.5mM Pi
Intensity
FRET
Time (s)

◀**Fig. 4.5** Real-time observation of processive RNA degradation by exoribonucleases. **a** Crystal structure of the archaeal Csl4 exosome (PDB code, 2BA1): top-to-bottom view (top left) and side view (top right); and experimental design showing reversible polymerization and degradation reactions. Archaeal exosome degrades RNA in the 3′ to 5′ direction in the presence of free inorganic phosphate (Pi), whereas it polymerizes RNA in the opposite direction in the presence of ADP or other NDPs. **b** FRET-time trajectories of degradation at various Pi concentrations. **c** FRET-time trajectories of polymerization at various ADP concentrations. **d** Both polymerization (blue) and degradation (red) reactions follow Michaelis–Menten kinetics, suggesting that there is no cooperative behavior among the three active sites. Superposition of both reactions generates a crossing point as a pseudo-equilibrium, where polymerization and degradation velocities are the same, at 6.5 mM ADP and 6.5 mM Pi. **e** A representative FRET-time trace at the pseudo-equilibrium, where the degradation and polymerization reactions are equally favorable, reveals a preference for one reaction over the other, suggesting a memory effect. **a**–**e** Reprinted with permission from Ref. [55]

The archaeal RNA exosome is a nine-subunit ring-shaped complex [110] that performs RNA polymerization and degradation reactions in a reversible manner without recruiting any additional subunits as in the eukaryote exosome. The enzyme is arranged into a trimer of dimers with each dimer having Rrp41 and Rrp42 in a ABABAB fashion, making it a hexameric ring (Fig. 4.5a). However, only Rrp41 contains an active site, so the complex contains three active sites. The enzyme does not use a water molecule to attach the phosphodiester bond for the cleavage reaction. Rather, it phosphorolytically degrades RNA substrates by an attack onto the phosphodiester bond using inorganic phosphate [112], producing 5′ nucleoside diphosphates (NDPs) rather than nucleoside monophosphates (NMPs) by hydrolysis. In contrast, the reverse polymerization of RNA occurs in the presence of DNPs. Thus, the reactions are reversible depending on either Pi or NDPs.

Single-molecule FRET (smFRET) technique has been also used to study the archaeal RNA degradation machine [55], demonstrating the real-time locomotion of the nanomachine in both directions. The reversible locomotion is dictated by either polymerization or degradation of ssRNA (Fig. 4.5a). First, degradation of ssRNA was monitored by a change in distance between two fluorophores. Monotonically decreasing FRET without pauses suggested processive activity of degradation (Fig. 4.5b). If the enzyme dissociates from the substrate during degradation, a sudden disappearance of FRET signal appears before the reaction has been finished. A fit to Michaelis–Menten kinetics yields a V_{max} of 2.8 nt/s and a K_m of 450 μM (Fig. 4.5d). Second, polymerization of the enzyme was also measured in the presence of ADP. Representative FRET-time trajectory showed gradually increasing FRET signals, suggesting that the polymerization is indeed processive as in degradation (Fig. 4.5c). A fit to Michaelis–Menten kinetics provides a V_{max} of 3.0 nt/s and a K_m of 1.3 Mm (Fig. 4.5d). Both polymerization and degradation data confirm that the extended binding of the surface along the central hole of the exosome would provide processivity. Despite the multimeric structure containing three active sites, the enzyme followed the Michaelis–Menten kinetics with a maximum speed of ~3 nucleotides per second, showing that the three catalytic cylinders fired independently (Fig. 4.5d). The formation of the double ring-structure does not provide any cooperative activity but rather offers processivity and specificity via threading of 3′ polyA tail onto the core.

Superposition of both polymerization and degradation creates a crossing point at which they have the same velocity by ~6.5 mM of Pi and ADP each, as a pseudo-equilibrium point (Fig. 4.5d). In other words, the enzyme prefers degradation below ~6.5 mM, whereas it favors polymerization above ~6.5 mM if the same amount of both cofactors is added (Fig. 4.5d). Cofactor preference was tested at this pseudo-equilibrium where the rates of both reactions were equal. Since the affinity of Pi and ADP onto the active site of the enzyme is the same at ~6.5 mM, both reactions might be balanced out by adding one nt and removing one nt with an equal probability. Unexpectedly, a series of different phases was observed where one of the two reactions is more overriding, as evidenced by spans of FRET increase and decrease that switch (Fig. 4.5e). The enzyme 'remembered' the previous reaction it catalyzed and stochastically switched between periods of favoring degradation and polymerization, as previously attributed to a memory effect [113]. What would be the molecular basis of this memory effect? It should be related to a change in structure that persists for a time period longer than the timescale of an enzymatic cycle. Then, this phenomenon can be accounted for by the following explanation. Since the pseudo-equilibrium is located at ~6.5 mM in a kinetically saturated region, association and dissociation of ADP and Pi into the active site are very quick but the structural relaxing from the cofactor-bound state (holo) to the cofactor-unbound state (apo) might be relatively slow. The relative cellular concentration between Pi and ADP would determine the direction between the two activities, but the exact mechanism of how the enzyme switches from one to the other in vivo is unknown.

This study [55] was the first single-molecule analysis of an exoribonuclease and the first example of a reversible and controllable biological motor with single-molecule precision and clarity. The measurements provided new and important information on the archaeal exosome's intrinsic properties, including enzymatic speed, processivity, cooperativity, and stochasticity. Further, the method can be directly applied to studies of other RNA processing enzymes that play crucial roles in regulating gene expression.

4.3.8 Asymmetric Inchworm Mechanism by Polynucleotide Phosphorylase (PNPase)

PNPase is a trimeric ring-structured exoribonuclease with a central channel, homologous to archaeal and eukaryotic exosomes [114]. Each unit of the trimer contains a KH and an S1 RNA-binding domain at the *C*-terminus, both of which are attached to the *N*-terminal core of the enzyme by flexible linkers [115]. These RNA-binding domains are equivalent to trimeric cap proteins in RNA exosomes, provide RNA targeting specificity, and help the RNA substrate to the central channel via a 'hands gripping a rope mechanism.'

Polynucleotide phosphorylase (PNPase) degrades various transcripts, including mRNA, rRNA, and structural RNA in bacterial cells, similar to the action of bacterial RNase R. Even though PNPase works with other cofactor proteins such as DEAD-box helicase RhlB as a degradosome to improve efficiency and processivity of degradation, it has been reported that this protein is capable of degrading structured RNA molecules in the absence of cofactor proteins. Thus, the mechanism PNPase uses to digest the structured regions of RNA without the aid of the other cofactors remains to be determined.

The unwinding and degradation mechanisms of PNPase were investigated by an optical-trapping assay that can measure a change in contour length during the structural conversion from dsRNA to ssRNA via processive degradation (Fig. 4.6a). This assay provided real-time trajectories, showing degradation rate, time delays due to the stability of structured RNA, and processivity with near-nucleotide resolution. dsRNA and DNA–RNA hybrid constructs of 155 bp were measured in the presence of PNPase under a 20-pN load by an optical force clamp (Fig. 4.6a). The enzyme degraded both dsRNA and DNA–RNA substrates on the average of ~23 nt as a processivity, and degradation rates were ~129 nt/s and ~121 nt/s, respectively, showing almost the same rates within error margins. When the 3′ upstream portion with 70% AU was increased from the first 26 bp to the first 50 bp, the enzyme processivity increased from ~23 to ~62 nt, while digesting at the similar rate of ~137 nt/s. Periodogram analysis (i.e., a Fourier analysis of the distribution of extension changes) revealed similar step sizes of ~6.1 nt and ~6.6 nt with the first 26 and 50 bp for both 70% AU substrates, respectively (Fig. 4.6b).

Next, KH and S1 RNA-binding domains of PNPase, known to be important for nucleolytic activity, were truncated to investigate their functional roles during degradation of structured RNA. The PNPase ΔKH-ΔS1 mutant showed a half of the processivity (~14 bp) on dsRNA compared to the WT PNPase, but its degradation rate remained the same as that of the WT PNPase. The reduced activity of the mutant was attributed to reduced binding affinity. Interestingly, the processivity of PNPase did not depend on the type of duplexes such as dsRNA and DNA–RNA hybrid, but it was strongly dependent on the % AU content of substrates, suggesting a correlation with the thermodynamic stability of the substrate [116].

Based on steps of ~six nucleotides, the researchers proposed an asymmetric inchworm mechanism (Fig. 4.6c), where the catalytic core degrades the unstructured substrate one nt at a time, progressively moving up to the bottom of KH/S1 domains, and KH/S1 domains leap off due to tension. Then, KH and/or S1 domains advance and bind to the 5′ side of the structured substrate, and melt six or seven base pairs via the binding free energy of KH and S1 domains, and the core degrades again in the 5′ direction to repeat the inchworm cycle. The degradation by the PNPase is a slow step while the melting event is a relatively fast event, since the assay could not detect a six or seven base pair melting step directly during the degradation reaction with the processivity of 62 nt. In addition, the fact that the ΔKH-ΔS1 mutant shows the same degradation rate as WT but it does not display an obvious delay in the melting step suggests that degradation is indeed rate-limiting (Fig. 4.6c).

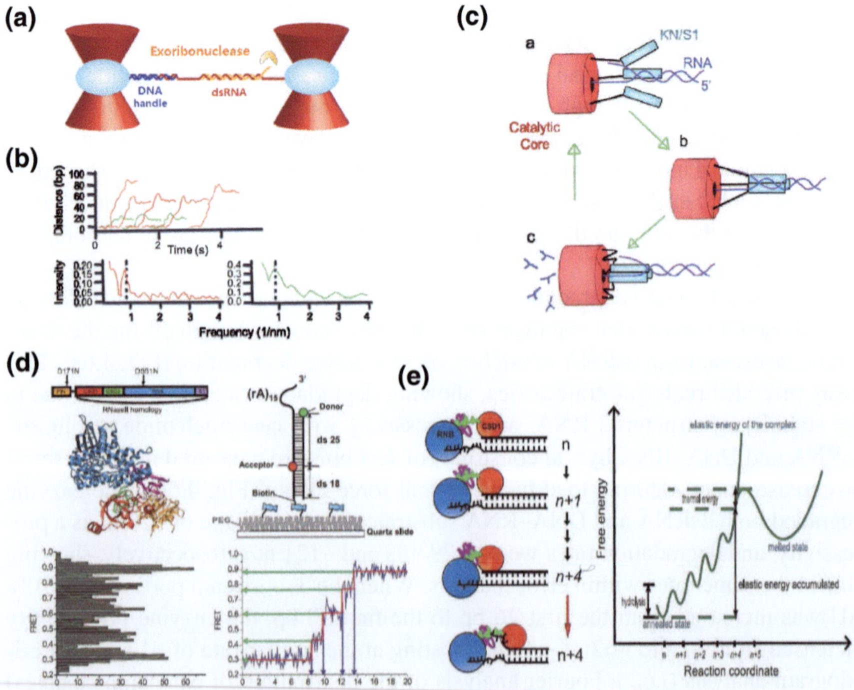

Fig. 4.6 Asymmetric inchworm mechanism for polynucleotide phosphorylase (PNPase) versus spring-loaded mechanism for Rrp44 nuclease. **a** Exoribonuclease assay via duel optical tweezers where a 155-bp segment of dsRNA is subjected under tension between two polystyrene beads. An RNA overhang containing a 3′ adenine-rich single-stranded stretch serves as a binding site for initiation and is degraded as an energy source for the exoribonuclease to mechanically unwind and processively digest the dsRNA in the 3′ to 5′ direction. As the PNPase degrades the RNA substrate, the distance between two trapped beads increases due to a structural conversion between ssRNA and dsRNA. **b** Single-molecule trajectories showing the process during which PNPase digests 155-bp DNA–RNA hybrid constructs. The nucleases processively degrade ~23 nt of the substrate in which the first 26 nt is composed of 70% AU (blue, middle panel), whereas they efficiently digest ~62 nt of the substrate in which the first 50 nt comprises 70% AU (red, middle panel). Periodogram analyses reveal a step size of ~6.6 nt for the construct with 50% AU content (red, bottom left panel) and that of ~6.1 nt for the construct with 70% AU content (blue, bottom right panel). **c** Asymmetric inchworm degradation model for the PNPase complex. PNPase consists of three KH/S1 domains (each one with elongated rod shape) and catalytic core ring (cylinder shape). Three KH/S1 domains bind and melt to the downstream of the dsRNA substrate ~six–seven base pairs, and the catalytic core digests as it advances in the 3′ to 5′ direction until it reaches the KH/S1 binding domains. When the core catches up, the binding domains are released and rebind further downstream along the RNA to continue another round of degradation. **d** Rrp44 unwinds RNA in four-nucleotide steps. The structure of Rrp44 and its domain composition (top left); DNA construct labeled with donor (Cy3) at the single strand and duplex junction and acceptor (Cy5) at 25 bp into the duplex (top right); histogram of FRET states stayed by Rrp44 during unwinding and degradation process (bottom left); and a representative FRET-time trajectory displaying FRET states and dwell times (bottom right). **e** Spring-loaded mechanism for duplex unwinding coupled with RNA degradation. Rrp44 transforms and combines a series of chemical energy, occurring during ~four cleave reactions, into a reservoir of elastic energy within the protein–RNA complex. The successive enzymatic actions allow Rrp44 to convert chemical energy released from RNA hydrolysis into accumulated elastic energy. **f** Elastic energy accumulated versus number of nucleotides digested with duplex unwinding. Adding thermal fluctuations to elastic energy accumulated triggers passing the energy barrier for melting. **a–c** Reprinted with permission from Ref. [55]. **d–f** Reprinted with permission from Ref. [56]

The asymmetric inchworm mechanism is different from the spring-loaded mechanism proposed in other enzyme systems [56, 71]. First, melting steps are ~six-seven bp and ~three-four bp for the asymmetric inchworm and the spring-loaded mechanisms, respectively. If melting/unwinding steps become small, the structured RNA can be completely melted into a single-stranded chain. Then, local degradation would be much faster. However, if melting steps are large, structured regions would remain locally and degradation would take longer. To rapidly degrade unwanted transcripts, an RNA helicase protein can be recruited into enzyme complexes to kinetically accelerate the degradation reaction, similar to the degradosome and exosome complex.

4.3.9 Chemo-mechanical Structured RNA Degradation by an Exoribonuclease

RNA degradation is essential for gene regulation and is performed by many different classes of ribonucleases [117]. It is a complex process involved in multiple pathways carried out by nuclease complexes with cofactors, since RNA tends to form many types of structures. A class of ribonuclease such as Rrp44 is able to carry out highly complex mechanical tasks in a processive and synchronized manner so that nucleases can degrade structured RNA. Rrp44 is the only catalytic active subunit of the yeast and human ten-subunit exosome (top left panel in Fig. 4.6d). This subunit is a 3′ to 5′ exonuclease that digests RNA substrates and moves in single-nucleotide increments. It also unwinds duplex RNA while steadily digesting the 3′ end of RNA [118]. It does not use ATP but instead burns the bridge behind the reaction, and thereby cannot move backward, unlike helicases.

To digest structured RNA, the enzyme must carry out coordinated RNA degradation and unwinding. How does Rrp44 couple its RNA degradation activity to its unwinding? Does it unwind double-stranded RNA in single-base steps, the same as degradation steps that occur one base at a time? A smFRET study investigated these questions using a duplex RNA with a 3′-polyA overhang (top right panel in Fig. 4.6d) and discovered a surprising mechanism of elastic coupling between RNA degradation and unwinding [56]. Unwinding occurred in ~four base pair steps even though the enzyme motion is fueled by RNA degradation in single-base steps (bottom panels in Fig. 4.6d). To coordinate this discrepancy between degradation and unwinding step sizes, unwinding should not occur until four nucleotides are degraded. Therefore, Rrp44 must store elastic energy released by hydrolysis during four steps of single-nucleotide degradation (Fig. 4.4e). The accumulated elastic energy triggers the unwinding of four base pairs in a burst. The unwinding reaction is not only rate-limiting, but also a thermally driven process (Fig. 4.6f). The concept that an enzyme stores elastic energy through a series of chemical hydrolysis events during cycles was also proposed for an NS2 helicase [71]. Biologically, such a mechanism would allow the enzyme to elastically accumulate energy and to overcome the free energy barriers that are too strong to overcome using the energy released from the hydrolysis of one nucleotide. Thus, the enzyme functions here as a chemo-mechanical machine that converts and combines a series of chemical energy releases from hydrolysis of the RNA chain into an accumulation of elastic energy for the unwinding reaction. From the viewpoint of proteins as 'nanomachines,' the data support the notion that proteins can behave like 'springs,' a component I often see in other man-made machines.

4.4 Conclusions

This review showcases how recent technical advances in single-molecule techniques have created new opportunities for studying detailed mechanistic elements of exonucleases, such as their catalytic and mechanical phases. The real power of these techniques is the ability to directly monitor not only the motions of enzymes but also structural changes in their substrates, in real time with unprecedented precision and clarity, which allows the understanding of spatiotemporal coordination of their activities. To date, single-molecule techniques have been successfully applied to study many fundamental aspects of underlying mechanisms, such as heterogeneity and stochasticity, transient intermediates, and multiple kinetic steps involved in hydrolytic reactions. However, significant advances have been made in vitro, which are different from physiological environments in vivo. An important future direction will be 'in vivo-like biochemistry' that closely mimics the actual cellular context in cells to trace biochemical pathways and the regulation of biological processes. Technical advances at the single-molecule level will provide the most accurate picture of dynamics and interactions in biological processes of living cells. There is little doubt that in vivo-like experiments will revolutionize our understanding of true dynamics and cellular mechanisms governing living cells. Advances in single-molecule imaging techniques will give rise to a new era that enables the biophysical motto: 'probing real-time dynamics of cellular processes in living cells.'

Acknowledgements This work was supported by the GIST Research Institute (GRI) and the National Research Foundation of Korea (NRF) grant funded by the Korea government (NRF-2016R1A2B4014556) and by the grant from the Korean Health Technology R&D Project, Ministry of Health and Welfare, Republic of Korea (HA17C0031: #1720050).

References

1. Fried, V. A., Smith, H. T., Hildebrandt, E., & Weiner, K. (1987). Ubiquitin has intrinsic proteolytic activity: Implications for cellular regulation. *Proceedings of the National Academy of Sciences of the United States of America, 84*(11), 3685–3689.
2. Barthelme, D., & Sauer, R. T. (2012). Identification of the Cdc48-20S proteasome as an ancient AAA+ proteolytic machine. *Science, 337*(6096), 843–846.
3. Levchenko, I., Seidel, M., Sauer, R. T., & Baker, T. A. (2000). A specificity-enhancing factor for the clpXP degradation machine. *Science, 289*(5488), 2354–2356.

4. Saffarian, S., Collier, I. E., Marmer, B. L., Elson, E. L., & Goldberg, G. (2004). Interstitial collagenase is a Brownian ratchet driven by proteolysis of collagen. *Science, 306*(5693), 108–111.
5. Bonneau, F., Basquin, J., Ebert, J., Lorentzen, E., & Conti, E. (2009). The yeast exosome functions as a macromolecular cage to channel RNA substrates for degradation. *Cell, 139*(3), 547–559.
6. Makino, D. L., Baumgartner, M., & Conti, E. (2013). Crystal structure of an RNA-bound 11-subunit eukaryotic exosome complex. *Nature, 494*(7439), 70–75.
7. Kovall, R., & Matthews, B. W. (1997). Toroidal structure of lambda-exonuclease. *Science, 277*(5333), 1824–1827.
8. Frazao, C., McVey, C. E., Amblar, M., Barbas, A., Vonrhein, C., Arraiano, C. M., et al. (2006). Unravelling the dynamics of RNA degradation by ribonuclease II and its RNA-bound complex. *Nature, 443*(7107), 110–114.
9. Xiang, S., Cooper-Morgan, A., Jiao, X., Kiledjian, M., Manley, J. L., & Tong, L. (2009). Structure and function of the $5' \rightarrow 3'$ exoribonuclease Rat1 and its activating partner Rai1. *Nature, 458*(7239), 784–788.
10. Doshi, U., McGowan, L. C., Ladani, S. T., & Hamelberg, D. (2012). Resolving the complex role of enzyme conformational dynamics in catalytic function. *Proceedings of the National Academy of Sciences of the United States of America, 109*(15), 5699–5704.
11. Klinman, J. P. (2013). Importance of protein dynamics during enzymatic C-H bond cleavage catalysis. *Biochemistry, 52*(12), 2068–2077.
12. Cornish, P. V., Ermolenko, D. N., Noller, H. F., & Ha, T. (2008). Spontaneous intersubunit rotation in single ribosomes. *Molecular Cell, 30*(5), 578–588.
13. Yin, Y. W., & Steitz, T. A. (2002). Structural basis for the transition from initiation to elongation transcription in T7 RNA polymerase. *Science, 298*(5597), 1387–1395.
14. Hodges, C., Bintu, L., Lubkowska, L., Kashlev, M., & Bustamante, C. (2009). Nucleosomal fluctuations govern the transcription dynamics of RNA polymerase II. *Science, 325*(5940), 626–628.
15. Marszalek, P. E., Lu, H., Li, H., Carrion-Vazquez, M., Oberhauser, A. F., Schulten, K., et al. (1999). Mechanical unfolding intermediates in titin modules. *Nature, 402*(6757), 100–103.
16. Nahas, M. K., Wilson, T. J., Hohng, S., Jarvie, K., Lilley, D. M. J., & Ha, T. (2004). Observation of internal cleavage and ligation reactions of a ribozyme. *Nature Structural and Molecular Biology, 11*(11), 1107–1113.
17. Laurence, T. A., Kong, X., Jager, M., & Weiss, S. (2005). Probing structural heterogeneities and fluctuations of nucleic acids and denatured proteins. *Proceedings of the National Academy of Sciences of the United States of America, 102*(48), 17348–17353.
18. Rothwell, P. J., Berger, S., Kensch, O., Felekyan, S., Antonik, M., Wöhrl, B. M., et al. (2003). Multiparameter single-molecule fluorescence spectroscopy reveals heterogeneity of HIV-1 reverse transcriptase: Primer/template complexes. *Proceedings of the National Academy of Sciences of the United States of America, 100*(4), 1655–1660.
19. Zhuang, X., Kim, H., Pereira, M. J. B., Babcock, H. P., Walter, N. G., & Chu, S. (2002). Correlating structural dynamics and function in single ribozyme molecules. *Science, 296*(5572), 1473–1476.
20. Onoa, B., Dumont, S., Liphardt, J., Smith, S. B., Tinoco, I., Jr., & Bustamante, C. (2003). Identifying kinetic barriers to mechanical unfolding of the *T. thermophila* ribozyme. *Science, 299*(5614), 1892–1895.
21. Myong, S., Rasnik, I., Joo, C., Lohman, T. M., & Ha, T. (2005). Repetitive shuttling of a motor protein on DNA. *Nature, 437*(7063), 1321–1325.
22. Zhuang, X., Bartley, L. E., Babcock, H. P., Russell, R., Ha, T., Herschlag, D., et al. (2000). A single-molecule study of RNA catalysis and folding. *Science, 288*(5473), 2048–2051.
23. Ha, T., Rasnik, I., Cheng, W., Babcock, H. P., Gauss, G. H., Lohman, T. M., et al. (2002). Initiation and re-initiation of DNA unwinding by the *Escherichia coli* Rep helicase. *Nature, 419*(6907), 638–641.

24. Tinoco, I., Jr., Li, P. T. X., & Bustamante, C. (2006). Determination of thermodynamics and kinetics of RNA reactions by force. *Quarterly Reviews of Biophysics, 39*(4), 325–360.
25. Keller, D., Swigon, D., & Bustamante, C. (2003). Relating single-molecule measurements to thermodynamics. *Biophysical Journal, 84*(2 I), 733–738.
26. Rief, M., Gautel, M., Oesterhelt, F., Fernandez, J. M., & Gaub, H. E. (1997). Reversible unfolding of individual titin immunoglobulin domains by AFM. *Science, 276*(5315), 1109–1112.
27. Zhao, Y., Terry, D. S., Shi, L., Quick, M., Weinstein, H., Blanchard, S. C., et al. (2011). Substrate-modulated gating dynamics in a Na^+-coupled neurotransmitter transporter homologue. *Nature, 474*(7349), 109–113.
28. Zhao, Y., Terry, D., Shi, L., Weinstein, H., Blanchard, S. C., & Javitch, J. A. (2010). Single-molecule dynamics of gating in a neurotransmitter transporter homologue. *Nature, 465*(7295), 188–193.
29. Lee, H. K., Yang, Y., Su, Z., Hyeon, C., Lee, T. S., Lee, H. W., et al. (2010). Dynamic Ca^{2+}-dependent stimulation of vesicle fusion by membrane-anchored synaptotagmin 1. *Science, 328*(5979), 760–763.
30. Yoon, T. Y., Okumus, B., Zhang, F., Shin, Y. K., & Ha, T. (2006). Multiple intermediates in SNARE-induced membrane fusion. *Proceedings of the National Academy of Sciences of the United States of America, 103*(52), 19731–19736.
31. Yoon, T. Y., Lu, X., Diao, J., Lee, S. M., Ha, T., & Shin, Y. K. (2008). Complexin and Ca^{2+} stimulate SNARE-mediated membrane fusion. *Nature Structural and Molecular Biology, 15*(7), 707–713.
32. Pandey, M., Syed, S., Donmez, I., Patel, G., Ha, T., & Patel, S. S. (2009). Coordinating DNA replication by means of priming loop and differential synthesis rate. *Nature, 462*(7275), 940–943.
33. Yardimci, H., Wang, X., Loveland, A. B., Tappin, I., Rudner, D. Z., Hurwitz, J., et al. (2012). Bypass of a protein barrier by a replicative DNA helicase. *Nature, 492*(7428), 205–209.
34. Finkelstein, I. J., Visnapuu, M. L., & Greene, E. C. (2010). Single-molecule imaging reveals mechanisms of protein disruption by a DNA translocase. *Nature, 468*(7326), 983–987.
35. Lee, J. B., Hite, R. K., Hamdan, S. M., Xie, X. S., Richardson, C. C., & Van Oijen, A. M. (2006). DNA primase acts as a molecular brake in DNA replication. *Nature, 439*(7076), 621–624.
36. Hamdan, S. M., Loparo, J. J., Takahashi, M., Richardson, C. C., & Van Oijen, A. M. (2009). Dynamics of DNA replication loops reveal temporal control of lagging-strand synthesis. *Nature, 457*(7227), 336–339.
37. Visnapuu, M. L., & Greene, E. C. (2009). Single-molecule imaging of DNA curtains reveals intrinsic energy landscapes for nucleosome deposition. *Nature Structural and Molecular Biology, 16*(10), 1056–1062.
38. Joo, C., McKinney, S. A., Nakamura, M., Rasnik, I., Myong, S., & Ha, T. (2006). Real-time observation of RecA filament dynamics with single monomer resolution. *Cell, 126*(3), 515–527.
39. Rothenberg, E., Grimme, J. M., Spies, M., & Ha, T. (2008). Human Rad52-mediated homology search and annealing occurs by continuous interactions between overlapping nucleoprotein complexes. *Proceedings of the National Academy of Sciences of the United States of America, 105*(51), 20274–20279.
40. Graneli, A., Yeykal, C. C., Robertson, R. B., & Greene, E. C. (2006). Long-distance lateral diffusion of human Rad51 on double-stranded DNA. *Proceedings of the National Academy of Sciences of the United States of America, 103*(5), 1221–1226.
41. Galletto, R., Amitani, I., Baskin, R. J., & Kowalczykowski, S. C. (2006). Direct observation of individual RecA filaments assembling on single DNA molecules. *Nature, 443*(7113), 875–878.
42. Bianco, P. R., Brewer, L. R., Corzett, M., Balhorn, R., Yeh, Y., Kowalczykowski, S. C., et al. (2001). Processive translocation and DNA unwinding by individual RecBCD enzyme molecules. *Nature, 409*(6818), 374–378.

43. Robertson, R. B., Moses, D. N., Kwon, Y., Chan, P., Chi, P., Klein, H., et al. (2009). Structural transitions within human Rad51 nucleoprotein filaments. *Proceedings of the National Academy of Sciences of the United States of America, 106*(31), 12688–12693.
44. Blainey, P. C., Van Oijen, A. M., Banerjee, A., Verdine, G. L., & Xie, X. S. (2006). A base-excision DNA-repair protein finds intrahelical lesion bases by fast sliding in contact with DNA. *Proceedings of the National Academy of Sciences of the United States of America, 103*(15), 5752–5757.
45. Gorman, J., Chowdhury, A., Surtees, J. A., Shimada, J., Reichman, D. R., Alani, E., et al. (2007). Dynamic basis for one-dimensional DNA scanning by the mismatch repair complex Msh2-Msh6. *Molecular Cell, 28*(3), 359–370.
46. Gorman, J., Wang, F., Redding, S., Plys, A. J., Fazio, T., Wind, S., et al. (2012). Single-molecule imaging reveals target-search mechanisms during DNA mismatch repair. *Proceedings of the National Academy of Sciences of the United States of America, 109*(45), E3074–E3083.
47. Abbondanzieri, E. A., Greenleaf, W. J., Shaevitz, J. W., Landick, R., & Block, S. M. (2005). Direct observation of base-pair stepping by RNA polymerase. *Nature, 438*(7067), 460–465.
48. Davenport, R. J., Wuite, G. J. L., Landick, R., & Bustamante, C. (2000). Single-molecule study of transcriptional pausing and arrest by *E. coli* RNA polymerase. *Science, 287*(5462), 2497–2500.
49. Bintu, L., Ishibashi, T., Dangkulwanich, M., Wu, Y. Y., Lubkowska, L., Kashlev, M., et al. (2013). Nucleosomal elements that control the topography of the barrier to transcription. *Cell, 151*(4), 738–749.
50. Larson, M. H., Greenleaf, W. J., Landick, R., & Block, S. M. (2008). Applied force reveals mechanistic and energetic details of transcription termination. *Cell, 132*(6), 971–982.
51. Neuman, K. C., Abbondanzieri, E. A., Landick, R., Gelles, J., & Block, S. M. (2003). Ubiquitous transcriptional pausing is independent of RNA polymerase backtracking. *Cell, 115*(4), 437–447.
52. Jeong, C., Cho, W. K., Song, K. M., Cook, C., Yoon, T. Y., Ban, C., et al. (2011). MutS switches between two fundamentally distinct clamps during mismatch repair. *Nature Structural and Molecular Biology, 18*(3), 379–385.
53. Cornish, P. V., Ermolenko, D. N., Staple, D. W., Hoang, L., Hickerson, R. P., Noller, H. F., et al. (2009). Following movement of the L1 stalk between three functional states in single ribosomes. *Proceedings of the National Academy of Sciences of the United States of America, 106*(8), 2571–2576.
54. Wen, J. D., Lancaster, L., Hodges, C., Zeri, A. C., Yoshimura, S. H., Noller, H. F., et al. (2008). Following translation by single ribosomes one codon at a time. *Nature, 452*(7187), 598–603.
55. Lee, G., Hartung, S., Hopfner, K. P., & Ha, T. (2010). Reversible and controllable nanolocomotion of an RNA-processing machinery. *Nano Letters, 10*(12), 5123–5130.
56. Lee, G., Bratkowski, M. A., Ding, F., Ke, A., & Ha, T. (2012). Elastic coupling between RNA degradation and unwinding by an exoribonuclease. *Science, 336*(6089), 1726–1729.
57. Maillard, R. A., Chistol, G., Sen, M., Righini, M., Tan, J., Kaiser, C. M., et al. (2011). ClpX(P) generates mechanical force to unfold and translocate its protein substrates. *Cell, 145*(3), 459–469.
58. Aubin-Tam, M. E., Olivares, A. O., Sauer, R. T., Baker, T. A., & Lang, M. J. (2011). Single-molecule protein unfolding and translocation by an ATP-fueled proteolytic machine. *Cell, 145*(2), 257–267.
59. Shin, Y., Davis, J. H., Brau, R. R., Martin, A., Kenniston, J. A., Baker, T. A., et al. (2009). Single-molecule denaturation and degradation of proteins by the AAA+ ClpXP protease. *Proceedings of the National Academy of Sciences of the United States of America, 106*(46), 19340–19345.
60. Sarkar, S. K., Marmer, B., Goldberg, G., & Neuman, K. C. (2012). Single-molecule tracking of collagenase on native type I collagen fibrils reveals degradation mechanism. *Current Biology, 22*(12), 1047–1056.

61. Funatsu, T., Harada, Y., Tokunaga, M., Saito, K., & Yanagida, T. (1995). Imaging of single fluorescent molecules and individual ATP turnovers by single myosin molecules in aqueous solution. *Nature, 374*(6522), 555–559.
62. Yildiz, A., Forkey, J. N., McKinney, S. A., Ha, T., Goldman, Y. E., & Selvin, P. R. (2003). Myosin V walks hand-over-hand: Single fluorophore imaging with 1.5-nm localization. *Science, 300*(5628), 2061–2065.
63. Veigel, C., Coluccio, L. M., Jontes, J. D., Sparrow, J. C., Milligan, R. A., & Molloy, J. E. (1999). The motor protein myosin-I produces its working stroke in two steps. *Nature, 398*(6727), 530–533.
64. Yasuda, R., Noji, H., Yoshida, M., Kinosita, K., Jr., & Itoh, H. (2001). Resolution of distinct rotational substeps by submillisecond kinetic analysis of F1-ATPase. *Nature, 410*(6831), 898–904.
65. Noji, H., Yasuda, R., Yoshida, M., & Kinosita, K., Jr. (1997). Direct observation of the rotation of F1-ATPase. *Nature, 386*(6622), 299–302.
66. van Oijen, A. M., Blainey, P. C., Crampton, D. J., Richardson, C. C., Ellenberger, T., & Xie, X. S. (2003). Single-molecule kinetics of lambda exonuclease reveal base dependence and dynamic disorder. *Science, 301*(5637), 1235–1238.
67. Hohng, S., Zhou, R., Nahas, M. K., Yu, J., Schulten, K., Lilley, D. M. J., et al. (2007). Fluorescence-force spectroscopy maps two-dimensional reaction landscape of the holliday junction. *Science, 318*(5848), 279–283.
68. Cecconi, G., Shank, E. A., Bustamante, C., & Marqusee, S. (2005). Biochemistry: Direct observation of the three-state folding of a single protein molecule. *Science, 309*(5743), 2057–2060.
69. Woodside, M. T., Anthony, P. C., Behnke-Parks, W. M., Larizadeh, K., Herschlag, D., & Block, S. M. (2006). Direct measurement of the full, sequence-dependent folding landscape of a nucleic acid. *Science, 314*(5801), 1001–1004.
70. Shi, J., Dertouzos, J., Gafni, A., Steel, D., & Palfey, B. A. (2006). Single-molecule kinetics reveals signatures of half-sites reactivity in dihydroorotate dehydrogenase A catalysis. *Proceedings of the National Academy of Sciences of the United States of America, 103*(15), 5775–5780.
71. Myong, S., Bruno, M. M., Pyle, A. M., & Ha, T. (2007). Spring-loaded mechanism of DNA unwinding by hepatitis C virus NS3 helicase. *Science, 317*(5837), 513–516.
72. Kim, S., Grant, R. A., & Sauer, R. T. (2011). Covalent linkage of distinct substrate degrons controls assembly and disassembly of DegP proteolytic cages. *Cell, 145*(1), 67–78.
73. Jiao, X., Xiang, S., Oh, C., Martin, C. E., Tong, L., & Kiledjian, M. (2010). Identification of a quality-control mechanism for mRNA 5′-end capping. *Nature, 467*(7315), 608–611.
74. Livak, K. J. (1999). Allelic discrimination using fluorogenic probes and the 5′ nuclease assay. *Genetic Analysis: Biomolecular Engineering, 14*(5–6), 143–149.
75. Li, J. J., Geyer, R., & Tan, W. (2000). Using molecular beacons as a sensitive fluorescence assay for enzymatic cleavage of single-stranded DNA. *Nucleic Acids Research, 28*(11).
76. Chemla, Y. R., Aathavan, K., Michaelis, J., Grimes, S., Jardine, P. J., Anderson, D. L., et al. (2005). Mechanism of force generation of a viral DNA packaging motor. *Cell, 122*(5), 683–692.
77. Moffitt, J. R., Chemla, Y. R., Aathavan, K., Grimes, S., Jardine, P. J., Anderson, D. L., et al. (2009). Intersubunit coordination in a homomeric ring ATPase. *Nature, 457*(7228), 446–450.
78. Aathavan, K., Politzer, A. T., Kaplan, A., Moffitt, J. R., Chemla, Y. R., Grimes, S., et al. (2009). Substrate interactions and promiscuity in a viral DNA packaging motor. *Nature, 461*(7264), 669–673.
79. Cheng, W., Arunajadai, S. G., Moffitt, J. R., Tinoco, I., Jr., & Bustamante, C. (2011). Single-base pair unwinding and asynchronous RNA release by the hepatitis C virus NS3 helicase. *Science, 333*(6050), 1746–1749.
80. Dumont, S., Cheng, W., Serebrov, V., Beran, R. K., Tinoco, I., Jr., Pyle, A. M., et al. (2006). RNA translocation and unwinding mechanism of HCV NS3 helicase and its coordination by ATP. *Nature, 439*(7072), 105–108.

81. Sun, B., Johnson, D. S., Patel, G., Smith, B. Y., Pandey, M., Patel, S. S., et al. (2011). ATP-induced helicase slippage reveals highly coordinated subunits. *Nature, 478*(7367), 132–135.
82. Qu, X., Wen, J. D., Lancaster, L., Noller, H. F., Bustamante, C., & Tinoco, I. (2011). The ribosome uses two active mechanisms to unwind messenger RNA during translation. *Nature, 475*(7354), 118–121.
83. Kaiser, C. M., Goldman, D. H., Chodera, J. D., Tinoco, I., Jr., & Bustamante, C. (2011). The ribosome modulates nascent protein folding. *Science, 334*(6063), 1723–1727.
84. Moffitt, J. R., Chemla, Y. R., Smith, S. B., & Bustamante, C. (2008). Recent advances in optical tweezers. *Annual Review of Biochemistry*, 205–228.
85. Neuman, K. C., & Nagy, A. (2008). Single-molecule force spectroscopy: Optical tweezers, magnetic tweezers and atomic force microscopy. *Nature Methods, 5*(6), 491–505.
86. Kim, S., Blainey, P. C., Schroeder, C. M., & Xie, X. S. (2007). Multiplexed single-molecule assay for enzymatic activity on flow-stretched DNA. *Nature Methods, 4*(5), 397–399.
87. Deniz, A. A., Dahan, M., Grunwell, J. R., Ha, T., Faulhaber, A. E., Chemla, D. S., et al. (1999). Single-pair fluorescence resonance energy transfer on freely diffusing molecules: Observation of Förster distance dependence and subpopulations. *Proceedings of the National Academy of Sciences of the United States of America, 96*(7), 3670–3675.
88. Talaga, D. S., Lau, W. L., Roder, H., Tang, J., Jia, Y., DeGrado, W. F., et al. (2000). Dynamics and folding of single two-stranded coiled-coil peptides studied by fluorescent energy transfer confocal microscopy. *Proceedings of the National Academy of Sciences of the United States of America, 97*(24), 13021–13026.
89. Ha, T. (2001). Single-molecule fluorescence resonance energy transfer. *Methods, 25*(1), 78–86.
90. Roy, R., Hohng, S., & Ha, T. (2008). A practical guide to single-molecule FRET. *Nature Methods, 5*(6), 507–516.
91. Hohng, S., Joo, C., & Ha, T. (2004). Single-molecule three-color FRET. *Biophysical Journal, 87*(2), 1328–1337.
92. Lee, J., Lee, S., Ragunathan, K., Joo, C., Ha, T., & Hohng, S. (2010). Single-molecule four-color FRET. *Angewandte Chemie - International Edition, 49*(51), 9922–9925.
93. Hohng, S., Lee, S., Lee, J., & Jo, M. H. (2014). Maximizing information content of single-molecule FRET experiments: Multi-color FRET and FRET combined with force or torque. *Chemical Society Reviews, 43*(4), 1007–1013.
94. Vaughan, J. C., Jia, S., & Zhuang, X. (2012). Ultrabright photoactivatable fluorophores created by reductive caging. *Nature Methods, 9*(12), 1181–1184.
95. Fazio, T., Visnapuu, M. L., Wind, S., & Greene, E. C. (2008). DNA curtains and nanoscale curtain rods: High-throughput tools for single molecule imaging. *Langmuir, 24*(18), 10524–10531.
96. Courtemanche, N., Lee, J. Y., Pollard, T. D., & Greene, E. C. (2013). Tension modulates actin filament polymerization mediated by formin and profilin. *Proceedings of the National Academy of Sciences of the United States of America, 110*(24), 9752–9757.
97. Fedor, M. J., & Williamson, J. R. (2005). The catalytic diversity of RNAs. *Nature Reviews Molecular Cell Biology, 6*(5), 399–412.
98. Nowotny, M., Gaidamakov, S. A., Crouch, R. J., & Yang, W. (2005). Crystal structures of RNase H bound to an RNA/DNA hybrid: Substrate specificity and metal-dependent catalysis. *Cell, 121*(7), 1005–1016.
99. Shevelev, I. V., & Hubscher, U. (2002). The 3′-5′ exonucleases. *Nature Reviews Molecular Cell Biology, 3*(5), 364–375.
100. Baumeister, W., Walz, J., Zühl, F., & Seemüller, E. (1998). The proteasome: Paradigm of a self-compartmentalizing protease. *Cell, 92*(3), 367–380.
101. Lee, G., Yoo, J., Leslie, B. J., & Ha, T. (2011). Single-molecule analysis reveals three phases of DNA degradation by an exonuclease. *Nature Chemical Biology, 7*(6), 367–374.
102. Radding, C. M. (1966). Regulation of lambda exonuclease. I. Properties of lambda exonuclease purified from lysogens of lambda T11 and wild type. *Journal of Molecular Biology, 18*(2), 235–250.

103. Little, J. W. (1967). An exonuclease induced by bacteriophage lambda. II. Nature of the enzymatic reaction. *Journal of Biological Chemistry, 242*(4), 679–686.
104. Perkins, T. T., Dalal, R. V., Mitsis, P. G., & Block, S. M. (2003). Sequence-dependent pausing of single lambda exonuclease molecules. *Science, 301*(5641), 1914–1918.
105. Kuo, T. L., Garcia-Manyes, S., Li, J., Barel, I., Lu, H., Berne, B. J., et al. (2010). Probing static disorder in Arrhenius kinetics by single-molecule force spectroscopy. *Proceedings of the National Academy of Sciences of the United States of America, 107*(25), 11336–11340.
106. Zhang, J., McCabe, K. A., & Bella, C. E. (2011). Crystal structures of lambda exonuclease in complex with DNA suggest an electrostatic ratchet mechanism for processivity. *Proceedings of the National Academy of Sciences of the United States of America, 108*(29), 11872–11877.
107. Young, B. A., Gruber, T. M., & Gross, C. A. (2002). Views of transcription initiation. *Cell, 109*(4), 417–420.
108. Marshall, R. A., Aitken, C. E., Dorywalska, M., & Puglisi, J. D. (2008). Translation at the single-molecule level. *Annual Review of Biochemistry*, 177–203.
109. Makino, D. L., Baumgärtner, M., & Conti, E. (2013). Crystal structure of an rna-bound 11-subunit eukaryotic exosome complex. *Nature, 495*(7439), 70–75.
110. Büttner, K., Wenig, K., & Hopfner, K. P. (2005). Structural framework for the mechanism of archaeal exosomes in RNA processing. *Molecular Cell, 20*(3), 461–471.
111. Lorentzen, E., & Conti, E. (2006). The exosome and the proteasome: Nano-compartments for degradation. *Cell, 125*(4), 651–654.
112. Wahle, E. (2007). Wrong PH for RNA degradation. *Nature Structural and Molecular Biology, 14*(1), 5–7.
113. Xie, X. S. (1998). Single-molecule enzymatic dynamics. *Science, 282*(5395), 1877–1882.
114. Lin-Chao, S., Chiou, N. T., & Schuster, G. (2007). The PNPase, exosome and RNA helicases as the building components of evolutionarily-conserved RNA degradation machines. *Journal of Biomedical Science, 14*(4), 523–532.
115. Shi, Z., Yang, W. Z., Lin-Chao, S., Chak, K. F., & Yuan, H. S. (2008). Crystal structure of Escherichia coli PNPase: Central channel residues are involved in processive RNA degradation. *RNA, 14*(11), 2361–2371.
116. Fazal, F. M., Koslover, D. J., Luisi, B. F., & Block, S. M. (2015). Direct observation of processive exoribonuclease motion using optical tweezers. *Proceedings of the National Academy of Sciences of the United States of America, 112*(49), 15101–15106.
117. Houseley, J., LaCava, J., & Tollervey, D. (2006). RNA-quality control by the exosome. *Nature Reviews Molecular Cell Biology, 7*(7), 529–539.
118. Lorentzen, E., Basquin, J., Tomecki, R., Dziembowski, A., & Conti, E. (2008). Structure of the active subunit of the yeast exosome core, Rrp44: Diverse modes of substrate recruitment in the RNase II nuclease family. *Molecular Cell, 29*(6), 717–728.

Chapter 5
Fitting in the Age of Single-Molecule Experiments: A Guide to Maximum-Likelihood Estimation and Its Advantages

Behrouz Eslami-Mosallam, Iason Katechis and Martin Depken

5.1 Introduction

Single-molecule (SM) experiments allow us to peer deep into the molecular dynamics that drive biology at the microscopic scale [4, 14]. Though observing the dynamics of a single molecule is an amazing feat in and of itself, the information gleaned is limited by the small number of observables that can be simultaneously tracked, and the resolution at which this can be done. Faced with such limitations, mechanistic modeling and parameter estimation are often used to extract as much quantitative information as possible.

Using SM fluorescence or Förster resonance energy transfer (FRET) [15], it is possible to generate time distributions for reactions, such as the unbinding-time distributions of ligands unbinding from a single receptor (Fig. 5.1). Such distributions are particularly useful when the pathway includes multiple steps, as they can be quite complex and information rich. Faced with systems exhibiting several characteristic times, least-squares (LS) fitting is often brought to bear on the problem. Though popular and often useful, there are situations in which standard LS approaches fail, and unfortunately often do so in quite non-obvious ways. To help the reader understand and avoid such pitfalls, we here explore some of these situations through the lens of ML estimation, an alternative approach that has become very popular in the physical sciences [1–3, 7, 8, 13, 16, 17, 24–26, 28].

As it is straight forward, adaptable, and well suited to SM experiments, we here provide a self-contained introduction to ML estimation. We heuristically show that ML estimation should generally outperform LS fitting and explicitly show this to be the case in relevant SM FRET examples. We close with a discussion of how to use bootstrapping to estimate the standard deviation of fit parameters. The presentation

B. Eslami-Mosallam · I. Katechis · M. Depken (✉)
Department of BioNanoScience, Kavli Institute of NanoScience, Delft University of Technology, 2629 HZ Delft, The Netherlands
e-mail: s.m.depken@tudelft.nl

C. Joo and D. Rueda (eds.), *Biophysics of RNA-Protein Interactions*, Biological and Medical Physics, Biomedical Engineering,
https://doi.org/10.1007/978-1-4939-9726-8_5

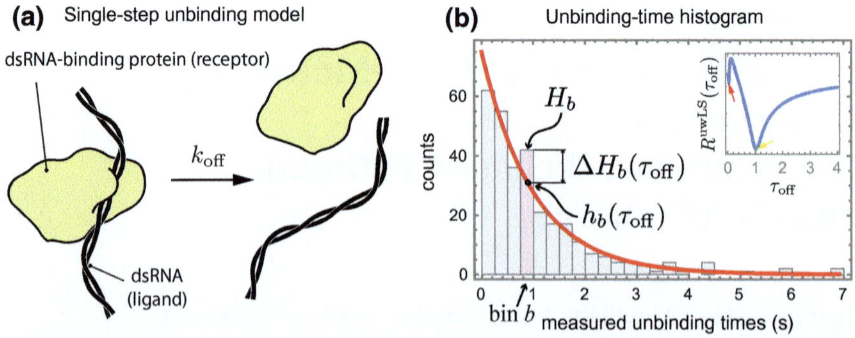

Fig. 5.1 **a** A single-step ligand-receptor unbinding model. A dsRNA-binding protein releases dsRNA at a characteristic rate k_{off}. For this model, we expect an exponential distribution of unbinding time, with the average unbinding time $\hat{\tau}_{\text{off}} = 1/k_{\text{off}}$. **b** A histogram (bars) formed from 300 unbinding times picked from an exponential distribution with the true average unbinding time $\hat{\tau}_{\text{off}} = 1$ s. The predicted bin counts for a model with average unbinding time $\tau_{\text{off}} = 1$ s are shown as a red curve, and the notation used in Eq. (5.1) is indicated for bin b (pink bar). In the inset, we show the unweighted LS residue $R^{\text{uwLS}}(\tau_{\text{off}})$ (in log-scale) as a function of the model parameter τ_{off}. The function displays a global minimum close to the true average unbinding time (yellow arrow), as well as a local minimum for short times (red arrow). Beware that local minima can sometimes trap numerical minimization algorithms, leading them to erroneously report the local minimum as the sought after global minimum

is intended for SM experimenters who find fitting data indispensable to their work, but might find the advantages/limitations/rationale of various approaches hard to ascertain.

5.2 Prerequisites

In an effort to be self-contained, we start by discussing LS fitting, as well as error estimation and some crucial concepts in probability theory. These sections can be skipped by the initiated reader.

5.2.1 LS Fitting and the Distance Between Model and Data

LS fitting comes in several flavors, depending on how statistical fluctuations in bin counts are accounted for. The fitting is generally performed by collecting the available data into bins $b = 1, 2, \ldots, B$, and finding the model parameter values that minimize the total square deviation between actual bin counts (H_b) and model predictions for bin counts (h_b) (Fig. 5.1b), normalized with the *true* standard deviation of the bin count (σ_b). We will refer to this approach as true LS (tLS) fitting. For unbinding times in the simple RNA-protein example of Fig. 5.1a, tLS fitting consists of finding the model parameter τ_{off} (the average unbinding time of the model) that minimizes the total residue

$$R^{\text{tLS}}(\tau_{\text{off}}) = \sum_{b=1}^{B} \left(\frac{\Delta H_b(\tau_{\text{off}})}{\sigma_b} \right)^2, \quad \Delta H_b(\tau_{\text{off}}) = H_b - h_b(\tau_{\text{off}}). \tag{5.1}$$

Minimizing the total residue R^{tLS} makes intuitive sense, as it penalizes parameter values that give large deviations between predictions and measurements, in a manner scaled by the size of statistical fluctuations in each bin. A perfect estimate in a bin ($H_b = h_b(\tau_{\text{off}})$) results in zero residue, while any positive (weighted) residue gives a measure of the "statistical distance" between model and data in that bin. By summing the residues in Eq. (5.1), we get a measure of the total distance between model and data; tLS fitting aims to minimize this distance.

Unfortunately, we do not often have access to the true standard deviation of counts in each bin, and various approximations to Eq. (5.1) must be deployed. For ease of presentation, we will here focus on two cases: In the first case, we assume that count fluctuations are almost constant over all bins, and we use unweighted LS (uwLS) residues by taking σ_b to be constant[1] (e.g., see inset in Fig. 5.1b); in the second case, we assume a fixed total number (N) of independent measurements, such that the count fluctuations in each bin are binomially distributed, with $\sigma_b = \sqrt{\langle H_b \rangle \left(1 - \langle H_b \rangle / N\right)} \approx \sqrt{\langle H_b \rangle}$. Here, the angle brackets represent the statistical average over a large number of experiments, and we have in the last step assumed bins to be small enough that no bin on average contains a large fraction of the total number of observations (i.e., $\langle H_b \rangle \ll N$ for all bins). With no better estimate at hand, the statistical average of bin counts is often approximated with the observed bin count by setting $\sigma_b \approx \sqrt{H_b}$ in Eq. (5.1). We will refer to this approach as weighted LS (wLS).

Both wLS and uwLS fitting can be problematic. Using uwLS, we assume fluctuations in bin counts to be uniform over bins. As we shall see, this is often a poor approximation for systems with multiple characteristic timescales. Using wLS, we instead use individual bin counts to estimate the standard deviation of counts in that bin. As individual bin counts can be small, relative fluctuations can be large, resulting in large approximation errors when using $\sigma_b \approx \sqrt{H_b}$ in Eq. (5.1).

5.2.2 Error Estimation, Variation, and Systematic Bias

For any estimation method applied to an experiment with a finite set of measurements, the estimated parameter value (τ) will deviate from the true value ($\hat{\tau}$). To compare

---

[1] Note that we do not need to know the actual constant value of σ_b, as it will not affect the position of the minimum of R^{uwLS}.

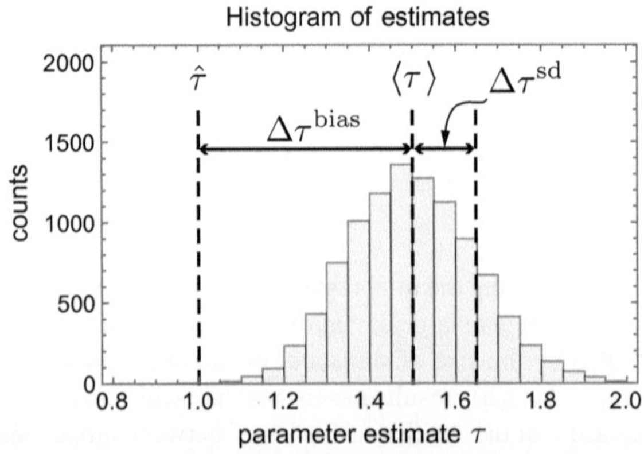

Fig. 5.2 A histogram of estimates for a hypothetical process with the true parameter value $\hat{\tau} = 1$. The systematic bias $\Delta\tau^{\text{bias}}$ and the typical size of fluctuations $\Delta\tau^{\text{sd}}$ around the average estimate $\langle\tau\rangle$ are indicated

two methods, we need to understand the distribution of parameter estimates that each approach would yield were it to be repeated many times. Over a large number of experiments, the typical error can be measured by the mean square error, MSE $= \left\langle(\tau - \hat{\tau}\,)^2\right\rangle$. To understand the nature of estimation errors, consider the bias $\Delta\tau^{\text{bias}} = \langle\tau\rangle - \hat{\tau}$, capturing how the average estimate deviates from the true parameter value, as well as the standard deviation $\Delta\tau^{\text{sd}} = \sqrt{\langle(\tau - \langle\tau\rangle)^2\rangle}$, capturing the typical spread of estimates around their average (Fig. 5.2). Conveniently, the bias and standard deviation add in quadrature to form the MSE [9]

$$\text{MSE} = \left(\Delta\tau^{\text{bias}}\right)^2 + \left(\Delta\tau^{\text{sd}}\right)^2.$$

The smaller the MSE the better, and we should seek to minimize both the bias and standard deviation as far as possible. A large bias can be introduced by the estimation method itself, while a large standard deviation typically results from a lack of data and/or accuracy of the measurements.

5.2.3 *Bayes' Equation and Observation Frequencies*

To explain the rationale behind ML estimation [9], we first introduce Bayes' equation by way of Venn diagrams and the frequentist interpretation of probability. According to this interpretation, probabilities can be seen as the asymptotic frequency of outcomes, recorded over a large number of repetitions [12]. For concreteness, imagine a steady rainfall with water drops hitting the yellow (event A)- and blue (event B)-striped shapes shown in Fig. 5.3. Further imagine keeping track of the number of raindrops that falls on the section with just yellow stripes (N_A), just blue stripes (N_B), both yellow and blue stripes ($N_{A\&B}$), or anywhere ($N_{\text{tot}} = N_A + N_B$). Among these various counts, the relationship

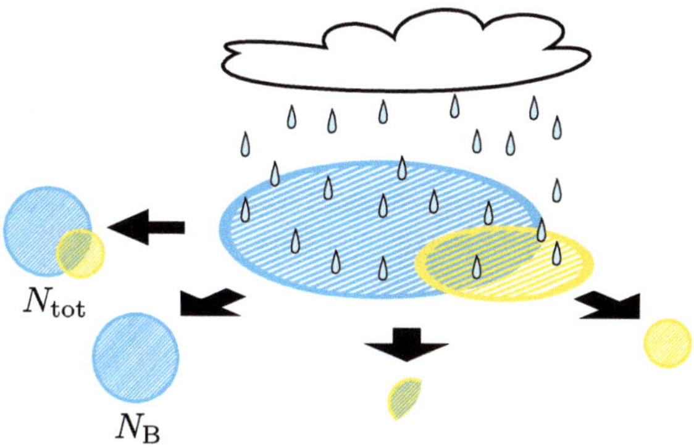

Fig. 5.3 Imagine exposing the blue and yellow shapes to rain, while keeping track of the number of raindrops that hit each differently striped area. If the rainfall is steady, we can use the frequentist interpretation of probability to relate the different fractions of raindrops landing on the various areas to probabilities. The trivial Eq. (5.2) then becomes Bayes' equation as expressed in Eq. (5.3)

$$\frac{N_{A,B}}{N_{\rm tot}} = \frac{N_{A\&B}}{N_B}\frac{N_B}{N_{\rm tot}} = \frac{N_{A\&B}}{N_A}\frac{N_A}{N_{\rm tot}} \tag{5.2}$$

holds trivially true, as can be seen by canceling the first denominator with the second numerator after each equal sign. If we collect enough raindrops, the fraction of raindrops that has so far landed on a particular section will approach the probability that also the next raindrop will land in that same section. Taking the frequentist approach, we can translate Eq. (5.2) into Bayes' equation for probabilities

$$P(A, B) = P(A|B)P(B) = P(B|A)P(A). \tag{5.3}$$

In the above, $P(A, B) = N_{A\&B}/N_{\rm tot}$ is the joint probability that both A and B occur, $P(A) = N_A/N_{\rm tot}$ is the probability that A occurs irrespective of whether B occurs or not, $P(A|B) = N_{A\&B}/N_B$ gives the conditional probability that A occurs, given that B occurs, and so on swapping A and B.

5.2.4 *Continuous Outcomes and Probability Densities*

We are ultimately interested in measurements that produce real numbers (such as unbinding times), while Bayes' equation (Eq. 5.3) is valid for probabilities of discrete events. For outcomes that can take a continuous value, the relevant concept is that of the probability density function (PDF). For two concurring continuous outcomes, denote recording the respective values in interval I_a and I_b, centered around a and b, as event A and B. For very short interval lengths ΔI_a and ΔI_b, the probability to end up in the interval (denoted by upper case P) is simply the relevant PDF (denoted by a lower case p) multiplied with the relevant interval length(s)

$$P(A, B) = p(a, b)\Delta I_a \Delta I_b,$$

$$P(A) = p(a)\Delta I_a,\ P(A|B) = p(a|b)\Delta I_a,\ P(B) = p(b)\Delta I_b,\ P(B|A) = p(b|a)\Delta I_b.$$

The above relations can be plugged into Bayes' equation for probabilities of discrete events (Eq. 5.3), giving the sought after Bayes' equation for PDFs of continuous outcomes

$$p(a, b) = p(a|b)p(b) = p(b|a)p(a). \tag{5.4}$$

With the prerequisites covered, we are now ready to address the rationale behind ML estimation and assess how it compares and relates to LS fitting.

5.3 Maximum Likelihood

To keep the discussion in general, consider an experiment where we collect N-independent measurements $\{t\}_N = \{t_1, t_2, \ldots, t_N\}$ and modeled it as process with M parameters $\{\tau\}_M = \{\tau_1, \tau_2, \ldots, \tau_M\}$. Based on our one experiment, we would like to determine the parameter values that gave rise to the data. As the stochasticity of the data makes it impossible to precisely determine the parameters exactly, our best bet would by definition be to find the most probable set of parameter values, *given* the data. In the language of conditional PDFs, this corresponds to finding the model parameters which maximize the PDF $p(\{\tau\}_M|\{t\}_N)$ of having a model with parameters $\{\tau\}_M$ given the measured data $\{t\}_N$ (for a lighthearted and instructive discussion of the meaning of the probability of a model, see [18]. Unfortunately, we do not have direct access to this conditional PDF. Still, we can make considerable progress by using Bayes' equation and introducing a few additional assumptions.

5.3.1 *The Most Likely Model*

Through Bayes' equation for PDFs (Eq. 5.4), we can relate the unknown PDF of interest to PDFs about which we do have some knowledge, or regarding which we can at least make some reasonable assumptions. Letting $a = \{t\}_N$ and $b = \{\tau\}_M$ in Eq. (5.4), we have[2]

$$p(\{\tau\}_M|\{t\}_N) = \frac{p(\{\tau\}_M)}{p(\{t\}_N)} p(\{t\}_N|\{\tau\}_M).$$

[2]A more intuitive way of writing this might be in the form $p(\text{model}|\text{data}) = \frac{p(\text{model})}{p(\text{data})} p(\text{data}|\text{model})$.

With the aim to maximize the left-hand side of the above expression with respect to the model parameters, we note that the denominator on the right-hand side does not depend on the model parameters and therefore will not influence which parameter value maximizes the left-hand side; we promptly ignore the denominator. The numerator can be interpreted as encoding what we knew of the correct parameter values before our experiments. If we assume little or no prior knowledge, it makes sense to also assume this prior PDF to be roughly uniform and thus largely independent of the model parameters[3]; we promptly ignore also the numerator. The last term on the right-hand side of the equation describes the PDF of a particular set of measurements, given the model parameters. This conditional PDF *can* be calculated if we have a model of the system!

Through the above argument, we conclude that by maximizing the *likelihood function* $p(\{t\}_N|\{\tau\}_M)$, we can find an estimate for the model parameter values that best describe the data. Equivalently, we could choose to minimize the *log-likelihood* function[4] $L^{\mathrm{ML}}(\{\tau\}_M) = -\ln p(\{t\}_N|\{\tau\}_M)$, which has a global minimum for the same parameter values as the likelihood function has a global maximum. As we assume *independent* measurements, the PDF of the whole experimental outcome $\{t\}_N$ can simply be written as the product of the PDFs for each measurement. The log-likelihood function then has the convenient property that it turns into a sum over measurements,

$$L^{\mathrm{ML}}(\{\tau\}_M) = -\ln\left(\prod_{n=1}^{N} p(t_n|\{\tau\}_M)\right) = -\sum_{n=1}^{N} \ln p(t_n|\{\tau\}_M). \tag{5.5}$$

Finding the parameter values that globally minimize Eq. (5.5) constitutes ML parameter estimation, and we now apply it to a few simple but illustrative examples to familiarize the reader with the approach.

5.3.2 *ML Estimation for an Exponential Process*

To demonstrate ML estimation in practice, we return to ligand–receptor unbinding. For simple unbinding kinetics, the unbinding times are exponentially distributed with the PDF $p(t|\tau_{\mathrm{off}}) = \mathrm{e}^{-t/\tau_{\mathrm{off}}}/\tau_{\mathrm{off}}$. Inserting this PDF into Eq. (5.5), we see that the log-likelihood function is given by

---

[3]There are subtleties here relating to variable changes [18], but these lie outside our present scope.

[4]It should be noted that as the logarithm takes a unit-less argument, while the PDF has units (inverse time in case of the unbinding experiments). Strictly, we therefore need to multiply the PDF with some constant that renders the argument of the logarithm unit less in the definition of $L^{\mathrm{ML}}(\{\tau\}_M)$. As the value of this constant does not affect the position of the minimum, we drop it for notational convenience.

$$L^{\mathrm{ML}}(\tau_{\mathrm{off}}) = N\left(\ln \tau_{\mathrm{off}} + \frac{\bar{t}}{\tau_{\mathrm{off}}}\right), \quad \bar{t} = \frac{1}{N}\sum_{n=1}^{N} t_n.$$

The ML estimate $\left(\tau_{\mathrm{off}}^{\mathrm{ML}}\right)$ is now arrived at by minimizing $L^{\mathrm{ML}}(\tau_{\mathrm{off}})$ with respect to τ_{off}. In this simple example, we can find the ML estimate analytically by using the zero-derivative test for finding an optimum,

$$0 = \frac{\partial L^{\mathrm{ML}}}{\partial \tau_{\mathrm{off}}}\left(\tau_{\mathrm{off}}^{\mathrm{ML}}\right) = N\left(\frac{1}{\tau_{\mathrm{off}}^{\mathrm{ML}}} - \frac{\bar{t}}{(\tau_{\mathrm{off}}^{\mathrm{ML}})^2}\right) \quad \Rightarrow \quad \tau_{\mathrm{off}}^{\mathrm{ML}} = \bar{t}. \tag{5.6}$$

Consequently, ML estimation confirms the well-known result that the characteristic time of an exponential process can be estimated by the average event time observed in the data; or simply, the off-rate estimate is $k_{\mathrm{off}}^{\mathrm{ML}} = 1/\bar{t}$. Note that we did not need to perform any binning to extract this estimate, which constitutes a clear advantage over standard LS fitting methods.

5.3.3 *ML Estimation for an Exponential Process with a Time Cutoff*

The simplest additional characteristic time to consider is possibly that introduced by photobleaching in FRET experiments. With photobleaching, the experimental signal in our unbinding example can, in addition to unbinding, also be lost due to the stochastic degradation of fluorophores over time. We can account for photobleaching by interpreting the estimated characteristic rate $(1/\tau_{\mathrm{off}}^{\mathrm{ML}})$ of the PDF (which is still exponential), not purely as the unbinding rate, but as the sum of the unbinding and bleaching rate. As the bleaching rate can usually be independently measured, we can often readily estimate the unbinding rate by subtracting the bleaching rate from the estimated total rate.

Next, consider having a hard cutoff time T_{cut} limiting the duration of each measurement. Slightly more complex than photobleaching, this scenario will serve to demonstrate that the ML approach often allows us to utilize extra information in a rational manner. Though we cannot know the precise duration for any binding event lasting longer than T_{cut}, there is information in the number of unbinding events that exceeded it. We start by noting that the simple ML recipe used in Eq. (5.6) does not work, as losing long unbinding times will clearly lead us to underestimate the characteristic unbinding time. Instead, we would like to keep the information regarding the number of measurements that exceeded the finite measurement time window. Combining the probability densities of the measured unbinding times $\left(\{t\}_{N_{\mathrm{rec}}}\right)$ with the probabilities of the missed times $\left(\{t'\}_{N_{\mathrm{cut}}}\right)$, the relevant likelihood function is

$$\underbrace{\prod_{n=1}^{N_{\text{rec}}} p(t_n|\tau_{\text{off}})}_{\substack{\text{PDF of the } N_{\text{rec}} \\ \text{recorded events}}} \underbrace{\prod_{n'=1}^{N_{\text{cut}}} P\left(t'_{n'} > T_{\text{cut}}|\tau_{\text{off}}\right)}_{\substack{\text{probability of the} \\ N_{\text{cut}} \text{ missed events}}} = \prod_{n=1}^{N_{\text{rec}}} p(t_n|\tau_{\text{off}}) \left(\int_{T_{\text{cut}}}^{\infty} \mathrm{d}t'\, p\left(t'|\tau_{\text{off}}\right)\right)^{N_{\text{cut}}}.$$

The corresponding log-likelihood function will now become a sum over both probability densities (for the N_{rec} recorded times) and probabilities (for the N_{cut} missed times)

$$\begin{aligned} L^{\text{ML}}(\tau_{\text{off}}) &= -\sum_{n=1}^{N_{\text{rec}}} \ln p(t_n|\tau_{\text{off}}) - N_{\text{cut}} \ln \int_{T_{\text{cut}}}^{\infty} \mathrm{d}t\, p(t|\tau_{\text{off}}) \\ &= N_{\text{rec}} \left(\ln \tau_{\text{off}} + \frac{\bar{t}}{\tau_{\text{off}}}\right) + \frac{N_{\text{cut}} T_{\text{cut}}}{\tau_{\text{off}}}. \end{aligned} \tag{5.7}$$

The ML estimate can once again be found analytically through the zero-derivative condition, yielding the simple formula

$$0 = \frac{\partial L^{\text{ML}}}{\partial \tau_{\text{off}}}\left(\tau_{\text{off}}^{\text{ML}}\right) \quad \Rightarrow \quad \tau_{\text{off}}^{\text{ML}} = \bar{t}\left(1 + \frac{T_{\text{cut}} N_{\text{cut}}}{\bar{t} N_{\text{rec}}}\right) \tag{5.8}$$

to correct for the cutoff-induced bias. Note that the correction only becomes significant when the lower bound of the total duration of cut events ($T_{\text{cut}} N_{\text{cut}}$) becomes comparable to the total time of recorded events ($\bar{t} N_{\text{rec}}$).

5.3.4 ML Estimation for a Double-Exponential Process

The unbinding process itself might have several characteristic times. We next consider the case where the model yields a double-exponential PDF of unbinding times and where the maximal measurement duration is large enough to be ignored. For the unbinding problem discussed above, such PDFs could originate in two interconvertible binding modes: a loose binding mode where the ligand first binds, and eventually unbinds from, and a tight binding mode from which the ligand cannot unbind directly (see Fig. 5.4a). Alternatively, it could result from two protein populations with different unbinding rates. The PDF for either system can be written as (Fig. 5.4b)

$$p(\mathrm{t}|\tau_1, \tau_2, P_1) = \frac{P_1}{\tau_1} \mathrm{e}^{-t/\tau_1} + \frac{1 - P_1}{\tau_2} \mathrm{e}^{-t/\tau_2} \tag{5.9}$$

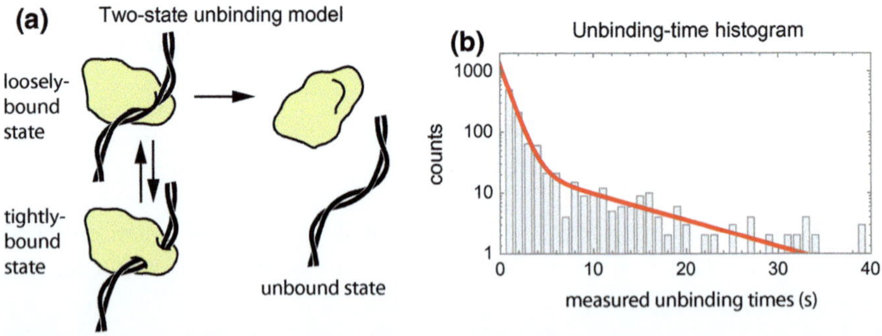

Fig. 5.4 **a** A dsRNA-binding protein exhibiting two bound states, resulting in a double-exponential PDF for the unbinding time. **b** Histogram (bars, in log-scale) formed by picking 1000 unbinding times from a double-exponential distribution with a PDF characterized by $\hat{\tau}_1 = 1$ s, $\hat{\tau}_2 = 5$ s, and $\hat{P}_1 = \hat{P}_2 = 0.5$. The predicted bin counts for a model with $\tau_1 = 1$ s, $\tau_2 = 5$ s, and $P_1 = P_2 = 0.5$ are shown as a red curve

where the characteristic times τ_1 and τ_2, as well as the population fraction P_1 associated with τ_1, can be directly related to the microscopic rates of the relevant system. Attempting to use the PDF of Eq. (5.9) to calculate the log-likelihood function according to Eq. (5.5), it quickly becomes clear that we can no longer find a simple analytic solution to the minimization problem. This is quite generally the case, and one has to perform the minimization numerically, as we will do when comparing LS and ML approaches on simulated data below.

5.3.5 *Coarse-Grained Likelihood*

Though ML estimation has the clear advantage of requiring no binning of the data, for large data sets, it often becomes computationally demanding to numerically minimize a log-likelihood function with as many terms as there are measurements (see sum in Eq. 5.5). The computational efficiency can be drastically increased by considering the likelihood over bins, which should be a reasonable approximation as long as we choose the bin size small enough for there to be little change in the PDF over each bin. The probability P_b of a particular measurement ending up in bin b can then be related to the model PDF and used to calculate the predicted bin count h_b as

$$h_b(\{\tau\}_M) = N P_b(\{\tau\}_M), \;\; P_b(\{\tau\}_M) = \int_{t_b-\Delta t_b/2}^{t_b+\Delta t_b/2} \mathrm{d}t \; p(t|\{\tau\}_M) \approx \Delta t_b p(t_b|\{\tau\}_M), \tag{5.10}$$

where the integral runs over the whole width Δt_b of bin b centered around t_b.

Splitting the sum over measurements in the definition of the log-likelihood function (Eq. 5.5) into a sum over bins and a sum over measurements in each bin, it can be approximated by the *coarse-grained* (cg) log-likelihood function

$$L^{\mathrm{ML}}(\{\tau\}_M) = -\sum_{b=1}^{B} \sum_{\substack{t_n \text{ in} \\ \text{bin } b}} \ln p(t_n|\{\tau\}_M) \approx -\sum_{b=1}^{B} H_b \ln h_b(\{\tau\}_M) = L^{\mathrm{cgML}}(\{\tau\}_M). \tag{5.11}$$

Here, the last equality is a definition, and we have dropped constant terms and factors not affecting the minimizing parameter values. Note that the results of using cgML estimation can always be made arbitrarily close to the original ML estimate by choosing the bin widths small enough.

5.3.6 The Connection Between LS and ML

We will now show that ML estimation can be seen as another approximation of tLS and, importantly, one that is generally expected to do better than both uwLS and wLS. The connection between LS and ML estimation has been studied for the case of independent and Gaussian-distributed data with equal variance [10], but in an effort to understand the differences in estimates more generally, we here employ a heuristic approach with wide applicability.

For any data set $\{t\}_N$ and model with parameter set $\{\tau\}_M$, we seek to compare tLS fitting to ML estimation. As the tLS scheme is based on binned data sets, we opt to compare it to equally binned cgML. The zero-derivative condition for finding the tLS parameter estimates $\left\{\tau^{\mathrm{tLS}}\right\}_M$ from Eq. (5.1) is

$$0 = \frac{\partial R^{\mathrm{tLS}}}{\partial \tau_m}\left(\left\{\tau^{\mathrm{tLS}}\right\}_M\right) \approx \sum_b \frac{\Delta H_b\left(\left\{\tau^{\mathrm{tLS}}\right\}_M\right)}{\langle H_b\rangle} \frac{\partial \Delta H_b\left(\left\{\tau^{\mathrm{tLS}}\right\}_M\right)}{\partial \tau_m}, \quad m = 1, \ldots, M. \tag{5.12}$$

Similarly, differentiating Eq. (5.11), and using the normalization of probabilities $\left(\sum_b h_b = N\right)$, the condition for finding the cgML estimate $\left\{\tau^{\mathrm{cgML}}\right\}_M$ can be written as

$$\begin{aligned} 0 &= -\frac{\partial L^{\mathrm{cgML}}}{\partial \tau_m}\left(\left\{\tau^{\mathrm{cgML}}\right\}_M\right) \\ &= \sum_b \frac{\Delta H_b\left(\left\{\tau^{\mathrm{cgML}}\right\}_M\right)}{h_b\left(\left\{\tau^{\mathrm{cgML}}\right\}_M\right)} \frac{\partial \Delta H_b\left(\left\{\tau^{\mathrm{cgML}}\right\}_M\right)}{\partial \tau_m}, \quad m = 1, \ldots, M. \end{aligned} \tag{5.13}$$

Interestingly, though the functions that are minimized during tLS (Eq. 5.1) and cgML (Eq. 5.11) estimation are quite different, their minima are located in close proximity. From above, it is clear that the cgML minimization condition (Eq. 5.13) can be seen as an approximation to the tLS minimization condition (Eq. 5.12) with $\langle H_b \rangle \approx h_b\left(\left\{\tau^{\mathrm{cgML}}\right\}_M\right)$.

The cgML approximation $\left(\langle H_b \rangle \approx h_b\left(\left\{\tau^{\mathrm{cgML}}\right\}_M\right)\right)$ should be compared to the wLS approximation ($\langle H_b \rangle \approx H_b$). The wLS approximation includes only the data of each bin when estimating the variance in each bin. The cgML approximation takes into account the data in all bins, since $\left\{\tau^{\mathrm{cgML}}\right\}_M$ is estimated from the whole data set by definition. As increasing the number of measurements generally reduces both the variance and systematic bias of estimates, we typically expect the cgML approach to outperform the wLS approach. It should be noted that the ML approach is *not* equivalent to setting $\sigma_b \approx \sqrt{h_b\left(\left\{\tau^{\mathrm{cgML}}\right\}_M\right)}$ already in Eq. (5.1), as we would then need to know the optimal parameters before we have minimized the residue to find them. ML estimation elegantly bypasses this problem by enforcing the same approximation, not on the function to be minimized but directly on the condition defining the minimum (Eq. 5.13).

Having argued that we should generally expect (cg)ML to outperform wLS, we explicitly compare their performance, together with that of uwLS, on the examples used above.

5.4 Comparing LS and ML Through Simulations

Having established that uwLS, wLS, and cgML can all be seen as tLS approximations of various severity, we here numerically explore the consequences of these approximations. By generating data with a known distribution, we can quantify the success of the different approaches at estimating known parameter values. We do not discuss the numerical minimization schemes we use when analytics fail, further than stating that it is implemented in Mathematica™, using a simulated-annealing algorithm [22] to minimize the risk of finding a local rather than global minimum (see inset in Fig. 5.1b, e.g., of a local (red arrow) and global (yellow arrow) minimum). There are many powerful software packages available with the required numerical optimization capabilities.

Without a sharp cutoff time for the measurements, we always expect many long-time bins to be empty in the tail end of the PDF. A zero count in any bin is catastrophic for wLS, as it gives a zero estimate for the standard deviation and so introduces infinite terms in Eq. (5.1). In an attempt to circumvent such issues, various re-binning procedures or reassignments of weights can be performed. Though such approaches avoid infinite terms in Eq. (5.1), they do change the details of the estimation method depending on the observed data, and so risk introducing a strong bias. For simplicity, we will here only consider the interval between the highest and lowest measured data points generated, and for wLS we choose the minimum constant bin size that leaves *no* empty bins in the intervening interval.

5.4.1 Method Comparison for an Exponential Process

Though trivial, we start with the simple exponential process. Using Eq. (5.10), we can calculate the predicted bin counts $h_b(\tau_{\text{off}})$ from the PDF. It should be noted that we could in principle estimate both N and τ_{off} by optimizing with respect to both in any LS or ML approach. Though this is often done, it is not advisable as it will increase the MSE compared to if we heed the fact that N is *known* and precisely dictates the translation from probability to histogram counts in Eq. (5.10).

In Fig. 5.5, we show the results of using uwLS, wLS, and ML estimation on 10,000 exponentially distributed data sets of 100 measurements each ($\hat{\tau}_{\text{off}} = 1$ s). Even after eliminating the zero bins for wLS (see above), the wLS estimate remains biased due to the unavoidable presence of the low-count bins [11, 19–21, 23, 27]. This bias has been shown to be inversely proportional to the average occupancy of the bins [11]. The fact that uwLS estimation introduces a much smaller—if not vanishing—bias compared to wLS estimation might seem strange, given that the latter estimates the standard deviations in bins based on the data, while the former ignores the data and assumes them all equal. The explanation can likely be found in that though the weighted approach clearly employs better approximation for bins

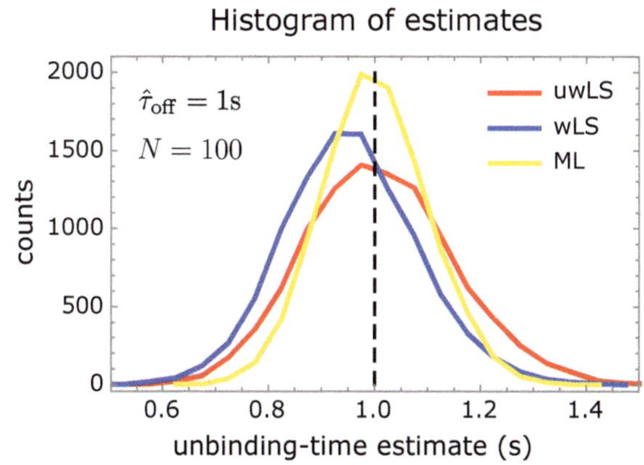

Fig. 5.5 Distribution of unbinding-time estimates from 10,000 exponentially distributed data sets containing 100 samples each. There is a clear bias for wLS estimation (−0.04 s), while little bias is apparent for uwLS (0.009 s) and ML (0.0002 s) estimation. The standard deviation of ML estimation (0.10 s) is less than for wLS estimation (0.13 s), which in turn is less than for uwLS estimation (0.14 s). Notwithstanding the larger absolute bias, the $\sqrt{\text{MSE}}$ for wLS estimation (0.13 s) outperforms that for uwLS estimation (0.14 s), while ML estimation outperforms both other methods (0.10 s)

with many counts, the relative errors in low-count bins can be very large, outstripping the error made when assuming the variance of counts to be equal in all bins. Among the three approaches, ML is clearly preferable as both bias and standard deviation are the smallest.

5.4.2 Method Comparison for an Exponential Process with a Cutoff

Next, we consider a measurement that is limited by a maximum measurement time T_{cut}. If this cutoff time is largely compared to the average unbinding time, we effectively have no cutoff, which we covered in the previous section. If we instead have a cutoff time that is comparable to the average unbinding time, there is information in the number of unbinding events that exceeded the maximal duration of the measurements. With a measurement cutoff time, the unbinding times are still exponentially distributed, but the number of experimental observations $N = N_{rec} + N_{cut}$ has to be split into the N_{rec} events where the time was recorded, and the N_{cut} events for which we know only that they lasted longer than T_{cut}. For both wLS and uwLS, we explicitly fit only the N_{rec} measurements falling within the observation window, while for ML estimation, we include also the information regarding the cut events, according to Eq. (5.8).

Though we lose data, introducing a short-time cutoff has the benefit of removing bins that are likely to have zero counts, and thus, we decrease the need to re-bin data for wLS estimation. For small data sets (Fig. 5.6a, b), the counts in each bin will still have large (relative) fluctuations, and it is not surprising that we see a substantial error in wLS estimation. This error decreases as the cutoff is lowered and progressively fewer low-count bins are included (c.f. Fig. 5.6a with b), even though a higher fraction of measurements falls outside the observation window. For the cutoff time close to the characteristic unbinding time, uwLS and ML estimation are comparable, as the variance in bin counts is roughly constant among bins below the cutoff time. This shows a scenario where uwLS outperforms wLS, though ML estimation consistently remains the better alternative.

As we increase the size of the data sets by a factor 100 (Fig. 5.6c, d), we expect the relative fluctuations around the predicted bin counts to decrease, bringing wLS estimation closer to ML estimation. This effect can be seen clearly seen in Fig. 5.6c, d. It is interesting to note that for these large data sets, the extra information regarding the cut measurements included in the ML estimation had little effect on the fit, as all fits roughly coincide in Fig. 5.6c, d.

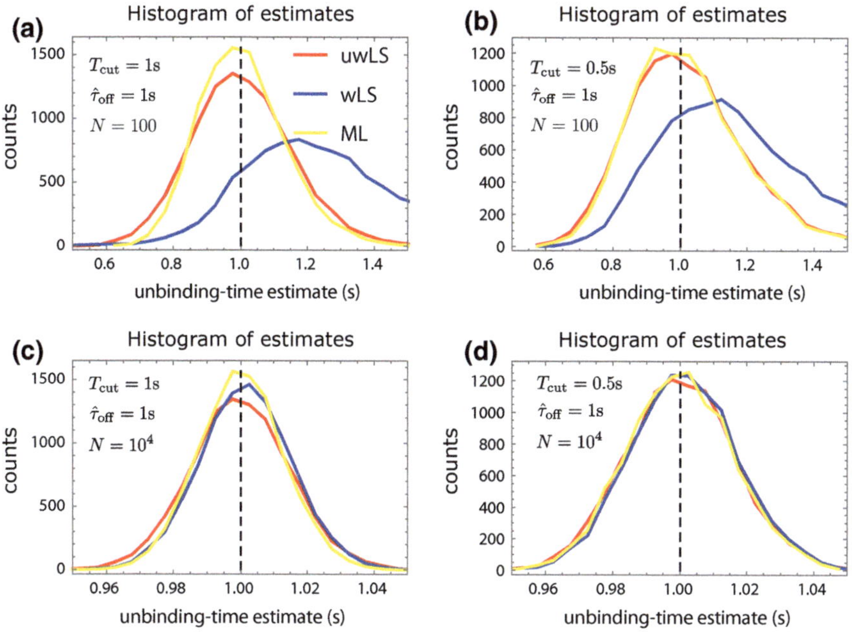

Fig. 5.6 Parameter estimation for 10,000 exponentially distributed data sets with a cutoff. **a** For sets with 100 measurements, and a $T_{cut} = 1$ s, we see a clear bias in wLS estimation, while uwLS estimation has a somewhat larger standard deviation than ML estimation. **b** For a lower $T_{cut} = 0.5$ s, the bias for wLS estimation decreases slightly, while uwLS approaches ML estimation. **c** Increasing the size of the data sets to 10,000 measurements and considering a moderate cutoff time, the difference between wLS estimation and ML estimation diminishes and both methods marginally outperform uwLS estimation. **c** For large data sets and a low cutoff time, all methods converge

5.4.3 Method Comparison for a Double-Exponential Process

For data distributed according to the double-exponential PDF of Eq. (5.9), we need to fit out two characteristic times ($\hat{\tau}_1$ and $\hat{\tau}_2$), together with the fraction of events belonging to each ($\hat{P}_1$, $\hat{P}_{12} = 1 - \hat{P}_{11}$). In Fig. 5.7, we show the results of 10,000 fits to data sets of size 10,000, for a process with moderately separated characteristic times ($\hat{\tau}_1 = 1$ s, $\hat{\tau}_2 = 3$ s) and for three different population fractions ($\hat{P}_1 = 0.1$ Fig. 5.7a–c, $\hat{P}_1 = 0.5$ Fig. 5.7d–e, $\hat{P}_1 = 0.9$ Fig. 5.7g–h). In each case, we report the $\sqrt{\text{MSE}}$/s within parenthesis in the legend.

The error in the short-timescale estimate (τ_1) is dominated by the variance around the average for all methods, and all methods perform better the larger the fraction of events corresponding to the shorter timescale are (Fig. 5.7a, d, and g). The error in the long-timescale estimates (τ_2) is also dominated by the variances, which is particularly large in uwLS estimation (Fig. 5.7b, e, and h). This can likely be traced back to the fact that the constant variance assumption of uwLS suppresses the relative influence of long timescales, introducing a relatively low penalty for variation here.

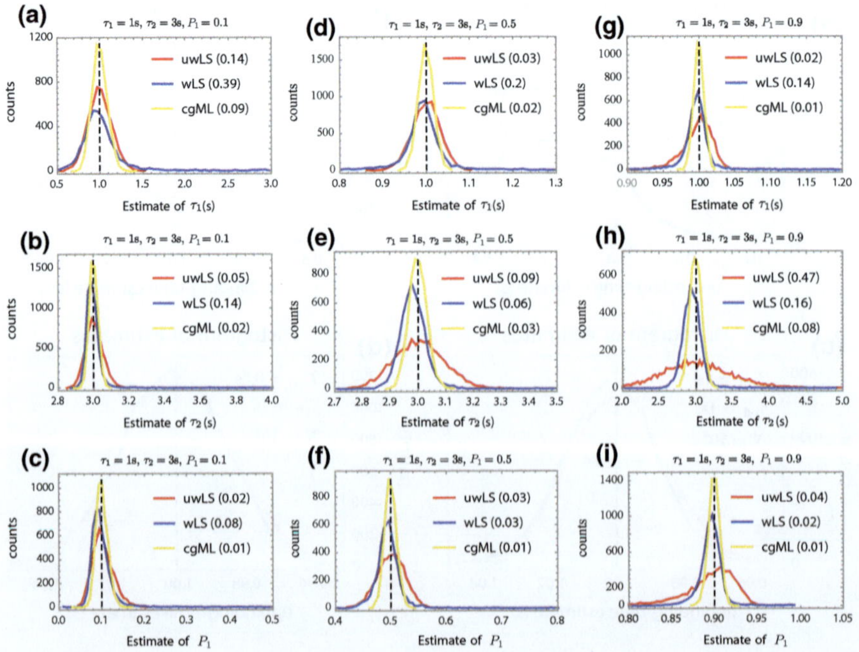

Fig. 5.7 Parameter estimation over 10,000 double-exponentially distributed data sets of size 10,000. Each column corresponds to parameter estimate distributions for a particular value of P_1, and each row corresponds to a particular model parameter. **a–c** $P_1 = 0.1$. **d, e** $P_1 = 0.5$. **f, h** $P_1 = 0.9$. In each case, we report the $\sqrt{\text{MSE}}$/s within parenthesis in the legend. In all considered situations, ML estimation is clearly the preferable choice as it has the lowest $\sqrt{\text{MSE}}$

The error in the estimation of the fraction of measurements belonging to the short timescale (P_1) is also dominated by the variance, and uwLS is particularly effected due to the poor accounting for the change in variance going from short to long timescales (Fig. 5.7c, f, and i). For all parameter values considered, cgML estimation again clearly outperforms the other methods as was expected from our theoretical developments.

5.5 Fitting Experimental Data

In the previous section, we have examined the performance of LS and ML estimation on well-specified data sets without experimental noise. Though a proper treatment of experimental noise is outside our present scope, it is still interesting to apply the three fitting methods on experimental data to see to what extent they agree. Considering experimental data will also give us the opportunity to comment on how to estimate the variance of parameter estimates through bootstrapping.

5.5.1 All Fits Different, but All Naively Plausible

Continuing with our RNA–protein unbinding example, we now analyze SM total internal reflection microscopy (TIRFM) data. The experiments measure the unbinding time of double-stranded (ds) RNA from viral RNA-binding proteins involved in protecting the viral genome from the hosts' RNA interference-based defenses [6]. The viral suppressors of RNA interference (VSR) proteins are immobilized on a glass surface, and the binding/unbinding of fluorescently tagged dsRNAs to the immobilized VSRs is followed (for more information on the biological aspects and the interpretation of the data, see [6]).

The unbinding-time data of 50 nucleotide dsRNA-binding VSR is fitted with uwLS, wLS, and cgML methods in Fig. 5.8a–c. In this particular system, and presumably due to the existence of weak and very strong binding modes, it is common to have a population of VSRs that unbind quickly, as well as a population that remain bound for the duration of the measurement. In the latter case, the apparent unbinding time will report on the photobleaching time of the fluorophores, as discussed previously. In such situations, the appropriate PDF is double exponential (Eq. 5.9), and the information regarding the number of molecules still bound and fluorescing at the end of the experiment (N_{cut}) can be incorporated in the ML estimation along

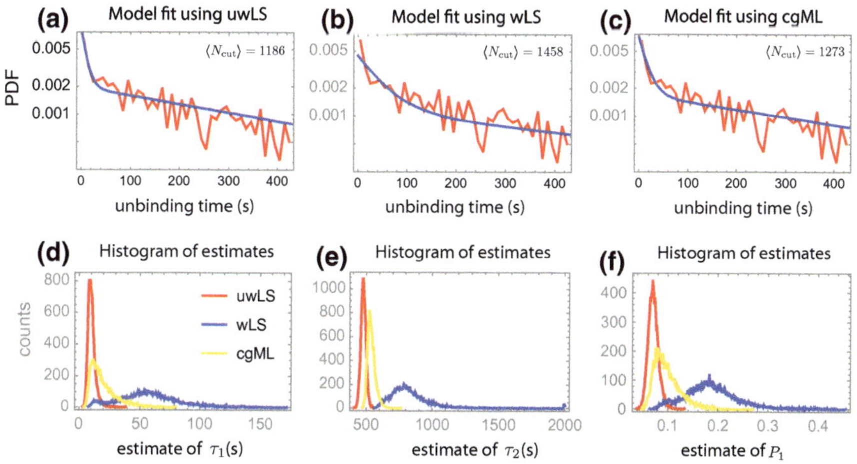

Fig. 5.8 **a** The measured distribution of unbinding times (red) together with the uwLS fit (blue). **b** The measured distribution of unbinding times (red) together with the wLS fit (blue). **c** The measured distribution of unbinding times (red) together with the cgML fit (blue). In **a–c**, the average number of measurements predicted to fall outside the observation window for the optimal fit is given as an inset. This should be compared to $N_{\text{cut}} = 1298$ in the fitted data set. **d** Histogram of estimates for the short timescale generated over 10,000 bootstrapped data sets. **e** Histogram of estimates for the long timescale generated over 10,000 bootstrapped data sets. **f** Histogram of estimates of the fraction of unbinding times originating in the short timescale, generated from 10,000 bootstrapped data sets. The parameter distributions vary significantly between data sets, even though all fits look plausible in **a–c**

the lines of Eq. (5.7)

$$L^{\mathrm{ML}}(\tau_1, \tau_2, P_1) = -\sum_{n=1}^{N_{\mathrm{rec}}} \ln\left(\frac{P_1}{\tau_1}\mathrm{e}^{-t_n/\tau_1} + \frac{1-P_1}{\tau_2}\mathrm{e}^{-t_n/\tau_2}\right) - N_{\mathrm{cut}} \ln\left(P_1 \mathrm{e}^{-T_{\mathrm{cut}}/\tau_1} + (1-P_1)\mathrm{e}^{-T_{\mathrm{cut}}/\tau_2}\right). \tag{5.14}$$

The information regarding N_{cut} is ignored in standard uwLS and wLS approaches, where the data is binned and fitted based on Eq. (5.9) only within the window capturing the N_{rec} unbinding times.

As can be seen in Fig. 5.8a–c, the three methods considered give very different results, all naively appearing to describe the data well. Lacking an objective way to evaluate the goodness of fit across scenarios, we can only point to the fact that our general developments and our numerical investigation suggest that the ML approach gives the best estimate of the fit parameters.

The insets in Fig. 5.8a–c report the average number of measurements that the best fit predicts should fall outside the measurement window. This average should be compared to the $N_{\mathrm{cut}} = 1298$ measurements that actually fell outside the observation window. From this, it is clear that the extra information included in the ML estimation regarding the cut data does increase its predictive capabilities in this case, which was not visibly the case for the fits in Fig. 5.6c, d.

5.5.2 Bootstrapping: Doing the Best We Can with Limited Resources

To determine the standard deviation of our parameter estimates, we would ideally like to establish their distribution by repeating the same experiment many times—much like we did in our earlier numerical comparison between estimation methods. A common practice is to report the standard deviation of fit estimates over a triplicate of identical experiments. However, not having a statistically significant sample can result in significant errors in estimating the standard deviation. Unfortunately, repeating the same experiment a sufficient number of times is often too time-consuming and costly, and we have to rely on other means.

If we could perform repeat experiments, we would in effect draw new unbinding times from the *true* PDF describing the unbinding kinetics. Instead of repeating the experiments by drawing from the true PDF, we here repeatedly draw from our best estimate of the true PDF: the original data set. This approach is called bootstrapping the data [5]. To generate each "new experiment," we randomly draw N unbinding times from our original data set (also of size N), *allowing for repeated draws* of the same data instance (this is known as random sampling with replacement). We then fit our bootstrapped data set in the same manner as we fit our original data sets. By repeating this process many times, we build up the desired distributions of fit parameters. In Fig. 5.8d–f, the distributions of the double-exponential fit parameters are plotted, using uwLS, wLS, and cgML methods over 10,000 bootstrapped data sets.

Contrary to the situation with our simulated data sets, we here do not know the true values of the model parameters and so cannot establish the bias nor the MSE and thus lack an objective metric by which to compare the different approaches. In light of this, it is important to stress that the fact that the standard deviation is consistently smallest for uwLS is not a good argument for why this approach should be preferable. Given the disparate results of the various methods—even though all fits naively look good (Fig. 5.8a–c)—it is clear that at least two of the three methods can go astray in very non-obvious ways, and that caution is warranted. Our heuristic arguments and simulations suggest ML estimation to be generally preferable.

5.6 Conclusion

We have provided an introduction to ML estimation as a powerful alternative to conventional LS fitting methods. Focusing on exponential distributions as examples, we showed how the ML method provides a general way to estimate the model parameters from stochastic data, in principle without the need for binning. We also showed that uwLS, wLS, and ML can all be thought of as approximations to tLS, utilizing various estimates for the a priori unknown standard deviation of bin counts. The main upshots of both our heuristic argument and numerical investigation are:

1. wLS becomes unreliable as soon as there are bins with low counts, as should always be expected in the tail end of distributions without a severe experimental cutoff time.
2. uwLS often outperforms wLS for processes with a single characteristic time, but for processes with multiple characteristic times, it becomes unreliable as it fails to appropriately weigh the contribution of data on different timescales.
3. (cg)ML consistently outperforms both wLS and uwLS by estimating bin-count variations from the whole data set, rather than ignoring them (uwLS) or estimating them on a bin-to-bin basis (wLS).

The two first points significantly limit the global applicability of both uwLS and wLS methods. The maximum-likelihood method is generally applicable though, needs no binning—but if binned, is not sensitive to empty bins—and outperforms both uwLS and wLS in all examples discussed. Although we focused on exponentially distributed data, our conclusions are general and should apply irrespective of the particular distribution describing the data. These advantages, together with the adaptability of the approach, have convinced the authors that ML estimation is the preferable choice for dealing with SM data; we hope our presentation has gone some way toward convincing the reader of the same.

Acknowledgements We thank Tao Ju (Thijs) Cui, Misha Klein, and Olivera Rakic for careful reading of the manuscript and thoughtful feedback. B. Eslami-Mosallam acknowledges financial support through the research program *Crowd management: the physics of genome processing in complex environments*, which is financed by the Netherlands Organisation for Scientific Research. I. Katechis acknowledges financial support from the Netherlands Organisation for Scientific Research, as part of the Frontiers in Nanoscience program.

References

1. Aartsen, M. G., Abraham, K., Ackermann, M., Adams, J., Aguilar, J. A., Ahlers, M., et al. (2015). A combined maximum-likelihood analysis of the high-energy astrophysical neutrino flux measured with icecube. *Astrophysical Journal, 809*(1), 1–15.
2. Avdis, E., & Wachter, J. A. (2017). Maximum likelihood estimation of the equity premium. *Journal of Financial Economics, 125*(3), 589–609.
3. Bahl, L. R., Jelinek, F., & Mercer, R. L. (1983). A maximum likelihood approach to continuous speech recognition. *IEEE Transactions on Pattern Analysis and Machine Intelligence, 5*(2), 179–190.
4. Dulin, D., Berghuis, B. A., Depken, M., & Dekker, N. H. (2015). Untangling reaction pathways through modern approaches to high-throughput single-molecule force-spectroscopy experiments. *Current Opinion in Structural Biology*.
5. Efron, B., & Tibshirani, R. (1994). *An introduction to the bootstrap*. Chapman & Hall.
6. Fareh, M., van Lopik, J., Katechis, I., Bronkhorst, A. W., Haagsma, A. C., van Rij, R. P., & Joo, C. (2018). Viral suppressors of RNAi employ a rapid screening mode to discriminate viral RNA from cellular small RNA. *Nucleic Acids Research* (March), 1–11.
7. Felsenstein, J. (1981). Evolutionary trees from DNA sequences: A maximum likelihood approach. *Journal of Molecular Evolution, 17*(6), 368–376.
8. Forney, G. D. (1972). Maximum-likelihood sequence estimation of digital sequences in the presence of intersymbol interference. *IEEE Transactions on Information Theory, 18*(3), 363–378.
9. Hastie, T., Tibsharani, R., & Friedman, J. (2009). *The elements of statistical learning. The mathematical intelligencer* (2nd ed.). New York: Springer.
10. Hauschild, T., & Jentschel, M. (2001). Comparison of maximum likelihood estimation and chi-square statistics applied to counting experiments. *Nuclear Instruments and Methods in Physics Research A, 457*(1–2), 384–401.
11. Humphrey, P. J., Liu, W., & Buote, D. A. (2009). χ2 and Poissonian data: BIASES even in the high-count regime and how to avoid them. *The Astrophysical Journal, 693*(1), 822–829.
12. Jaynes, E. T. (2003). *Probability theory: The logic of science*. Cambridge: Cambridge University Press.
13. Johansen, S., & Juselius, K. (1990). Maximum likelihood estimation and inference on cointegration—With applications to the demand for money. *Oxford Bulletin of Economics and Statistics, 52*(2), 169–210.
14. Joo, C., Balci, H., Ishitsuka, Y., Buranachai, C., & Ha, T. (2008). Advances in single-molecule fluorescence methods for molecular biology. *Annual Review of Biochemistry, 77*, 51–76.
15. Joo, C., & Ha, T. (2012). Single-molecule FRET with total internal reflection microscopy. *Cold Spring Harbor Protocols, 7*(12), 1223–1237.
16. Leggetter, C. J., & Woodland, P. C. (1995). Maximum likelihood linear regression for speaker adaptation of continuous density hidden markov models. *Computer Speech & Language, 9*(2), 171–185.
17. Murshudov, G. N., Vagin, A. A., & Dodson, E. J. (1997). Refinement of macromolecular structures by the maximum-likelihood method. *Acta Crystallographica. Section D, Biological Crystallography, 53*(3), 240–255.

18. Nelson, P. (2015). *Physical models of living systems*. New York: W. H. Freeman.
19. Nishimura, G., & Tamura, M. (2005). Artefacts in the analysis of temporal response functions measured by photon counting. *Physics in Medicine & Biology, 50*(6), 1327–1342.
20. Nørrelykke, S. F., & Flyvbjerg, H. (2010). Power spectrum analysis with least-squares fitting: Amplitude bias and its elimination, with application to optical tweezers and atomic force microscope cantilevers. *Review of Scientific Instruments, 81*(7).
21. Nousek, J. A., & Shue, D. R. (1989). Chi-squared and C statistic minimization for low count per bin data. *Astrophysical Journal, 342,* 1207–1211.
22. Press, W., Teukolsky, S., Vetterling, W., Flannery, B., Ziegel, E., Press, W., et al. (2007). *Numerical recipes: The art of scientific computing* (3rd ed.). Cambridge: Cambridge University Press.
23. Santra, K., Zhan, J., Song, X., Smith, E. A., Vaswani, N., & Petrich, J. W. (2016). What is the best method to fit time-resolved data? A comparison of the residual minimization and the maximum likelihood techniques as applied to experimental time-correlated, single-photon counting data. *Journal of Physical Chemistry B, 120*(9), 2484–2490.
24. Scholten, T. L., & Blume-Kohout, R. (2018). Behavior of the maximum likelihood in quantum state tomography. *New Journal of Physics, 20,* 023050.
25. Stamatakis, A. (2006). RAxML-VI-HPC: Maximum likelihood-based phylogenetic analyses with thousands of taxa and mixed models. *Bioinformatics, 22*(21), 2688–2690.
26. Trifinopoulos, J., Nguyen, L. T., von Haeseler, A., & Minh, B. Q. (2016). W-IQ-TREE: A fast online phylogenetic tool for maximum likelihood analysis. *Nucleic Acids Research, 44*(W1), W232–W235.
27. Turton, D. A., Reid, G. D., & Beddard, G. S. (2003). Accurate analysis of fluorescence decays from single molecules in photon counting experiments. *Analytical Chemistry, 75*(16), 4182–4187.
28. Whelan, S., & Goldman, N. (2001). A general empirical model of protein evolution derived from multiple protein families using a maximum-likelihood approach. *Molecular Biology and Evolution, 18*(5), 691–699.

Part II
Transcription and Translation

Chapter 6
A Single-Molecule View on Cellular and Viral RNA Synthesis

Eugen Ostrofet, Flavia Stal Papini, Anssi M. Malinen and David Dulin

6.1 Introduction

Ribonucleic acid (RNA) mediates the genetic information to protein synthesis and is thus an essential component of the central dogma in molecular biology. Additionally, specific RNAs serve as structural and catalytic constituents in ribozymes, such as ribosome and spliceosome, or participate in the regulation of gene expression. Furthermore, RNA uniquely both stores and mediates the genetic information in RNA viruses [4]. RNA polymerase enzymes make RNA by joining together the ribonucleotides building blocks in a template-guided reaction. On the one hand, if the RNA polymerase uses deoxyribonucleic acid (DNA) as a template to build the complementary RNA strand, the RNA polymerase is classified as DNA-dependent RNA polymerases (DdRp) [4]. All cellular RNA polymerases (RNAPs) belong to this class. On the other hand, if the template is RNA, as in RNA virus genome replication, the RNA polymerase is called RNA-dependent RNA polymerase (RdRp) [134].

This essay focuses on the multisubunit cellular DdRps, also coined RNAPs, and specifically, bacterial RNAP, archaeal RNAP, and three distinct eukaryotic RNAPs (Pol I, Pol II, and Pol III) [4]. The tight regulation of RNAP activity is key to gene expression, as it provides the cell the means to respond quickly to environmental changes. Cellular transcription is regulated by a myriad of factors, e.g., DNA promoter sequence, transcribed sequence, the secondary structures of the RNA transcript, protein transcription factors interacting with the DNA or RNAP itself, nucleoproteins, and RNAP trafficking [7, 12, 74, 113, 120, 126, 146]. As a consequence,

E. Ostrofet · F. Stal Papini · D. Dulin (✉)
Junior Research Group 2, Interdisciplinary Center for Clinical Research, Friedrich Alexander University Erlangen-Nürnberg (FAU), Cauerstr. 3, 91058 Erlangen, Germany
e-mail: david.dulin@uk-erlangen.de

A. M. Malinen
Department of Biochemistry, University of Turku, Tykistökatu 6 A, Biocity 6th floor, 20520 Turku, Finland

C. Joo and D. Rueda (eds.), *Biophysics of RNA-Protein Interactions*, Biological and Medical Physics, Biomedical Engineering,
https://doi.org/10.1007/978-1-4939-9726-8_6

the RNAP transcriptional activity is very dynamic, with bursts of RNA synthesis being interrupted by pauses that can vary in length and nature [74, 86, 126]. The overlapping layers of RNAP regulation make the transcription process stochastic and noisy, which eventually influences protein production levels and the phenotype of an organism [40, 76, 112].

The RdRp class contains two evolutionary non-related families: Cellular RdRps are present in plants and produce short RNAs that are important in the developmental regulation, genome integrity maintenance and defense against pathogens [3, 71, 156], while the second family of RdRp is one of the key components of replication of RNA viruses [26, 111, 134]. We discuss here only the latter, as no cellular RdRps single-molecule study has been published to date. Viral RdRps not only replicate the viral genome but also allow it to evolve rapidly because of their very high error rate, originating from nucleotide misincorporations [51, 68]. Though high, the error rate is tightly controlled to simultaneously provide an evolutionary advantage over the host immune system and a robust production of infectious virions [92, 132]. RdRps replication activity is highly regulated by different factors, e.g., viral genome secondary structures, viral-associated proteins and host factors. Furthermore, because of their central role in virus proliferation and their conserved catalytic structural domains, RdRps also represent a target for broad-spectrum antiviral drugs, such as nucleotide analogs [72].

The development of single-molecule techniques has offered a unique view of the action of enzymes, including that of RNA polymerases [37, 104]. The world of a biomolecule is strikingly different from the one we experience as humans: The characteristic length of molecules is ~1–10 nm, inertial forces are negligible, and cellular processes are driven by diffusion and low-energy activations in the order of few $k_B T$ (k_B being the Boltzmann constant and T the absolute temperature). For example, a polymerase that moves forward by a length of a DNA base pair (bp) (0.34 nm) against a force of 12 pN (10^{-12} N) generates a mechanical work of ~1 $k_B T$, which equals the thermal energy provided by the environment. To give some perspective, a ~25 pN hindering force is necessary to stall a bacterial RNAP [145]. As it translocates with 0.34 nm DNA base-pair (bp) steps, it performs up to ~2 $k_B T$ mechanical work ($E = F \times \delta$, where E is the mechanical work, F the applied force, and δ the step size of the biomolecular motor), while the Φ29 DNA packaging motor, that translocates by 2.3 bp step size, withstands a force up to ~57 pN and performs up to ~11 $k_B T$ mechanical work [99]. The kinesin withstands forces up to 8 pN, but translocates forward with ~8 nm steps, and therefore performs up to ~16 $k_B T$ mechanical work [143]. In the context of a long DNA sequence and ignoring the nearest neighbors influence, breaking a DNA A/T base pair costs in average ~1.5 $k_B T$, while breaking a G/C base pair costs ~3 $k_B T$ [10]. Because the energies involved are low, positions and conformations fluctuate constantly rendering the catalytic activity of enzymes "noisy" or stochastic. Furthermore, enzymes often explore several kinetic states, with different interconversion rates and not necessarily catalytically active, making their catalytic activity like a journey through a very bumpy road. Observing these different kinetic states is difficult—if not impossible—with classic ensemble approaches, as it averages out the heterogeneity of the sample, extracting only the average behavior of a population of biomolecules [137]. By monitoring enzyme catalysis at the single-molecule level, it is possible to uncover complex and parallel kinetic pathways, where transient and rare events play important roles in the overall activity of the enzyme [41, 151]. In vitro single-molecule techniques can be divided in two main categories: (i) force spectroscopy techniques, where the change in the extension of a tethered nucleic acids upon a single-enzyme activity is monitored under an applied force, (ii) fluorescence spectroscopy, where dye-labeled biomolecules are localized and followed using high-end microscopy to monitor inter- or intra-molecule conformational changes [37, 104]. Both approaches have led to many important discoveries on the molecular mechanism of RNA synthesis in all domains of life.

In Figs. 6.1 and 6.2, we present the single-molecule techniques used in in vitro experiments on cellular RNAPs or viral RdRps. These figures mainly serve to "visualize" the experimental assay used in the described studies, and we advise the reader to look into the numerous specialized reviews that treat the techniques from a more technical viewpoint of single-molecule biophysics [14, 27, 37, 67, 75, 84, 104, 105, 110, 121]. In the following parts of the book chapter, we discuss the recent literature of the in vitro single-molecule enzymology studies of RNA synthesis by cellular RNAPs and viral RdRps.

6.2 In Vitro Single-Molecule Studies of Cellular RNAPs

Cellular transcription performed by large multisubunit RNAPs can be divided into three different phases, called initiation, elongation, and termination. The RNAP finds and initiates RNA synthesis at a promoter—a specifically recognized DNA sequence preceding every gene or operon (a set of adjacent co-regulated genes). The RNAP subsequently moves into the coding region of the gene and elongates the nascent RNA until it meets a termination signal at the end of the gene or operon. These three phases of transcription have been extensively studied at the single-molecule level using the RNAP from *Escherichia coli* (*E. coli*) as a model system, and more recently also using structurally more complex eukaryotic and archaeal RNAPs. We highlight here some of the key studies.

6.2.1 Initiation

6.2.1.1 Bacterial Transcription Machinery

During the initiation phase, the RNAP must (1) recognize the promoter, (2) form the transcription bubble, i.e., open the double-stranded DNA, (3) initiate the synthesis of RNA, (4) stably hold the short nascent RNA in the active site, and (5) break interactions with the promoter and transcription initiation factor(s) on the way to the elongation phase [11, 12, 122, 124]. However, the RNAP may fail at each of

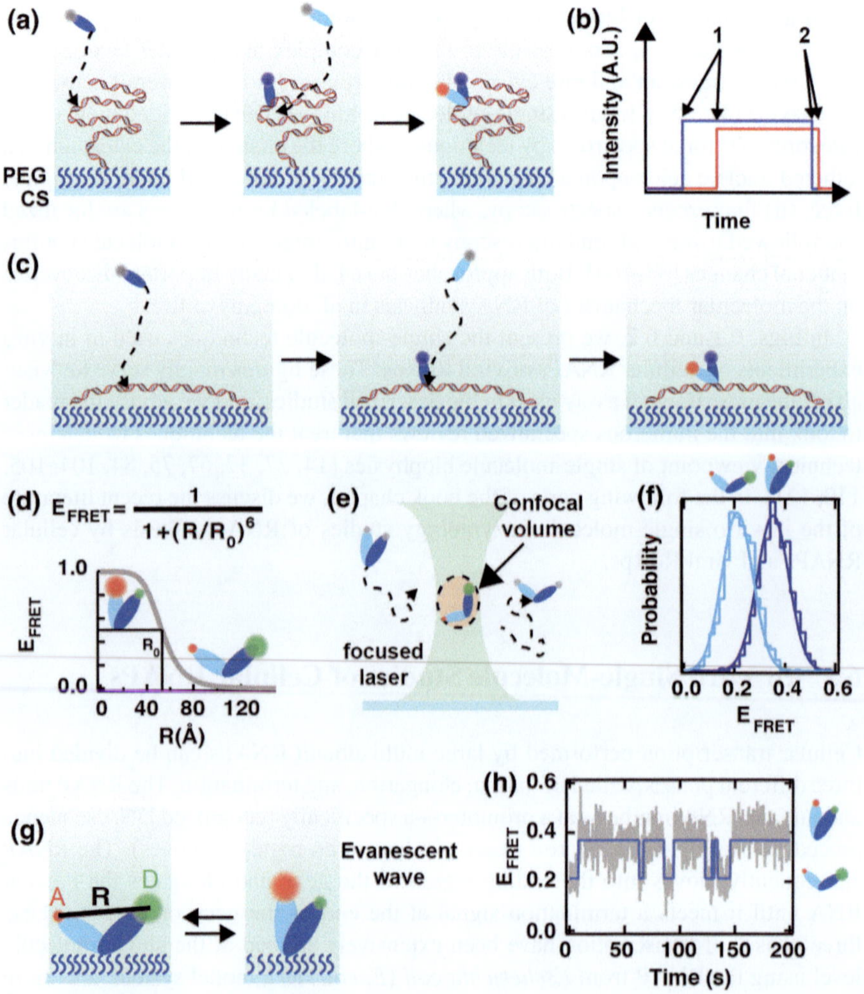
(a)
PEG
CS
(b)
Intensity (A.U.)
1
2
Time
(c)
(d) E_FRET = 1 / (1+(R/R_0)^6)
1.0
E_FRET
R_0
0.0
0
40
80
120
R(Å)
(e)
Confocal volume
focused laser
(f)
Probability
0.0
0.2
0.4
0.6
E_FRET
(g)
A
R
D
Evanescent wave
(h)
0.6
0.4
0.2
0.0
E_FRET
0
50
100
150
200
Time (s)

◂**Fig. 6.1** In vitro single-molecule fluorescence spectroscopy techniques in transcription studies. **a** Single-molecule fluorescence co-localization microscopy monitors in real-time binding–unbinding events of interacting molecules [18, 48, 130]. Total internal reflection fluorescence microscopy (TIRFM) is used to image surface-attached biomolecules [5]. In TIRFM, the excitation laser is reflected at the glass–water interface, generating an evanescent wave (green shade) above the coverslip (CS) that excites dyes within ~100 nm from the coverslip top surface. Non-specific adsorption of the labeled biomolecules is blocked with specific surface chemistry, e.g., polyethylene glycol coating [16]. **b** Observed variation in fluorescence at a specific location where two biomolecules successively bind (1) and dissociate (2) from the nucleic acid molecule as depicted in panel (**a**). **c** By stretching the nucleic acids, as in flow stretching [53] or in DNA curtains [20], one is able to localize the DNA binding molecules with ~100 nm accuracy. **d** In single-molecule Forster Resonant Energy Transfer (smFRET), a fraction of the excitation energy of the donor dye (green) is transferred non-radiatively to the acceptor dye (red) with an efficiency E_{FRET} that decreases when the distance between the two dyes increases [59]. Using this molecular ruler, distances ranging from 2 to 10 nm can be determined [67]. **e** In confocal microscopy, dye-labeled biomolecules at low concentration diffuse freely in the solution and a short burst of photons is detected when a labeled molecule crosses the confocal detection volume. To increase sensitivity, photons emitted outside the confocal volume are spatially filtered out in the imaging optical path. smFRET combined with alternative-laser excitation (ALEX) in confocal microscopy is a powerful tool to access conformational equilibrium and complex formation by biomolecules in solution [67, 79]. **f** Confocal smFRET provides the prevalence and the number of different conformations, which are revealed by the amplitude and the width of the normal distributions describing the E_{FRET} histogram, respectively; however, solution smFRET cannot obtain the rates of conformational change. **g** To observe the conformational changes happening in an individual biomolecule, it has to be immobilized to the surface of a coverslip. **h** E_{FRET} (grey) time trace, typically obtained using TIRFM-based smFRET, shows the biomolecule to interconvert between two distinct conformations. The kinetic constants defining the stabilities of the two states are recovered using a hidden Markov model (blue) [140]. Single-molecule FRET can also determine accurate distances between the donor and acceptor dyes; the obtained distances help to model the 3D structure of the biomolecule [6, 77]

these steps, making transcription initiation a highly stochastic and heterogeneous process. Single-molecule techniques are therefore particularly suitable to uncover the determinants of the transitions from one step to the next [82].

The bacterial transcription initiation complex—formed by the association of the core RNAP with a σ initiation factor as a holoenzyme—is the most studied transcription complex at the single-molecule level. The holoenzyme has to respond to a large variety of biochemical signals to control gene expression. The holoenzyme also needs to negotiate substantial variation in the promoter sequences that take place ~100 bp upstream and ~20 bp downstream of the transcription start site (TSS, locates at position +1); there are ~3000 different promoters in *E. coli* each imposing a unique set of parameter (rates, stabilities) to the substages of initial transcription [11, 12]. At the beginning of gene transcription, the holoenzyme finds the promoter and forms the RNAP–promoter closed (RP_C) complex by establishing interactions with specific elements of the promoter. The promoter search mechanism was investigated using a single-molecule fluorescence co-localization assay in combination with total internal reflection microscopy (TIRFM) (Fig. 6.1a, b) [48]. The data consists of holoenzyme binding dwell times on the DNA promoter as well as the dwell times separating two binding events to the same promoter; further variables included holoenzyme

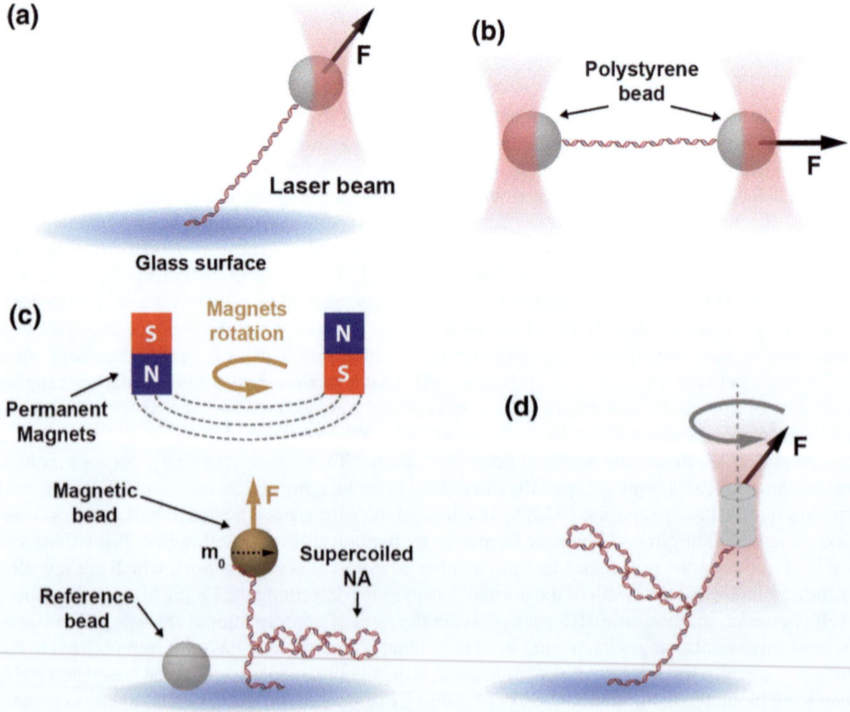

Fig. 6.2 Single-molecule force spectroscopy techniques applied to transcription studies. **a** In an optical tweezers assay, one possible configuration is to attach a nucleic acid molecule (NA) from one end to the glass coverslip of the flow chamber and from the other end to a polystyrene bead trapped in a focused laser beam. The bead position can be controlled in three dimensions by moving the position of the laser focal point. Displacement of the bead from the equilibrium position, i.e., the center of the trap, increases linearly the force F experienced by the NA tether. F ranges from ~0.1 pN to hundreds of pN [109]. **b** Most modern optical tweezers utilize two optical traps to pull the NA from both ends. The configuration produces a signal with a smaller drift artifact and thus a resolution high enough to distinguish translocation steps at ~0.34 nm/s velocity [57]. **c** In a magnetic tweezers assay, double- or single-stranded NA is attached from one end to the coverslip and from the other end to a magnetic bead. The force F (from ~10 fN to ~1 nN) is generated by pulling the bead with a pair of magnets located above the flow chamber. A reference bead on the coverslip surface is used to correct for the mechanical vibrations caused drift in the position of the sample bead [142]. Camera-based detection allows the simultaneous tracking of hundreds of beads at near base-pair resolution. The magnetization (m_0) of the bead permits its rotation, by rotating the magnetic field originating from the magnets. Rotation of the bead adds supercoiling to the torsionally constrained double-stranded NA, which eventually leads to the formation of plectonemes. Torque-dependent behavior of protein–NA interactions can thus be studied [27]. **d** In an optical torque wrench, a birefringent particle (here a cylinder) is trapped by a polarized laser beam. By rotating the polarization of the laser, the birefringent cylinder rotates and induces supercoiling in the NA [85, 125]. The "sticky" ends of the NA, necessary in optical tweezers and magnetic tweezers assays, are typically generated with biotin- and digoxigenin-labeled nucleotides that bind very stably to streptavidin/neutravidin and antidigoxigenin antibody, respectively, coating the bead or coverslip surface [73]

concentration and the total length of DNA construct [49]. The results were consistent with a search process being dominated by three-dimensional diffusion instead of one-dimensional sliding along the DNA [49]. A similar conclusion was reached using a DNA curtains approach, where stretched bacteriophage λ-DNA containing several *E. coli* RNAP–promoters was used to observe binding/unbinding RNAPs eventually converting into productive transcribing complexes [144] (Fig. 6.1c). The contacts formed between the holoenzyme and promoter were recently studied using an optical tweezers assay in a dumbbell configuration (Fig. 6.2b), where the promoter is encoded in the stem of a hairpin (Fig. 6.3a) [103]. Using this elegant experimental configuration, it is possible to determine the contact points between the promoter and the holoenzyme, even the most transient ones, by opening the hairpin with a linear increase in the force [149] (Fig. 6.3b). By comparing the hairpin opening profiles of the free promoter and holoenzyme-bound promoter, the authors found that, besides the well-known contacts formed with the −35 and −10 elements of the promoter, the holoenzyme forms strong contacts with the spacer element—located between the −35 and −10 elements—and that the various contacts are released in a non-sequential fashion during promoter escape [103].

After recognizing the promoter, the holoenzyme opens ~12 bp stretch of the double-stranded DNA (dsDNA) to form the RNAP open–promoter complex (RP_O). dsDNA melting is monitored using the topological changes taking place in a torsionally constrained DNA molecule [115, 123]. DNA has a fixed amount of twist between each nucleotide, which results in a helical pitch of ~10.5 bp/turn. When adding twist to a DNA molecule, e.g., by adding turns, and for a stretching force below ~0.5 pN, the DNA molecule eventually buckles and forms plectonemes (Fig. 6.2c), and therefore adds writhe, i.e., the DNA molecule crossing over itself. The sum of writhe and twist is conserved in a torsionally constrained DNA molecule [17], and therefore, the change in twist eventually leads to a change in writhe to compensate. In other words, for a negatively (or positively) supercoiled DNA molecule that has passed the buckling transition and has formed plectonemes, the number of writhes decreases (or increases) upon promoter melting, i.e., decrease in twist, leading to an increase (or decrease) in the end-to-end extension of the DNA molecule. Using this property in combination with a torque spectroscopy technique, i.e., optical torque wrench (Fig. 6.2d) or magnetic tweezers (Fig. 6.3c), it is therefore possible to monitor promoter melting. Strick and co-workers used a magnetic tweezers assay to monitor the RP_O dynamics on a consensus and ribosomal promoters, which, respectively, make stable and unstable contacts with the holoenzyme (Fig. 6.3c, d). The authors showed that the addition of torque in a torsionally constrained DNA molecule affects the formation and the stability of the RP_O: The addition of negative supercoil assists promoter opening by lowering the DNA melting energy penalty and promotes the formation of a stable RP_O, whereas the addition of positive supercoil increases the DNA melting energy penalty, hinders promoter opening, and reduces the RP_O lifetime [114]. A recent study using single-molecule Förster resonance energy transfer (smFRET) with TIRFM (Fig. 6.1d, g, h) showed that in the absence of promoter supercoiling, the RP_O is very dynamic, fluctuating in millisecond timescale between the open and the closed DNA conformations; the contacts formed by the $\sigma_{3.2}$ domain of

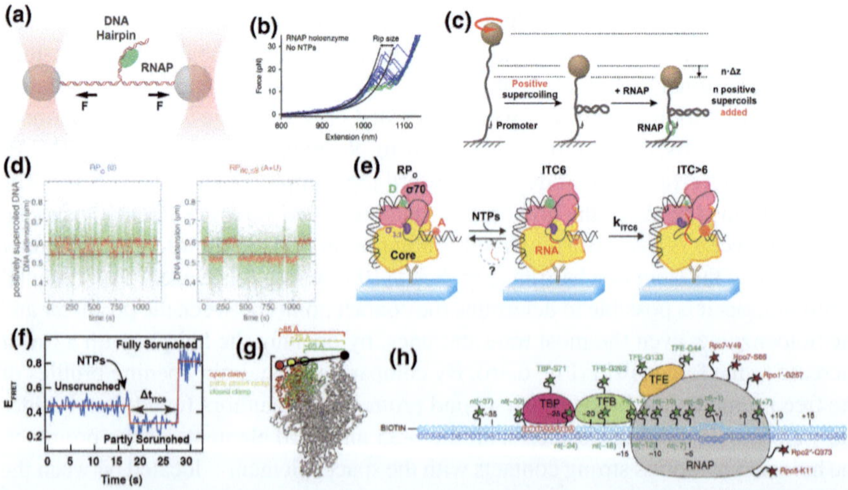

Fig. 6.3 In vitro single-molecule studies of transcription initiation by multisubunit RNA polymerases. **a** The free energy landscape of the RNAP–DNA promoter interaction is manipulated and characterized by progressively unwinding the promoter sequence present in the DNA hairpin by moving the optical traps further apart. **b** Force extensions curves from the experiments presented in (**a**) without (green) and with RNAP (blue) bound to the promoter in the hairpin stem. Figure adapted from Ref. [103]. **c** Monitoring transcription initiation using DNA supercoiled by magnetic tweezers. When the RNAP forms the transcription bubble on a supercoiled DNA, it changes the total amount of twist, which is thus compensated by a change in writhe, therefore in z-axis magnetic bead position, i.e. a decrease in end-to-end tether extension for a positively supercoiled DNA molecule. **d** Conformational dynamics of RP_O (left) and abortive synthesis of 8-mer RNAs by the initially transcribing complex (ITC, right) was monitored by magnetic tweezers. Adapted from Ref. [116]. **e** TIRFM-based smFRET assay can record the dynamics of initial transcription. RP_O complex is formed with a promoter containing a donor dye (green sphere) and an acceptor dye (red sphere) upstream and downstream of the transcription bubble, respectively. The RP_O is immobilized to the coverslip surface with an antibody. RNA synthesis is coupled to the promoter scrunching and movement of the downstream DNA toward the RNAP, leading to the change in the dye pair distance and E_{FRET}. **f** E_{FRET} was continuously measured with the assay described in (**f**) to monitor the magnitude of promoter scrunching and thus the progress of initial transcription. **f**, **g** are adapted from Ref. [34]. **g** Model of alternative clamp positions in the bacterial RNAP was developed based on confocal microscopy smFRET experiments. Red, yellow, and green indicate the positions explored by the β′ clamp domain and the donor dye (spheres), respectively. The black sphere indicates the immobile position of the acceptor dye on the opposite side of the DNA binding cleft. Adapted from Ref. [15]. **h** Nano-Positioning System has been used to map the structure of archaeal pre-initiation complex by determining multiple distances between the transcription initiation factors (TBP, TFB, and TFE) and the RNAP as well as between the non-template DNA strand and RNAP. The inter-dye distances were calculated from the measured E_{FRET} values. The donor and acceptor dye locations used in the study are indicated with green and red stars, respectively. Adapted from Ref. [107]

σ^{70} (the housekeeping initiation factor of *E. coli*) with the template DNA stabilize the open form of RP_O [30]. Noteworthy, TIRFM-smFRET and magnetic tweezers have temporal resolutions of ~10 ms and ~1 s, respectively. Two studies indirectly assessed the submillisecond dynamics of the RP_O using confocal smFRET data (Fig. 6.1e) in combination with the signal burst variance analysis [118] or the photon-by-photon hidden Markov modeling [95]. Both studies showed that the RP_O explores different transcription bubble sizes by opening more downstream DNA, which have been suggested to determine the transcription start site (TSS). The mechanistic basis for TSS selection was recently further explored using magnetic tweezers (Fig. 6.3CD) and DNA–protein photocrosslinking experiments, showing that the energetics of the transcription bubble size eventually regulates the TSS selection, in practice limiting its range to the positions −1, +1, and + 2 [153].

In the presence of NTPs, the RP_O quickly engages in the synthesis of the nascent RNA, forming an initially transcribing complex (ITC). Early biochemical experiments showed that the (average) position of the upstream position of the ITC does not change during the addition of the first 9–11 nucleotides to the RNA. Three different mechanisms were proposed to describe how the RNAP manages these constraints: (1) transient excursion (RNAP diffuses back and forth between subsequent abortive initiations), (2) inchworming (flexible RNAP body containing the active site stretches forward at each nucleotide incorporation), and (3) downstream DNA scrunching (the DNA bubble is extended inside and on the surface of the RNAP). Elegant confocal smFRET and magnetic tweezers studies showed that the scrunching model was the correct one (Figs. 6.1d, e and 6.3c, d, respectively), where only the downstream DNA region of the promoter is moving relative to the holoenzyme during initial transcription [80, 116].

The initial transcription leads either to successful promoter escape and synthesis of the full-length RNA or the release of short (up to ~11 bases) aborted RNA products and reversion to the initial RP_O state [12]. The overall efficiency of transcription initiation is determined at two distinct phases: either at the stage of RP_O formation if the promoter, e.g. *rrnB*, forms an unstable DNA bubble, or during initial transcription if the promoter, e.g. *lacUV5*, forms a stable RP_O. The former case was explored by Strick and co-workers and described above [114]. The latter case was recently explored by several TIRFM-smFRET, confocal smFRET, and magnetic tweezer-based studies [29, 35, 94]. It has become evident that initial transcription is interrupted by two types of pauses: a short pause (half-life ~10 s) occurring after the synthesis of a 6-mer RNA, originating from the clash of the nascent RNA and the σ factor blocking the RNA exit channel (Fig. 6.3e) [29], and a long pause (~100–1000 s) involving a stable backtracked complex [94]. The latest of the three studies (Fig. 6.3e, f) found that these two pauses are actually connected via a branched mechanism, where a fraction of the initially transcribing complexes pauses after the synthesis of a 6-mer RNA and isomerizes to a long-lived backtracked pause state [35]. The backtracked fraction increases with the strength of the pause at a 6-mer RNA, which in turn depends on the initially transcribed sequence and the NTP concentration. This study additionally showed that promoter unscrunching does not always require the release of the abortive RNA, thereby expanding the earlier models that assumed the two processes to be tightly connected [102, 116].

The initiation factor σ^{70}, the housekeeping σ factor in *E. coli*, was thought to impact only the initiation phase and be released upon transition to the elongation phase. However, a confocal smFRET (Fig. 6.1e) study showed that σ^{70} is indeed retained during elongation, at least in vitro [81]. A more recent study from Harden and co-workers using TIRFM single-molecule co-localization (Fig. 6.2a, b) confirmed the retention of σ^{70} and found that σ^{70} influences the progress of the transcript elongation hundreds of bp downstream the promoter region by binding to and inducing pauses at -10 element-like DNA sequence [61].

The bacterial RNAP resembles structurally a crab claw, with the two "pincers," formed by the β and the β' subunits, defining the walls of the primary DNA binding cleft of the polymerase. The pincer of the β' subunit is called the clamp, which according to the crystal structures adopts multiple conformations, including the open and closed conformations that differ at most by a 20° swinging motion of the clamp from a hinge at the base of the clamp. Chakraborty et al. employed unnatural amino acid mutagenesis to specifically attach fluorophores to both pincers, thus generating a FRET ruler to monitor the clamp positions (Fig. 6.3g) [15]. Using confocal smFRET (Fig. 6.1d–f), the clamp conformation in different phases of transcription was observed. The authors found that the clamp of DNA-free core enzyme, RNAP-σ^{70} holoenzyme, and RNAP-σ^{54} holoenzyme have three distinct conformations at equilibrium, assigned as the open, closed, and collapsed clamp, respectively. The authors further extended the study by trapping the structural intermediates on the σ^{54}-dependent open complex formation pathway. In this experimental configuration, the clamp remains predominantly open until the promoter DNA melts and forms the transcription bubble in the open complex (RP_O); the clamp remains consistently closed in the initially transcribing complex and in elongation complex. The clamp state in the holoenzyme is modulated by antibiotics myxopyronin, corallopyronin, and ripostatin, as well as by bacteriophage T7 protein Gp2, with all of them depopulating the open clamp conformation. However, the confocal smFRET experiments did not have access to the kinetics of the clamp conformations. Duchi et al. [31] thus performed further experiments using TIRFM-smFRET (Fig. 6.1g, h) and similarly labeled RNAP. This experimental configuration allowed to set strict selection criteria for the homogeneity of the analyzed molecules, which lead to the re-assignment of the holoenzyme clamp states as open, partly closed, and closed. A significant fraction of the holoenzymes has a dynamic clamp that interconverts between these states in a timescale of ~0.1–1 s. The binding of stringent-response alarmone ppGpp stabilizes the partly closed clamp state of the RNAP. By combining cryo-EM based structural information with TIRFM-smFRET based data on clamp dynamics [96], it was possible to uncover the mechanism of RNAP inhibition by lipiarmycin, an antibiotic clinically used to treat *Clostridium difficile* infection. Lipiarmycin dramatically modifies the clamp by locking it in the open conformation and thereby prevents the isomerization of RP_C to RP_O.

6.2.1.2 Archaeal Transcription Machinery

The archaeal transcription machinery has many similarities with the eukaryotic Pol II machine as the RNAP and many associated factors are homologous [147]. Because the archaeal transcription machinery is less complex and can be readily reconstituted from recombinant proteins [129, 148], it constitutes, in addition to its inherent value, a good model system for understanding the mechanism of eukaryotic transcription. Archaeal RNAP and eukaryotic Pol II require two additional proteins for the basal level of transcription initiation: the TATA-binding protein (TBP) and the transcription factor B (known as TFB in archaea and TFIIB in eukaryotes). TBP bends the DNA promoter, associates with TFB and subsequently recruits the RNAP to form the pre-initiation complex (PIC). A TIRFM-smFRET (Fig. 6.1d, g, h) study found that the archaeal TBP dynamically bends and unbends the promoter, whereas its eukaryotic counterpart bends the promoter into two stable populations with different bending angle; the less bended population is eventually converted into the more bended conformation upon the addition of TFIIB [54]. Nagy and co-workers used TIRFM-smFRET (Fig. 6.1g, h) and Nano-Positioning System analysis [106] (Fig. 6.3h) to provide a structural model of the archaeal RP_O by determining the positions of the promoter DNA, RNAP and the transcription initiation factors TBP, TFB, and the transcription factor E (TFE) [107]. In a separate study, the RNAP binding sites of TFE and transcription elongation factor Spt4/5 were shown to overlap, which makes the two factors competing against each other to bind RNAP; this competition likely has implication in the regulation of transcription initiation and elongation. Schultz et al. used TIRFM-smFRET to analyze the clamp conformation in archaeal RNAP labeled with two fluorophores on the opposite sides of the DNA binding cleft [128]. The authors found that most (~80%) of the DNA-free RNAPs adopts a closed clamp conformation with a smaller amount of RNAPs having an open clamp (~20%). In contrast to the bacterial RNAP study [32], the authors did not observe real-time interconversions between the two clamp states. The opening of the transcription bubble upon RP_O formation shifted the clamp equilibrium toward the open state, whereas, interestingly, the exact opposite has been observed with the bacterial transcription system [15]. Consistent with the mutual dependence of the transcription bubble and clamp opening in the archaeal system, they showed that the transcription initiation factor TFE, which stimulates DNA opening, increases the fraction of RNAPs with an open clamp. Because the open clamp in the transcription elongation complex is also promoted by the binding of a correct (templated) nucleotide to the active site and elongation factor Spt4/5, the catalytically competent, highly processive archaeal RNAP may require a relatively open clamp conformation.

6.2.1.3 Eukaryotic Transcription Machineries

Eukaryotic transcription is far more complex than its bacterial counterpart and has therefore only recently been investigated. For example, the yeast Pol II system assembles a total of 32 proteins when it forms a pre-initiation complex (PIC) on the promoter. Recently, Galburt et al. [139] adapted the magnetic tweezers assay pioneered by the Strick lab (Fig. 6.3c, d) [114] to measure real-time promoter melting events by the Pol II and the distributions of DNA-bubble sizes generated during different phases of initiation. They found that promoter opening is in fact a two-steps process: First, the Ssl2 subunit of transcription initiation factor TFIIH pumps the downstream DNA toward the Pol II in the process generating torsional and mechanical stress that leads to the formation of an initial ~6-bp bubble; second, Pol II synthesizes the initial RNA at the transcription start site which expands the bubble to its final ~13-bp size. On the other hand, an earlier study by the Block lab using high-resolution optical tweezers (Fig. 6.4a) had predicted the formation of much a bigger transcription bubble (~85-bp) by the action of Ssl2 during initial transcription [44]. Clearly, a lot of work remains to be done by single-molecule biophysicists to bring the understanding of the mechanistic of the Pol II transcription initiation complex anywhere close to the bacterial one.

Transcription initiation by Pol III depends heavily on the transcription factor IIIB (TFIIIB), which is a complex formed by Brf1 (or Brf2), TBP, and Bdp1. A recent study combined the use of X-ray crystallography, TIRFM-smFRET (Fig. 6.1d, g, h), and biochemical analysis to provide structural and functional insights into the assembly process of TFIIIB on the U6 snRNA promoter DNA [55]. In particular, smFRET provided the means to monitor the TBP-mediated bending of the promoter and thus the binding dynamics of TFIIIB and its subcomplex to the promoter.

6.2.2 Elongation

6.2.2.1 Bacterial Transcription Machinery

Early single-molecule experiments following bacterial transcription elongation concluded that RNAP is a strong motor that withstands hindering force up to ~25 pN [145] and that single RNAPs progress at similar average rates [127, 152]. However, improvements to the optical tweezers spatiotemporal resolution [87, 150] (Fig. 6.3a, b) have allowed more detailed studies and have demonstrated that the steady transcription elongation is halted by ~10–100 s duration pauses at specific sequence sites [25, 46]. Because the probability to enter these long-lived pauses is force dependent, it has been suggested that the pauses originated from RNAP backtracking, i.e., backward sliding of the polymerase on the DNA that drives the 3′-RNA end out of register to the NTP entry channel. Having improved the optical tweezers assay sufficiently, backtracking was finally directly observed by Shaevitz et al. They showed that the backtracked pauses (Fig. 6.4c) have long lifetimes, ranging from 20 s to above 30 min [131]. They also observed that the pause duration is significantly reduced by the addition of GreA and GreB transcription factors, which bind the RNAP at the NTP entry channel (also known as secondary channel) and restore the elongation competent translocation register by stimulating the cleavage of the overhanging 3′-end of the RNA. However, all pauses did not originate from backtracked RNAP; these shorter

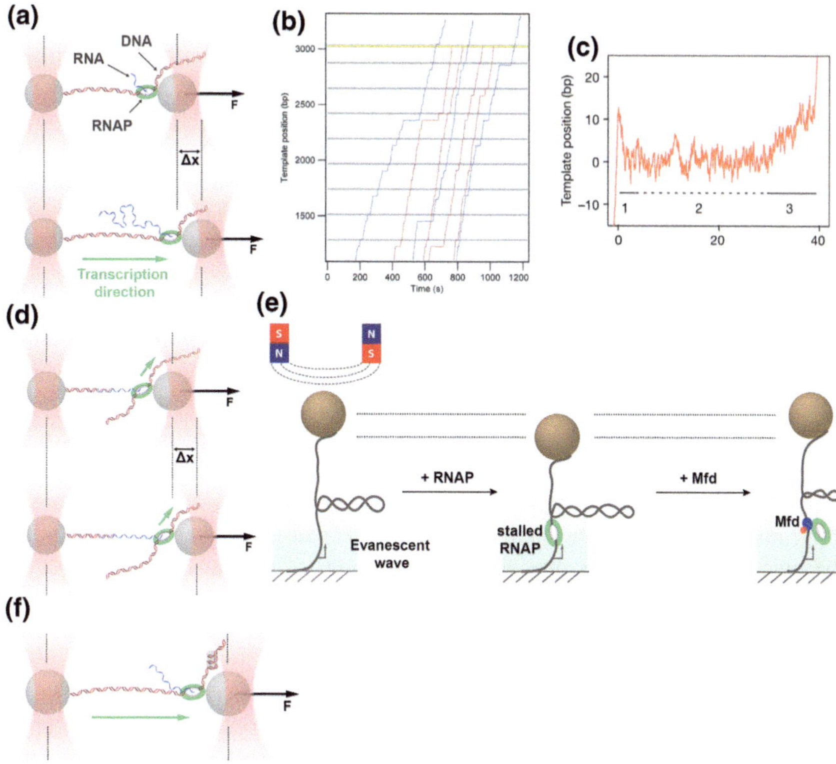

Fig. 6.4 Optical and magnetic tweezers assays to study transcription elongation and termination by multisubunit RNA polymerases. **a** Optical tweezers based assay allows to subject the RNAP either to assisting or hindering force depending on which direction the transcription is designed to progress on the template DNA. **b** Transcription activity traces for individual RNAPs obtained with the high-resolution optical tweezers assay depicted in (**a**). Adapted from Ref. [63]. **c** A close-up of a transcription activity trace shows the RNAP to backtrack on the template DNA in the optical tweezers assay. Adapted from Ref. [131]. **d** Optical tweezers transcription assay where the assisting force is applied to the nascent RNA. **e** Magnetic tweezers were combined with TIRFM to study transcription-coupled repair in Ref. [43, 56]. The bead position is affected by the size of the transcription bubble thus transmitting information on the RNAP occupancy on the DNA and the stage of transcription. The binding of dye-labeled Mfd to the DNA or DNA-bound RNAP is inferred from the sudden appearance of a fluorescent spot on the coverslip surface. Because the strength of evanescent field decays exponentially with the distance from the surface, the position of Mfd on the DNA can be extracted from the intensity of the fluorescent spot. **f** Schematic of the optical tweezers assay monitoring Pol II transcription past a nucleosome

(1–10 s) force independent pauses known as ubiquitous pauses (or elemental pauses) interrupt the progress of RNAP at any sequence position, while the probability to pause at any particular position is low [2, 108]. After the spatiotemporal resolution of optical tweezers has been pushed to its limit [57, 105], the 1 bp step translocation of the RNAP was monitored at low NTPs concentration (up to ~10 μM) [1, 117]. This type of data led to the formulation of a Brownian ratchet model of RNAP translocation during nucleotide addition cycles [1]. The Block lab further scrutinized the RNAP pausing and found that the ubiquitous pauses are indeed sequence dependent, similar to the long-known sequence encoded *his* and *ops* pauses (Fig. 6.4a, b) [63]. In combination with biochemical experiments, a new model was developed [86], where the RNAP has a certain sequence-dependent probability to isomerize into a catalytically inactive conformation, the ubiquitous pause state, today more widely known as the elemental pause state. The short-lived elemental pause state (typical half-life ~2 s) may further isomerize into a more stable (longer) pause by backtracking or a conformational change triggered when a RNA-hairpin forms in the RNA exit channel of RNAP [58, 78]. The sequence code (consensus: $G_{-10}Y_{-1}G_{+1}$; Y standing for pyrimidine) imposing the elemental/ubiquitous pause has been unraveled when it became possible to determine the exact locations of the paused-RNAPs on the transcribed genes by massive parallel sequencing of the nascent RNAs 3′-ends, as RNAPs are enriched at pause sites. Single-molecule optical tweezers assays as well as biochemical approaches provided detailed mechanistic dissection of the consensus pause sequence [89]. Interestingly, the pauses, elongation rate, and processivity of the RNAP are not affected when the RNA transcript is pulled by a force up to 30 pN, i.e. twice stronger as is typically needed to melt RNA secondary structures [23], showing that RNA structure has little influence on ubiquitous transcriptional pausing.

The pausing behavior of bacterial RNAP is also influenced by external transcription factors. For example, NusA increases the probability and lifetime of the elemental (short-lived) and RNA hairpin stabilized (long-lived) pauses [155], while NusG has an opposite effect [64]. The RNAP elongation rate is also affected by the amount of supercoil generated by the polymerase when it transcribes a torsionally constrained DNA molecule, e.g. the bacterial chromosome. Using optical microscopy, it was first observed that the RNAP generates torque during transcription elongation [60]. Ma and collaborators used an optical torque wrench (Fig. 6.2d) to control the amount of torque applied to the DNA and showed that the bacterial RNAP generates and sustains a torque up to ~11 pN nm, i.e. enough to melt DNA; if the RNAP is stalled by excessive resisting torque, the complex eventually resumes elongation when the torque is released [100].

Bulky DNA lesions on the template strand stall RNAP. Bacteria have evolved a mechanism to utilize this RNAP property to detect the harmful DNA lesions and guide the start of the repair process. When RNAP stalls on the DNA lesion, it is recognized by Mfd, an ATP-dependent DNA translocase. Mfd dissociates the RNAP from DNA and recruits the UvrABC endonuclease to cleave off and repair the damaged DNA. Recently, several single-molecule studies have dissected this transcription-coupled DNA repair pathway. Howan and co-workers used magnetic tweezers (Figs. 6.2c and 6.3c, d) to observe the dynamic interactions of Mfd with the stalled RNAP [69]. They found that Mfd binds and dissociates the RNAP in an ATP-dependent process, reaching an intermediate state where Mfd is simultaneously bound to the DNA and the RNAP. Subsequently, Mfd dissociates the RNAP from the DNA in another ATP-dependent step of a remarkable duration (~6 min). However, it remained unclear whether the RNAP dislodged by Mfd from the DNA lesion site remained bound to the Mfd/DNA and whether the RNAP retained the RNA transcript. To answer these questions, Graves and co-workers combined smTIRFM with their magnetic tweezers assay to simultaneously monitor both the real-time composition (using fluorescence) and the transcription bubble size (using the magnetic tweezers) of the transcription-coupled repair machinery (Fig. 6.4d) [56]. They showed that the RNAP releases the RNA transcript when the Mfd dislodges the RNAP from the DNA lesion. Interestingly, the formed RNAP–Mfd complex is stable and translocates thousands of base pairs on the DNA [56]. The studies from the Strick lab did not investigate how Mfd translocated along the DNA molecule before finding a stall RNAP. Using an optical tweezers DNA hairpin assay (Fig. 6.3a), where the hairpin is progressively opened upon Mfd progression, further revealed that the Mfd independently translocates along the DNA at ~7 bp/s; this rate is too slow to follow a normally transcribing RNAP, but enough to catch up with a stalled RNAP [93]. Collectively, the locomotive action of the Mfd assists the RNAP to either overcome translocation arrest on, e.g. a strong pause site, or to remove and terminate transcription for an RNAP stalled on an insurmountable obstacle [93].

On highly expressed genes, e.g. ribosomal RNA genes, multiple RNAPs transcribe simultaneously the same gene. If the leading RNAP encounters an obstacle (e.g. a pause or DNA-bound protein), the trailing RNAPs will catch up the stalled leading RNAP, push it forward, and rescue it into active transcription [42]. To investigate whether the rescue of the leading RNAP by the trailing RNAPs could be linked to transcriptional bursting, Fujita and co-workers derived a smTIRFM assay that allows both monitoring the production of messenger RNAs by single-molecule fluorescence in situ hybridization (smFISH) and locating quantum dot labeled individual RNAPs on the template DNA [50]. Their mathematical modeling of the observed transcription dynamics supports the assumption that a significant amount of transcriptional bursting simply stems from the arrest of the leading RNAP and its rescue by the trailing RNAPs.

6.2.2.2 Eukaryotic Transcription Machineries

In eukaryotic transcription, the elongation phase has been first studied with Pol II [98]. Seminal work from Galburt and co-workers used a high-resolution optical tweezers assay (Fig. 6.4a) to show that Pol II molecules ceased to transcribe and were unable to recover from backtracks (Fig. 6.4c) at a force of ~8 pN, only one-third of the *E. coli* RNAP stalling force [52]. Most Pol II pauses were explained by backtracking. TFIIS—a eukaryotic analog of the bacterial Gre factors—rescues Pol II from backtrack by stimulating the cleavage of the protruding 3′-RNA end and thus allows Pol II to work against a two-fold higher hindering force. The authors suggested that there exists a full layer of transcription elongation regulation that depends on transcription factors modifying the mechanical performance of Pol II. Lisica and coworkers investigated further the mechanism of backtrack recovery by Pol I and Pol II using a similar optical tweezers assay (Fig. 6.4a) [97]. Backtracking was enforced by pulling the polymerase backward with a rapidly spiked, strong hindering force. Analysis showed that the recovery from shallow backtrack takes place via 1D diffusion of the RNAP, while recovery from deeper backtracks depends on RNA cleavage. Many transcription factors are expected to affect the elongating Pol II. For example, TFIIF—a transcription initiation factor involved in the recruitment phase of the Pol II-PIC—was shown to be also active during elongation, reducing the backtracking propensity of Pol II [70]. The only elongation factor found in eukaryotes, archaea, and bacteria—Spt4/5 (Spt5 is homologous to bacterial NusG)—on the other hand appears to regulate Pol II transcription through the nucleosome [22]. The GC content of the nucleic acids has also been shown to influence Pol II pausing dynamic. Indeed, it was reported that Pol II transcription elongation encounters less and shorter pauses when the DNA template is GC-rich than when it is AT-rich [154]. The authors suggest that the strong and bulky secondary structures, preferentially formed in the GC-rich RNA transcript, prevent Pol II from backtracking. Consistently, RNase-mediated degradation of the transcript abolishes the GC-content dependent pausing bias.

During eukaryotic transcription, polymerases have to pass through nucleosomes, i.e., a 146 bp stretch of DNA wrapped around a bundle of eight histone proteins. Nucleosomes form a mechanical barrier to transcription, and consequently, the accessibility of the DNA for transcription is also regulated by histone chemical modifications and ATP-dependent nucleosome remodeling enzymes. How an elongating Pol II bypasses a nucleosome that is placed on its path was investigated using optical tweezers [66]. The authors derived a mathematical model for the observed Pol II dynamics in the vicinity of the nucleosome and concluded that the polymerase, instead of actively separating the DNA from the histones, waits for fluctuations that locally unwrap the nucleosome and allow the Pol II to advance. In a follow-up study, the roles of various nucleosomal elements were investigated as a function of the strength and location of the barrier to transcription [9]. Specifically, the authors determined, how the trajectories of individual Pol II complexes transcribing past nucleosomes responds to the modifications in specific histone–DNA interactions or histone tails. They observed that the DNA unwrapped and rewrapped faster around the tails-free histones, which favors Pol II movement closer to the nucleosome. In addition, they noted that point mutations compromising the DNA–histone interactions at the center of the nucleosome (dyad) decrease the local rewrapping rate of the DNA and thus remove a barrier for Pol II to translocate forward and that the nucleosomes amplify Pol II sequence-dependent pausing. The Block lab investigated the fundamental steps of the nucleotide addition cycle—substrate selection, catalysis, and translocation—using Pol II mutants with altered trigger loop function (Fig. 6.4a) [90]. The trigger loops are mobile and conserved RNA polymerase subdomain that stabilizes substrate NTPs in the active site. The global fits to the force–velocity curves they extracted converge with a branched Brownian ratchet model for elongation, where the incoming NTP binds either the expected post-translocated state or the pre-translocated state, similar to what was previously proposed for the bacterial RNAP [1]. The latter binding mode is expected to take place in the pre-insertion site of the RNAP and does not require the NTP to form interactions with the active site bound template base. Furthermore, the trigger loop was suggested to control the transitions between the pre- and post-translocated states. Another study by the Bustamante lab utilized nucleosomes as specific barriers to forward translocation—a trick that allows to determine separately both the forward and reverse translocation rates of Pol II and further estimate all main kinetics parameters involved in the nucleotide addition and pausing phases of transcription elongation [24]. In contrast to the earlier studies that had assumed the polymerase to reach fast equilibrium between the pre- and post-translocated states prior and after each RNA elongation step [1, 90], the authors found that the forward translocation rate (88 s^{-1}) of Pol II is actually similar to the RNA extension rate (35 s^{-1}). Therefore, the translocation and RNA extension together constitute the rate-limiting steps in the nucleotide addition cycle. From these findings the authors proposed a simpler linear Brownian ratchet model of transcription elongation, where the incoming NTP binds only to the post-translocated state, which is consistent with biochemical evidences.

6.2.3 Termination

6.2.3.1 Bacterial Transcription Machinery

Larson et al. [88] investigated the molecular mechanism of transcription termination at the single-molecule level using optical tweezers. The authors characterized three different terminators (*his*, t500, and tR2) each consisting of a GC-rich hairpin followed by a U-rich tract. Two distinct termination mechanisms for these intrinsic terminators were observed, namely termination by forward hypertranslocation or RNA: DNA hybrid shearing. When observing the forward translocation strategy, the authors found that the RNAP hypertranslocates forward by ~1.5 bp leading to shorter RNA–DNA hybrid, which destabilizes the complex enough for subsequent termination. In the RNA:DNA hybrid shearing strategy, the U-rich tract forms a weak RNA–DNA hybrid, which is easily further destabilized by the folding of an upstream RNA hairpin. By pulling the hairpin from the 5′-RNA end, two modes of action were observed: (i) at a force larger than the hairpin melting force, the termination efficiency increases because of the shearing of the RNA–DNA hybrid; (ii) at a force lower than the hairpin melting force, the termination efficiency also increases, though here because the pulling force modulates the balance of termination hairpin and other secondary structures in the RNA. The authors concluded that the most frequent cause of termination failure is the folding of the RNA into one of the competing secondary structures. The termination by hypertranslocation likely dominates in the sites where the energy penalty of shearing the RNA–DNA hybrid is higher compared to RNAP forward hypertranslocation.

Frieda and Block developed an optical trap-based assay to monitor the co-transcriptional folding of the *pbuE* riboswitch [47]. This riboswitch regulates the concentration of adenine in the cell by forming an adenine binding aptamer. The folding of the aptamer, which is stabilized by the binding of an adenine, prevents the formation of the competing terminator hairpin and thus allows the expression of the downstream genes. The authors determined that aptamer-dependent RNAP termination is a function of the adenine concentration and the applied force to the RNA (Fig. 6.4d); they also identified the folding signature of the riboswitch. The termination versus read-through outcome turns out to be kinetically controlled indicating that the riboswitch-based regulation of gene expression is mechanistically tightly linked with the transcription elongation kinetics and the regulatory layer that controls the elongation kinetics in the cell.

6.3 In Vitro Single-Molecule Studies of Viral RNA-Dependent RNA Polymerases

RNA viruses are particularly remarkable for their diverse genome replication strategies. The genome of RNA viruses may be positive (+), negative (−), or double-stranded (ds). The protein synthesis machinery of the host cell directly employs the positive virus genome as the template whereas the negative genome must first be copied into a complementary strand. The RNA viruses rely on the viral polymerase, formally called the RNA-dependent RNA polymerase (RdRp), to replicate and transcribe their genomes [134]. The process of genome replication and transcription is divided into two main phases—initiation and elongation. Viruses have developed many different strategies to initiate replication and transcription. For example, Φ6 bacteriophage and flaviviruses, e.g. dengue, employ de novo initiation (i.e. on a free 3′-RNA end) to replicate their genomes. Influenza virus initiates genome transcription by primer-extension but employs de novo initiation to begin the replication of its (−) RNA genome. Poliovirus primes replication of its (+) RNA genome using a VPg (viral protein genome-linked) attached at the 5′-end of the viral genome [45]. The initiation phase is critical for viral survival because its specificity and efficiency must ensure sufficient synthesis of viral RNA by the RdRp to meet the demand of both viral proteins synthesis and virion assembly. After a successful initiation, the RdRp enters the elongation phase. This phase is equally important for the RNA virus because the full-length genomes are necessary for correct translation and virion assembly. Viral RdRps have a relatively high misincorporation rate that serves to increase the genome diversity of the virus population, thus helping the virus to evade the host immune response [92].

Single-molecule techniques have only recently been applied to study the genome replication/transcription of RNA viruses. Here, we describe the results from single-molecule experiments that have shed light on the RdRp initiation mechanism on influenza A virus (IAV) and the elongation dynamics of Φ6 P2 and poliovirus RdRps.

6.3.1 *Replication and Transcription Initiation in Influenza A Virus*

IAV is a segmented (−) RNA virus, meaning that its genome is divided into eight segments of viral single-stranded RNA (vRNA). The RNA strands form a ribonucleoprotein complex with viral nucleoproteins, and the partially complementary 3′- and 5′-RNA ends are bound by a single copy of IAV RdRp. The IAV RdRp is formed by three individual polypeptides called PB1, PB2, and PA. Each of them has a separate task in IAV genome processing: PB1 is the core polymerase, PB2 is the cap-binding domain, and PA is the metal ion-dependent endonuclease [133]. This heterotrimeric complex replicates and transcribes the vRNA. The 3′ and 5′ ends of the vRNA are highly conserved and hybridize to form a partially double-stranded panhandle structure that takes the shape of a corkscrew and specifically binds to the RdRp [135]. Though this structure has first been suggested using functional studies, a structural confirmation was lacking. In an elegant study using confocal smFRET (Fig. 6.1d–f), Tomescu and co-workers have mapped the structure of the hybridized termini bound to the RdRp [138]. The authors have studied the FRET efficiency for different FRET pair locations along the RNA and determined the inter-dye distances from the measured FRET efficiencies. The measured distances—determined separately for the free and RdRp-bound RNA—are consistent with the RNA corkscrew structure model when RdRp is bound to the panhandle RNA structure (Fig. 6.5a). Robb and co-workers have recently expanded this work [119] by characterizing the 3′-RNA end structure. They showed that the 3′-end of the vRNA takes two alternative conformations upon RdRp binding, one bound on the RdRp surface in the

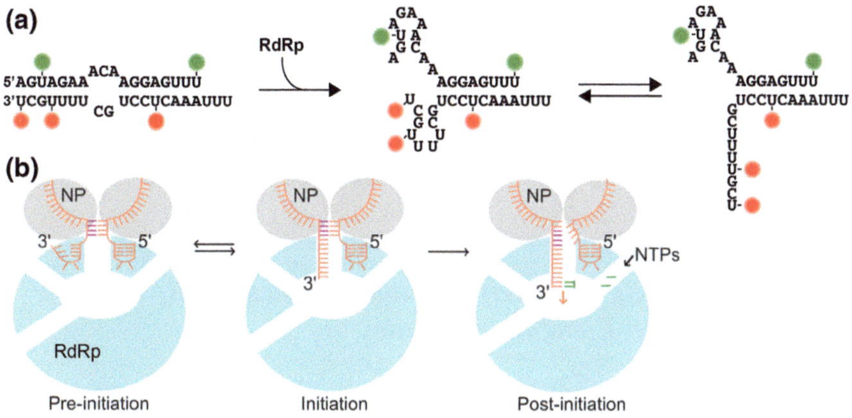

Fig. 6.5 Solution smFRET studies of de novo replication initiation by influenza virus RNA-dependent RNA polymerase. **a** Different configurations observed for the nucleic acid scaffold mimicking the 3′ and 5′ ends of the influenza RNA genome. The green and red spheres indicate the donor and acceptor dye positions, respectively. **b** Model for de novo replication initiation by influenza RdRp. Adapted from Ref. [119]

pre-initiation state and another bound to the active site in the initiation-competent state. Both conformations are present at equilibrium in the absence of NTPs, while the initiation state is favored in the presence of a dinucleotide that mimics the state of the complex after the synthesis of a 2-bp RNA (Fig. 6.5b).

Once complexed with the panhandle structure of the viral genome termini, the RdRp starts either transcription or replication. In the transcription mode, the RdRp captures and cleaves a capped host 5′-mRNA to prime the mRNA synthesis from the viral genome. In the replication mode, two steps are needed to produce a new copy of the vRNA. In the first step, known as terminal initiation, a complementary RNA (cRNA) intermediate is produced from the vRNA when the RdRp initiates de novo by joining together the NTPs complementary to the first two residues of the 3′-vRNA end. In the second step known as internal initiation, the initiation takes place at the cRNA positions 4 and 5 leading to the synthesis of pppAG dinucleotide, which subsequently realigns with the positions 1 and 2 and is elongated by the RdRp [135]. Viral RdRps that initiate replication de novo generally contain a priming loop domain, which stacks the 3′-RNA end of the template strand and the first nucleotide of the product strand [135]. The IAV RdRp supports both primer-dependent and de novo initiation; however, it was unknown whether the priming loop is involved in the terminal and internal phases of replication initiation or in transcription initiation. Te Velthuis and co-workers used in vitro and in vivo ensemble biochemical assays together with confocal smFRET to investigate the question [136]. They showed that the priming loop is indeed needed to support de novo terminal replication initiation but not the internal replication initiation or primer-dependent transcription initiation. Interestingly, the priming loop actually represents an obstacle to transcription, its removal being the rate-limiting step for primer-dependent transcription initiation.

The above studies used confocal smFRET to pave the way for understanding the inter- and intramolecular conformational changes occurring during IVA RdRp mediated replication and transcription initiation, thereby complementing X-ray and cryo-EM based structural studies. Future work using TIRFM-smFRET (Fig. 6.1d, g, h) will allow the observation in real time of the full trajectories of individual RdRp complexes engaged in the initiation of replication or transcription, which will provide the detailed dynamics of the IAV initiation mechanisms.

6.3.2 *Φ6 P2 RdRp Transcription and Poliovirus RdRp Replication Elongation Kinetics*

The RdRps from RNA viruses have to replicate or transcribe the ~5–30 kb long viral RNA in order to produce new viral genomes to be incorporated into the new generation of virions or to provide templates for translation [134]. The elongation phase of replication is also very important for the viral evolution because the nucleotide misincorporations made by the RdRp are the main source of genetic diversity in the virus population [92]. Typical RdRp error incorporation rate, ~10^{-3}–10^{-4} per a nucleotide incorporation cycle, is one of the highest in all replication machineries [68]. However, the high error rate bears a fitness cost and must therefore be tightly balanced: An error rate too low would leave the virus unable to adapt to the host immune defense, while an error rate too high would be detrimental for the production of a sufficient amount of active virions [132]. The kinetics of nucleotide incorporation and misincorporation have been heavily studied using fast mixing enzyme kinetics assay, such as quenched flow and stopped flow [13]. These approaches offer an exquisite resolution, i.e., single-nucleotide additions at millisecond timescale; however, this resolution is only attainable for short templates, typically less than 10 nucleotides. Even though the misincorporation rate of RNA viruses is high, it still remains a rare event and can only be observed in bulk assays in the absence of the correct, templated nucleotide. The viral replication also represents an important target for antiviral therapeutic strategies, currently taking advantage from the large library of antiviral nucleotide analogs. Nucleotide analog incorporation studies, similar to the nucleotide misincorporation studies, suffer from the limitation that these events are rare when observed in the presence of natural nucleotides. Therefore, an experimental approach compatible with the use of natural length templates, i.e., few kilobases, and discrimination power capable of distinguishing rare misincorporation or nucleotide analog incorporation events in the background of normal replication/transcription would greatly benefit the mechanistic studies of viral RdRp-mediated RNA elongation.

Single-molecule force spectroscopy approaches, especially optical tweezers and magnetic tweezers, come close to fulfill the specific technical requirements of RdRp elongation studies by offering the possibility to observe the RNA synthesis by individual RdRps on kilobase(s) length RNA templates at ~10–100 ms temporal and near base-pair spatial resolution. However, to observe events as rare as 10^{-3} per nucleotide incorporation cycle, highly multiplexed approach is needed. Unlike fluorescence spectroscopy techniques, force spectroscopy techniques have historically suffered from poor throughput. However, this limitation has been recently overcome with the development of several high-throughput techniques [36, 65]. One of the high-throughput enabling solutions involved upgrading magnetic tweezers apparatus with the latest generation of large sensor CMOS cameras with a real-time image analysis algorithm, capable of tracking hundreds of individual molecules in parallel [8, 19]. The real-time high-throughput magnetic tweezers were first applied to characterize the nucleotide incorporation dynamics of RdRps from bacteriophage Φ6 P2 (has dsRNA genome) and human poliovirus [(+) RNA] [33, 38, 39].

To study the viral RdRps, a double-stranded RNA tether is used to attach the magnetic bead to the surface of a microscope coverslip. When RdRp employs the dsRNA as a template for RNA synthesis, it gradually displaces the template RNA strand leaving the bead anchored to the surface via a single-stranded RNA. The progress of the RdRp action is monitored in real time as the movement of the bead further away from the surface (Fig. 6.6a). Large data sets of Φ6 P2 RdRp transcription activity were acquired at different NTP concentrations and applied force. Interestingly, Φ6 P2 RdRp shows fast bursts of nucleotide additions that are interrupted by pauses of 1–1000 s duration (Fig. 6.6b). Previously developed data analysis approach charac-

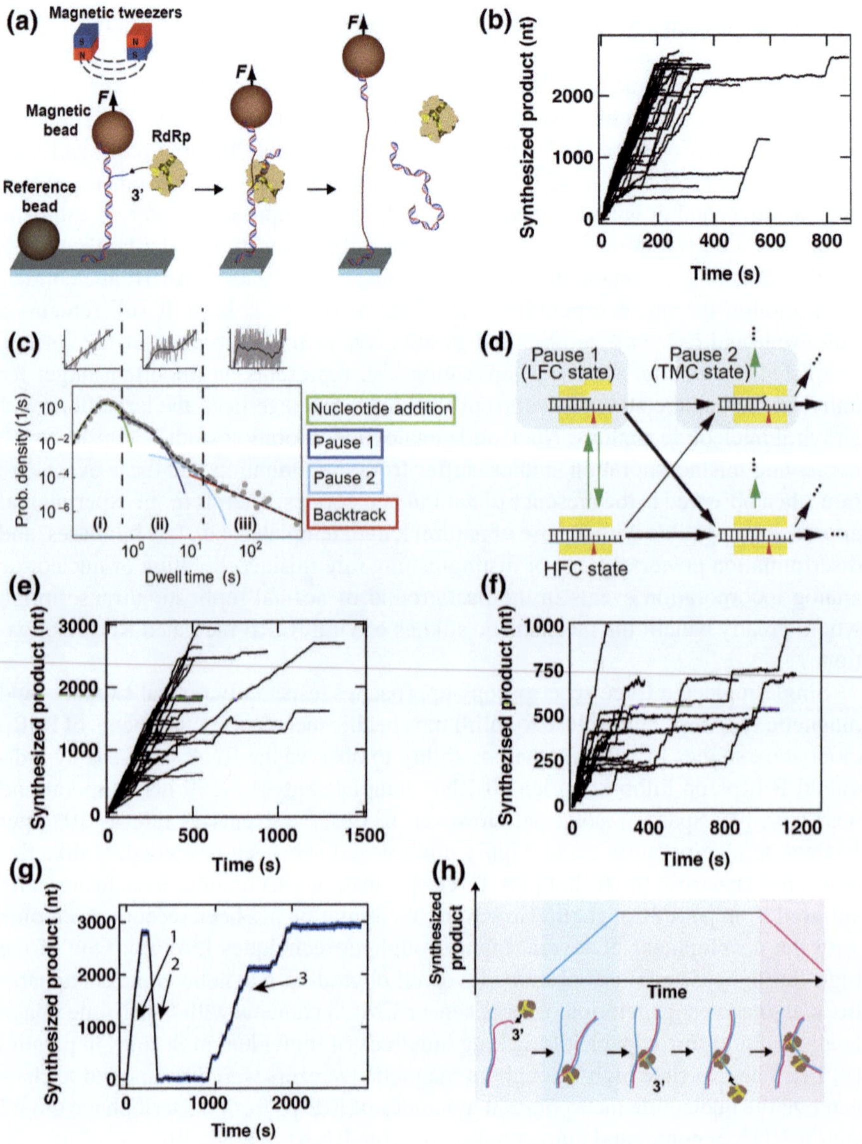
(a)
Magnetic tweezers
Magnetic bead
RdRp
Reference bead
3'
F
(b)
Synthesized product (nt)
Time (s)
(c)
Prob. density (1/s)
Dwell time (s)
(i)
(ii)
(iii)
Nucleotide addition
Pause 1
Pause 2
Backtrack
(d)
Pause 1
(LFC state)
Pause 2
(TMC state)
HFC state
(e)
Synthesized product (nt)
Time (s)
(f)
Synthezised product (nt)
Time (s)
(g)
Synthesized product (nt)
Time (s)
1
2
3
(h)
Synthesized product
Time
3'

◀**Fig. 6.6** In vitro single-molecule studies of RdRp transcription elongation. **a** Magnetic tweezers assay can be used to study the dynamics of RdRp transcription elongation. The magnetic bead is tethered to the coverslip surface by a double-stranded RNA that experiences a constant force. A short non-hybridized segment of the RNA template presents a free 3′-end for RdRps to perform de novo initiation. To study primer dependent initiating RdRps, such as poliovirus RdRp, the 3′-end of the template RNA is modified to contain a short priming hairpin. Following successful initiation, the RdRp elongates the RNA product strand, unwinding the template strand and converting the tether to ssRNA. In the process, the end-to-end distance of the tether changes, thus reporting on the RdRp activity. **b** 52 traces of transcribing Φ6 P2 RdRps were acquired in a single experiment using high-throughput magnetic tweezers [19]. Adapted from Ref. [39]. **c** Probability density distribution of the dwell times corresponding to the synthesis of ten consecutive nucleotides stretches of RNA. Four distinct dwell time distributions are fitted; these correspond to the pause-free nucleotide incorporation (nucleotide addition, green), short pauses (Pause 1, dark blue), intermediate pauses (Pause 2, light blue), and long pauses caused by polymerase backtracking (backtrack, red). Example trace snapshots above illustrate each dwell time type. Adapted from Ref. [38]. **d** Nucleotide error incorporation model explains the dwell time distribution of the Φ6 P2 RdRp. The model details are explained in the main text. HFC, the high-fidelity catalytic state; LFC, low-fidelity catalytic state; TMC, terminal mismatched catalytic state. **e** Poliovirus RdRp replication traces in the presence of 100 μM of NTPs and **f** 100 μM of NTPs with 10 μM of antiviral nucleotide analog T1106-triphosphate. **e, f** are adapted from Ref. [33]. **g** A fraction of Φ6 P2 RdRp transcription traces displayed "reversal" activity (arrows). **h** The reversal activity originates from a backtracked RdRp that presents a protruding 3′-end of the product RNA strand, which is used by another RdRp as a template of transcription. The second RdRp pushes back the first RdRp resulting in the rehybridization of the original template and non-template strands. The shortening of the end-to-end distance of the tether is thus detected as a "reversal" trace. **g, h** are adapted from Ref. [39]

terizes separately the nucleotide addition and the pause kinetics by picking the pauses out of the traces. However, it is impossible to distinguish pauses shorter than ~1 s because of the finite spatiotemporal resolution of tweezers assay, and therefore, the nucleotide addition kinetics is polluted by the missed short pauses [38]. To overcome this issue, a dwell time analysis combined with a maximum likelihood estimation (MLE) fit has been developed to extract the elongation kinetics parameters, i.e., rates and probabilities, at once without sorting the pauses out of the traces (Fig. 6.6c) [36]. Using this dwell time analysis, the probability and average interconversion rates of the catalytic and non-catalytic states as well as the nucleotide addition rates are recovered. The dwell time distribution of Φ6 P2 RdRp synthesizing each consecutive 10 nt stretch of RNA (a limit set by the resolution of the used magnetic tweezers assay) was measured and found to be composed of four subdistributions (Fig. 6.6c). The shortest dwell times (<1 s at saturating NTP concentration) is assigned to the pause-free nucleotide addition rate ((i) in Fig. 6.6c). Intermediate dwell times (~1–10 s) are split into two populations exponentially distributed, representing pauses of short (Pause 1) and intermediate durations (Pause 2) ((ii) in Fig. 6.6c). Finally, the distribution of the longest dwell times (>20 s) is best described by a power law ($t^{-3/2}$), suggesting a backtrack state for the polymerase [28]. Furthermore, Pause 1 and Pause 2 probabilities and lifetimes are surprisingly dependent on the NTP concentration, and Pause 2 probability is affected by inosine triphosphate incorporations. These findings suggest that Pause 1 and Pause 2 are intimately linked with nucleotide misincorporation (Fig. 6.6d). It was formerly believed that RdRp misincorporation events were a rare incident happening along the same catalytic pathway as the correct nucleotide incorporation. The new model derived from the single-molecule data, in contrast, suggests that Φ6 P2 RdRp has two catalytic conformations: a high-fidelity catalytic (HFC) state and a low-fidelity catalytic (LFC) state, respectively (Fig. 6.6d). Majority of the RNA synthesis by RdRp takes place on the HFC pathway and leads to the rapid incorporation of the correct nucleotides to the RNA. However, the RdRp has a certain probability to isomerize into the LFC conformation, which leads to a slow nucleotide addition, i.e. Pause 1 (Fig. 6.6c). The LFC has also an elevated probability (though still low in absolute terms) to elongate the RNA product strand with a wrong nucleotide. Upon misincorporation, Φ6 P2 RdRp enters an even slower catalytic state, i.e. Pause 2 or the terminal mismatched catalytic (TMC) state (Fig. 6.6c, d), as the catalytic activity is further compromised by the mismatched 3′-RNA end.

A follow-up study focusing on poliovirus RdRp (Fig. 6.6e) revealed very similar elongation kinetics compared to the Φ6 P2 RdRp; the coexistence of high- and low-fidelity catalytic conformations emerges thus as a general property of viral RdRps [33]. The mechanistic model defining the intermediate pauses as misincorporation events was further corroborated by a set of experiments performed with an error-prone poliovirus RdRp mutant. Specifically, the mutator RdRp had a threefold increased probability to enter Pause 2, an increase similar to what was determined using deep sequencing [83]. Taking advantage of the high-throughput capability of the magnetic tweezers, the effects of five nucleotide analogs on the replication activity of poliovirus RdRp were investigated with a physiological concentration of NTP (100 μM, saturating condition relative to the K_m). The tested compounds included the mutagenic nucleotide analog ribavirin triphosphate (RTP), inosine triphosphate (ITP), obligatory chain terminator 3′-deoxy ATP, non-obligatory chain terminator 2′-C-met-ATP and T1106-triphosphate (T1106-TP) whose mechanism of action was unclear until then. As expected from the misincorporation–pause model, RTP and ITP specifically increase the occurrence of Pause 2. Also as expected, the chain terminators 2′-C-met-ATP and 3′-deoxy ATP decrease the processivity of the replicating RdRp, e.g., the median processivity drops from ~1200 to ~400 nt when 100 μM 3′-deoxy ATP is added to the 100 μM of natural NTPs. However, this result also demonstrates how strongly the poliovirus RdRp selects against the nucleotide analogs, i.e., ~400 correct nucleotide incorporations take place before one 3′-dATP is added to the elongated RNA. Finally, the data revealed that the addition of T1106 to the RNA chain unexpectedly triggers the RdRp to enter a unique long-lived pause, seemingly backtrack-related (Fig. 6.6e, f). In conclusion, high-throughput magnetic tweezers have provided new insights into the mechanisms of viral RdRps replication activity and antiviral nucleotide analogs function.

The backtracking activity of Φ6 P2 RdRp has also been characterized by high-throughput magnetic tweezers (Fig. 6.6g) [39]. The probability of Φ6 P2 RdRp to enter long-lived backtracked states decreases with the increase in the applied force. Because the force destabilizes the ds-ssRNA junction in the front of the RdRp, the dominant factor determining the RdRp backtracking appears to be the base pairing energy at the dsRNA fork. Surprisingly, it was also found that the extensively backtracked 3′-RNA end of the newly synthesized product strand may be used as a template for de novo initiation by another RdRp. Eventually, the second RdRp pushes the first RdRp backward, all the way to the upstream end of the product strand, which produces the "reversal" traces shown in Fig. 6.6g, h. One possible biological function of the reversal mechanism could be assisting viral RNA recombination—another important evolutionary pathway. The ~two-fold higher rate of the reversal transcription compared to the forward transcription [39] may also provide the virus with a more efficient viral RNA production pathway in the host cell.

6.4 Perspective

Single-molecule techniques have offered a complete new angle on the understanding of the molecular mechanisms of cellular and viral RNA polymerases. The unique power of single-molecule approaches largely arises from its ability to resolve individual steps in complex reaction pathways, competing reaction pathways and multiple coexisting conformations. The force spectroscopy methods additionally allow nanomanipulation of the biological molecules (pushing, pulling, and twisting), thus creating a versatile experimental tool that can be used to steer the energy landscape of biomolecular reactions. Technical improvements in observation parallelization [36, 65] and spatiotemporal resolution [91], or the combination of fluorescence and force spectroscopy [21, 62, 101, 141] will allow to monitor the activities and structural dynamics of individual RNA polymerase molecules in ever more accurate and complex settings.

Acknowledgements DD was supported by the Interdisciplinary Center for Clinical Research (IZKF) at the University Hospital of the University of Erlangen-Nuremberg. A. M. M. was supported by the Academy of Finland (grant number: 307775).

References

1. Abbondanzieri, E. A., Greenleaf, W. J., Shaevitz, J. W., Landick, R., & Block, S. M. (2005). Direct observation of base-pair stepping by RNA polymerase. *Nature, 438,* 460–465.
2. Adelman, K., La Porta, A., Santangelo, T. J., Lis, J. T., Roberts, J. W., & Wang, M. D. (2002). Single molecule analysis of RNA polymerase elongation reveals uniform kinetic behavior. *Proceedings of the National Academy of Sciences of the United States of America, 99,* 13538–13543.
3. Ahlquist, P. (2002). RNA-dependent RNA polymerases, viruses, and RNA silencing. *Science (New York, N.Y.), 296*, 1270–1273.
4. Alberts, B., Johnson, A., Lewis, J., Raff, M., Roberts, K., & Walter, P. (2002). *Molecular biology of the cell* (4th ed.). New York: Garland Science.
5. Axelrod, D. (2001). Total internal reflection fluorescence microscopy in cell biology. *Traffic, 2,* 764–774.

6. Beckers, M., Drechsler, F., Eilert, T., Nagy, J., & Michaelis, J. (2015). Quantitative structural information from single-molecule FRET. *Faraday Discussions.*
7. Belogurov, G. A., & Artsimovitch, I. (2015). Regulation of transcript elongation. *Annual Review of Microbiology, 69,* 49–69.
8. Berghuis, B. A., Dulin, D., Xu, Z. Q., van Laar, T., Cross, B., Janissen, R., et al. (2015). Strand separation establishes a sustained lock at the Tus-Ter replication fork barrier. *Nature Chemical Biology, 11,* 579–585.
9. Bintu, L., Kopaczynska, M., Hodges, C., Lubkowska, L., Kashlev, M., & Bustamante, C. (2011). The elongation rate of RNA polymerase determines the fate of transcribed nucleosomes. *Nature Structural & Molecular Biology, 18,* 1394–1399.
10. Bockelmann, U., Essevaz-Roulet, B., & Heslot, F. (1998). DNA strand separation studied by single molecule force measurements. *Physical Review E, 58,* 2386–2394.
11. Browning, D. F., & Busby, S. J. (2004). The regulation of bacterial transcription initiation. *Nature Reviews Microbiology, 2,* 57–65.
12. Browning, D. F., & Busby, S. J. (2016). Local and global regulation of transcription initiation in bacteria. *Nature Reviews Microbiology, 14,* 638–650.
13. Cameron, C. E., Moustafa, I. M., & Arnold, J. J. (2016). Fidelity of nucleotide incorporation by the RNA-dependent RNA polymerase from poliovirus. *Enzymes, 39,* 293–323.
14. Chakraborty, A., Meng, C. A., & Block, S. M. (2017). Observing single RNA polymerase molecules down to base-pair resolution. *Methods in Molecular Biology, 1486,* 391–409.
15. Chakraborty, A., Wang, D., Ebright, Y. W., Korlann, Y., Kortkhonjia, E., Kim, T., et al. (2012). Opening and closing of the bacterial RNA polymerase clamp. *Science (New York, N.Y.), 337,* 591–595.
16. Chandradoss, S. D., Haagsma, A. C., Lee, Y. K., Hwang, J. H., Nam, J. M., & Joo, C. (2014). Surface passivation for single-molecule protein studies. *Journal of Visualized Experiments: JoVE.*
17. Charvin, G., Allemand, J. F., Strick, T. R., Bensimon, D., & Croquette, V. (2004). Twisting DNA: Single molecule studies. *Contemporary Physics, 45,* 383–403.
18. Churchman, L. S., & Spudich, J. A. (2012). Colocalization of fluorescent probes: Accurate and precise registration with nanometer resolution. *Cold Spring Harbor Protocols, 2012,* 141–149.
19. Cnossen, J. P., Dulin, D., & Dekker, N. H. (2014). An optimized software framework for real-time, high-throughput tracking of spherical beads. *The Review of Scientific Instruments, 85,* 103712.
20. Collins, B. E., Ye, L. F., Duzdevich, D., & Greene, E. C. (2014). DNA curtains: Novel tools for imaging protein-nucleic acid interactions at the single-molecule level. *Methods in Cell Biology, 123,* 217–234.
21. Comstock, M. J., Ha, T., & Chemla, Y. R. (2011). Ultrahigh-resolution optical trap with single-fluorophore sensitivity. *Nature Methods, 8,* 335–340.
22. Crickard, J. B., Lee, J., Lee, T. H., & Reese, J. C. (2017). The elongation factor Spt4/5 regulates RNA polymerase II transcription through the nucleosome. *Nucleic Acids Research, 45,* 6362–6374.
23. Dalal, R. V., Larson, M. H., Neuman, K. C., Gelles, J., Landick, R., & Block, S. M. (2006). Pulling on the nascent RNA during transcription does not alter kinetics of elongation or ubiquitous pausing. *Molecular Cell, 23,* 231–239.
24. Dangkulwanich, M., Ishibashi, T., Liu, S., Kireeva, M. L., Lubkowska, L., Kashlev, M., & Bustamante, C. J. (2013). Complete dissection of transcription elongation reveals slow translocation of RNA polymerase II in a linear ratchet mechanism. *eLife, 2,* e00971.
25. Davenport, R. J., Wuite, G. J., Landick, R., & Bustamante, C. (2000). Single-molecule study of transcriptional pausing and arrest by *E. coli* RNA polymerase. *Science (New York, N.Y.), 287,* 2497–2500.
26. de Farias, S. T., Dos Santos Junior, A. P., Rego, T. G., & Jose, M. V. (2017). Origin and evolution of RNA-dependent RNA polymerase. *Frontiers in Genetics, 8,* 125.

27. De Vlaminck, I., & Dekker, C. (2012). Recent advances in magnetic tweezers. *Annual Review of Biophysics, 41,* 453–472.
28. Depken, M., Galburt, E. A., & Grill, S. W. (2009). The origin of short transcriptional pauses. *Biophysical Journal, 96,* 2189–2193.
29. Duchi, D., Bauer, D. L. V., Fernandez, L., Evans, G., Robb, N., Hwang, L. C., et al. (2016). RNA polymerase pausing during initial transcription. *Molecular Cell, 63,* 939–950.
30. Duchi, D., Gryte, K., Robb, N. C., Morichaud, Z., Sheppard, C., Brodolin, K., et al. (2018). Conformational heterogeneity and bubble dynamics in single bacterial transcription initiation complexes. *Nucleic Acids Research, 46,* 677–688.
31. Duchi, D., Mazumder, A., Malinen, A. M., Ebright, R. H., & Kapanidis, A. N. (2018). The RNA polymerase clamp interconverts dynamically among three states and is stabilized in a partly closed state by ppGpp. *Nucleic Acids Research.*
32. Duchi, D., Mazumdera, A., Malinen, A. M., Ebright, R. H., & Kapanidis, A. N. (2018). The RNA polymerase clamp interconverts dynamically among three states and is stabilized in a partly closed state by ppGpp. *BioRxiv.*
33. Dulin, D., Arnold, J. J., van Laar, T., Oh, H. S., Lee, C., Perkins, A. L., et al. (2017). Signatures of nucleotide analog incorporation by an RNA-dependent RNA polymerase revealed using high-throughput magnetic tweezers. *Cell Reports, 21,* 1063–1076.
34. Dulin, D., Bauer, D. L. V., Malinen, A. M., Bakermans, J. J. W., Kaller, M., Morichaud, Z., et al. (2017). Pausing controls branching between productive and non-productive pathways during initial transcription. *BioRxiv.*
35. Dulin, D., Bauer, D. L. V., Malinen, A. M., Bakermans, J. J. W., Kaller, M., Morichaud, Z., et al. (2018). Pausing controls branching between productive and non-productive pathways during initial transcription in bacteria. *Nature Communications, 9,* 1478.
36. Dulin, D., Berghuis, B. A., Depken, M., & Dekker, N. H. (2015). Untangling reaction pathways through modern approaches to high-throughput single-molecule force-spectroscopy experiments. *Current Opinion in Structural Biology, 34,* 116–122.
37. Dulin, D., Lipfert, J., Moolman, M. C., & Dekker, N. H. (2013). Studying genomic processes at the single-molecule level: Introducing the tools and applications. *Nature Reviews Genetics, 14,* 9–22.
38. Dulin, D., Vilfan, I. D., Berghuis, B. A., Hage, S., Bamford, D. H., Poranen, M. M., et al. (2015). Elongation-competent pauses govern the fidelity of a viral RNA-dependent RNA polymerase. *Cell Reports, 10,* 983–992.
39. Dulin, D., Vilfan, I. D., Berghuis, B. A., Poranen, M. M., Depken, M., & Dekker, N. H. (2015). Backtracking behavior in viral RNA-dependent RNA polymerase provides the basis for a second initiation site. *Nucleic Acids Research.*
40. Elowitz, M. B., Levine, A. J., Siggia, E. D., & Swain, P. S. (2002). Stochastic gene expression in a single cell. *Science (New York, N.Y.), 297,* 1183–1186.
41. English, B. P., Min, W., van Oijen, A. M., Lee, K. T., Luo, G., Sun, H., et al. (2006). Ever-fluctuating single enzyme molecules: Michaelis-Menten equation revisited. *Nature Chemical Biology, 2,* 87–94.
42. Epshtein, V., & Nudler, E. (2003). Cooperation between RNA polymerase molecules in transcription elongation. *Science (New York, N.Y.), 300,* 801–805.
43. Fan, J., Leroux-Coyau, M., Savery, N. J., & Strick, T. R. (2016). Reconstruction of bacterial transcription-coupled repair at single-molecule resolution. *Nature, 536,* 234–237.
44. Fazal, F. M., Meng, C. A., Murakami, K., Kornberg, R. D., & Block, S. M. (2015). Real-time observation of the initiation of RNA polymerase II transcription. *Nature, 525,* 274–277.
45. Fields, B. N., Knipe, D. M., & Howley, P. M. (2013). *Fields virology* (6th ed.). Philadelphia: Wolters Kluwer Health/Lippincott Williams & Wilkins.
46. Forde, N. R., Izhaky, D., Woodcock, G. R., Wuite, G. J., & Bustamante, C. (2002). Using mechanical force to probe the mechanism of pausing and arrest during continuous elongation by *Escherichia coli* RNA polymerase. *Proceedings of the National Academy of Sciences of the United States of America, 99,* 11682–11687.

47. Frieda, K. L., & Block, S. M. (2012). Direct observation of cotranscriptional folding in an adenine riboswitch. *Science (New York, N.Y.), 338*, 397–400.
48. Friedman, L. J., Chung, J., & Gelles, J. (2006). Viewing dynamic assembly of molecular complexes by multi-wavelength single-molecule fluorescence. *Biophysical Journal, 91,* 1023–1031.
49. Friedman, L. J., Mumm, J. P., & Gelles, J. (2013). RNA polymerase approaches its promoter without long-range sliding along DNA. *Proceedings of the National Academy of Sciences of the United States of America, 110,* 9740–9745.
50. Fujita, K., Iwaki, M., & Yanagida, T. (2016). Transcriptional bursting is intrinsically caused by interplay between RNA polymerases on DNA. *Nature Communications, 7,* 13788.
51. Gago, S., Elena, S. F., Flores, R., & Sanjuan, R. (2009). Extremely high mutation rate of a hammerhead viroid. *Science (New York, N.Y.), 323*, 1308.
52. Galburt, E. A., Grill, S. W., Wiedmann, A., Lubkowska, L., Choy, J., Nogales, E., et al. (2007). Backtracking determines the force sensitivity of RNAP II in a factor-dependent manner. *Nature, 446,* 820–823.
53. Geertsema, H. J., Duderstadt, K. E., & van Oijen, A. M. (2015). Single-molecule observation of prokaryotic DNA replication. *Methods in Molecular Biology, 1300,* 219–238.
54. Gietl, A., Holzmeister, P., Blombach, F., Schulz, S., von Voithenberg, L. V., Lamb, D. C., et al. (2014). Eukaryotic and archaeal TBP and TFB/TF(II)B follow different promoter DNA bending pathways. *Nucleic Acids Research, 42,* 6219–6231.
55. Gouge, J., Guthertz, N., Kramm, K., Dergai, O., Abascal-Palacios, G., Satia, K., et al. (2017). Molecular mechanisms of Bdp1 in TFIIIB assembly and RNA polymerase III transcription initiation. *Nature Communications, 8,* 130.
56. Graves, E. T., Duboc, C., Fan, J., Stransky, F., Leroux-Coyau, M., & Strick, T. R. (2015). A dynamic DNA-repair complex observed by correlative single-molecule nanomanipulation and fluorescence. *Nature Structural & Molecular Biology, 22,* 452–457.
57. Greenleaf, W. J., Woodside, M. T., Abbondanzieri, E. A., & Block, S. M. (2005). Passive all-optical force clamp for high-resolution laser trapping. *Physical Review Letters, 95,* 208102.
58. Guo, X., Myasnikov, A. G., Chen, J., Crucifix, C., Papai, G., Takacs, M., et al. (2018). Structural basis for NusA stabilized transcriptional pausing. *Molecular Cell, 69*(816–827), e814.
59. Ha, T., Enderle, T., Ogletree, D. F., Chemla, D. S., Selvin, P. R., & Weiss, S. (1996). Probing the interaction between two single molecules: Fluorescence resonance energy transfer between a single donor and a single acceptor. *Proceedings of the National Academy of Sciences of the United States of America, 93,* 6264–6268.
60. Harada, Y., Ohara, O., Takatsuki, A., Itoh, H., Shimamoto, N., & Kinosita, K., Jr. (2001). Direct observation of DNA rotation during transcription by *Escherichia coli* RNA polymerase. *Nature, 409,* 113–115.
61. Harden, T. T., Wells, C. D., Friedman, L. J., Landick, R., Hochschild, A., Kondev, J., et al. (2016). Bacterial RNA polymerase can retain sigma70 throughout transcription. *Proceedings of the National Academy of Sciences of the United States of America, 113,* 602–607.
62. Heller, I., Sitters, G., Broekmans, O. D., Farge, G., Menges, C., Wende, W., et al. (2013). STED nanoscopy combined with optical tweezers reveals protein dynamics on densely covered DNA. *Nature Methods, 10,* 910–916.
63. Herbert, K. M., La Porta, A., Wong, B. J., Mooney, R. A., Neuman, K. C., Landick, R., et al. (2006). Sequence-resolved detection of pausing by single RNA polymerase molecules. *Cell, 125,* 1083–1094.
64. Herbert, K. M., Zhou, J., Mooney, R. A., Porta, A. L., Landick, R., & Block, S. M. (2010). *E. coli* NusG inhibits backtracking and accelerates pause-free transcription by promoting forward translocation of RNA polymerase. *Journal of Molecular Biology, 399,* 17–30.
65. Hill, F. R., Monachino, E., & van Oijen, A. M. (2017). The more the merrier: High-throughput single-molecule techniques. *Biochemical Society Transactions, 45,* 759–769.
66. Hodges, C., Bintu, L., Lubkowska, L., Kashlev, M., & Bustamante, C. (2009). Nucleosomal fluctuations govern the transcription dynamics of RNA polymerase II. *Science (New York, N.Y.), 325*, 626–628.

67. Hohlbein, J., Craggs, T. D., & Cordes, T. (2014). Alternating-laser excitation: Single-molecule FRET and beyond. *Chemical Society Reviews, 43,* 1156–1171.
68. Holmes, E. C. (2010). Evolution in health and medicine Sackler colloquium: The comparative genomics of viral emergence. *Proceedings of the National Academy of Sciences of the United States of America, 107*(Suppl 1), 1742–1746.
69. Howan, K., Smith, A. J., Westblade, L. F., Joly, N., Grange, W., Zorman, S., et al. (2012). Initiation of transcription-coupled repair characterized at single-molecule resolution. *Nature, 490,* 431–434.
70. Ishibashi, T., Dangkulwanich, M., Coello, Y., Lionberger, T. A., Lubkowska, L., Ponticelli, A. S., et al. (2014). Transcription factors IIS and IIF enhance transcription efficiency by differentially modifying RNA polymerase pausing dynamics. *Proceedings of the National Academy of Sciences of the United States of America, 111,* 3419–3424.
71. Iyer, L. M., Koonin, E. V., & Aravind, L. (2003). Evolutionary connection between the catalytic subunits of DNA-dependent RNA polymerases and eukaryotic RNA-dependent RNA polymerases and the origin of RNA polymerases. *BMC Structural Biology, 3,* 1.
72. Jacome, R., Becerra, A., Ponce de Leon, S., & Lazcano, A. (2015). Structural analysis of monomeric RNA-dependent polymerases: Evolutionary and therapeutic implications. *PLoS ONE, 10,* e0139001.
73. Janissen, R., Berghuis, B. A., Dulin, D., Wink, M., van Laar, T., & Dekker, N. H. (2014). Invincible DNA tethers: Covalent DNA anchoring for enhanced temporal and force stability in magnetic tweezers experiments. *Nucleic Acids Research, 42,* e137.
74. Jonkers, I., & Lis, J. T. (2015). Getting up to speed with transcription elongation by RNA polymerase II. *Nature Reviews. Molecular Cell Biology, 16,* 167–177.
75. Joo, C., Balci, H., Ishitsuka, Y., Buranachai, C., & Ha, T. (2008). Advances in single-molecule fluorescence methods for molecular biology. *Annual Review of Biochemistry, 77,* 51–76.
76. Kaern, M., Elston, T. C., Blake, W. J., & Collins, J. J. (2005). Stochasticity in gene expression: From theories to phenotypes. *Nature Reviews Genetics, 6,* 451–464.
77. Kalinin, S., Peulen, T., Sindbert, S., Rothwell, P. J., Berger, S., Restle, T., et al. (2012). A toolkit and benchmark study for FRET-restrained high-precision structural modeling. *Nature Methods, 9,* 1218–1225.
78. Kang, J. Y., Mishanina, T. V., Bellecourt, M. J., Mooney, R. A., Darst, S. A., & Landick, R. (2018). RNA polymerase accommodates a pause RNA hairpin by global conformational rearrangements that prolong pausing. *Molecular Cell, 69*, 802–815, e801.
79. Kapanidis, A. N., Lee, N. K., Laurence, T. A., Doose, S., Margeat, E., & Weiss, S. (2004). Fluorescence-aided molecule sorting: Analysis of structure and interactions by alternating-laser excitation of single molecules. *Proceedings of the National Academy of Sciences of the United States of America, 101,* 8936–8941.
80. Kapanidis, A. N., Margeat, E., Ho, S. O., Kortkhonjia, E., Weiss, S., & Ebright, R. H. (2006). Initial transcription by RNA polymerase proceeds through a DNA-scrunching mechanism. *Science (New York, N.Y.), 314,* 1144–1147.
81. Kapanidis, A. N., Margeat, E., Laurence, T. A., Doose, S., Ho, S. O., Mukhopadhyay, J., et al. (2005). Retention of transcription initiation factor sigma70 in transcription elongation: Single-molecule analysis. *Molecular Cell, 20,* 347–356.
82. Kapanidis, A. N., & Strick, T. (2009). Biology, one molecule at a time. *Trends in Biochemical Sciences, 34,* 234–243.
83. Korboukh, V. K., Lee, C. A., Acevedo, A., Vignuzzi, M., Xiao, Y., Arnold, J. J., et al. (2014). RNA virus population diversity, an optimum for maximal fitness and virulence. *Journal of Biological Chemistry, 289,* 29531–29544.
84. Kriegel, F., Ermann, N., & Lipfert, J. (2017). Probing the mechanical properties, conformational changes, and interactions of nucleic acids with magnetic tweezers. *Journal of Structural Biology, 197,* 26–36.
85. La Porta, A., & Wang, M. D. (2004). Optical torque wrench: Angular trapping, rotation, and torque detection of quartz microparticles. *Physical Review Letters, 92,* 190801.

86. Landick, R. (2006). The regulatory roles and mechanism of transcriptional pausing. *Biochemical Society Transactions, 34,* 1062–1066.
87. Lang, M. J., Asbury, C. L., Shaevitz, J. W., & Block, S. M. (2002). An automated two-dimensional optical force clamp for single molecule studies. *Biophysical Journal, 83,* 491–501.
88. Larson, M. H., Greenleaf, W. J., Landick, R., & Block, S. M. (2008). Applied force reveals mechanistic and energetic details of transcription termination. *Cell, 132,* 971–982.
89. Larson, M. H., Mooney, R. A., Peters, J. M., Windgassen, T., Nayak, D., Gross, C. A., et al. (2014). A pause sequence enriched at translation start sites drives transcription dynamics in vivo. *Science (New York, N.Y.), 344,* 1042–1047.
90. Larson, M. H., Zhou, J., Kaplan, C. D., Palangat, M., Kornberg, R. D., Landick, R., et al. (2012). Trigger loop dynamics mediate the balance between the transcriptional fidelity and speed of RNA polymerase II. *Proceedings of the National Academy of Sciences of the United States of America, 109,* 6555–6560.
91. Laszlo, A. H., Derrington, I. M., & Gundlach, J. H. (2017). Subangstrom measurements of enzyme function using a biological nanopore, SPRNT. *Methods in Enzymology, 582,* 387–414.
92. Lauring, A. S., Frydman, J., & Andino, R. (2013). The role of mutational robustness in RNA virus evolution. *Nature Reviews Microbiology, 11,* 327–336.
93. Le, T. T., Yang, Y., Tan, C., Suhanovsky, M. M., Fulbright, R. M., Jr., Inman, J. T., et al. (2018). Mfd dynamically regulates transcription via a release and catch-up mechanism. *Cell, 172*(344–357), e315.
94. Lerner, E., Chung, S., Allen, B. L., Wang, S., Lee, J., Lu, S. W., et al. (2016). Backtracked and paused transcription initiation intermediate of *Escherichia coli* RNA polymerase. *Proceedings of the National Academy of Sciences of the United States of America, 113,* E6562–E6571.
95. Lerner, E., Ingargiola, A., & Weiss, S. (2018). Characterizing highly dynamic conformational states: The transcription bubble in RNAP-promoter open complex as an example. *The Journal of Chemical Physics, 148,* 10.
96. Lin, W., Das, K., Degen, D., Mazumder, A., Duchi, D., Wang, D., et al. (2018). Structural basis of transcription inhibition by fidaxomicin (lipiarmycin A3). *Molecular Cell, 70*(60–71), e15.
97. Lisica, A., Engel, C., Jahnel, M., Roldan, E., Galburt, E. A., Cramer, P., et al. (2016). Mechanisms of backtrack recovery by RNA polymerases I and II. *Proceedings of the National Academy of Sciences of the United States of America, 113,* 2946–2951.
98. Lisica, A., & Grill, S. W. (2017). Optical tweezers studies of transcription by eukaryotic RNA polymerases. *Biomolecular Concepts, 8,* 1–11.
99. Liu, S., Chistol, G., Hetherington, C. L., Tafoya, S., Aathavan, K., Schnitzbauer, J., et al. (2014). A viral packaging motor varies its DNA rotation and step size to preserve subunit coordination as the capsid fills. *Cell, 157,* 702–713.
100. Ma, J., Bai, L., & Wang, M. D. (2013). Transcription under torsion. *Science (New York, N.Y.), 340,* 1580–1583.
101. Madariaga-Marcos, J., Hormeno, S., Pastrana, C. L., Fisher, G. L. M., Dillingham, M. S., & Moreno-Herrero, F. (2018). Force determination in lateral magnetic tweezers combined with TIRF microscopy. *Nanoscale, 10,* 4579–4590.
102. Margeat, E., Kapanidis, A. N., Tinnefeld, P., Wang, Y., Mukhopadhyay, J., Ebright, R. H., et al. (2006). Direct observation of abortive initiation and promoter escape within single immobilized transcription complexes. *Biophysical Journal, 90,* 1419–1431.
103. Meng, C. A., Fazal, F. M., & Block, S. M. (2017). Real-time observation of polymerase-promoter contact remodeling during transcription initiation. *Nature Communications, 8,* 1178.
104. Miller, H., Zhou, Z., Shepherd, J., Wollman, A. J. M., & Leake, M. C. (2018). Single-molecule techniques in biophysics: A review of the progress in methods and applications. *Reports on Progress in Physics, 81,* 024601.
105. Moffitt, J. R., Chemla, Y. R., Smith, S. B., & Bustamante, C. (2008). Recent advances in optical tweezers. *Annual Review of Biochemistry, 77,* 205–228.

106. Muschielok, A., Andrecka, J., Jawhari, A., Bruckner, F., Cramer, P., & Michaelis, J. (2008). A nano-positioning system for macromolecular structural analysis. *Nature Methods, 5,* 965–971.
107. Nagy, J., Grohmann, D., Cheung, A. C., Schulz, S., Smollett, K., Werner, F., et al. (2015). Complete architecture of the archaeal RNA polymerase open complex from single-molecule FRET and NPS. *Nature Communications, 6,* 6161.
108. Neuman, K. C., Abbondanzieri, E. A., Landick, R., Gelles, J., & Block, S. M. (2003). Ubiquitous transcriptional pausing is independent of RNA polymerase backtracking. *Cell, 115,* 437–447.
109. Neuman, K. C., & Block, S. M. (2004). Optical trapping. *Review of Scientific Instruments, 75,* 2787–2809.
110. Neuman, K. C., & Nagy, A. (2008). Single-molecule force spectroscopy: optical tweezers, magnetic tweezers and atomic force microscopy. *Nature Methods, 5,* 491–505.
111. Ng, K. K., Arnold, J. J., & Cameron, C. E. (2008). Structure-function relationships among RNA-dependent RNA polymerases. *Current Topics in Microbiology and Immunology, 320,* 137–156.
112. Raser, J. M., & O'Shea, E. K. (2004). Control of stochasticity in eukaryotic gene expression. *Science (New York, NY), 304,* 1811–1814.
113. Ray-Soni, A., Bellecourt, M. J., & Landick, R. (2016). Mechanisms of bacterial transcription termination: All good things must end. *Annual Review of Biochemistry, 85,* 319–347.
114. Revyakin, A., Ebright, R. H., & Strick, T. R. (2004). Promoter unwinding and promoter clearance by RNA polymerase: detection by single-molecule DNA nanomanipulation. *Proceedings of the National Academy of Sciences of the United States of America, 101,* 4776–4780.
115. Revyakin, A., Ebright, R. H., & Strick, T. R. (2005). Single-molecule DNA nanomanipulation: Improved resolution through use of shorter DNA fragments. *Nature Methods, 2,* 127–138.
116. Revyakin, A., Liu, C., Ebright, R. H., & Strick, T. R. (2006). Abortive initiation and productive initiation by RNA polymerase involve DNA scrunching. *Science (New York, N.Y.), 314,* 1139–1143.
117. Righini, M., Lee, A., Canari-Chumpitaz, C., Lionberger, T., Gabizon, R., Coello, Y., Tinoco, I., Jr., & Bustamante, C. (2018). Full molecular trajectories of RNA polymerase at single base-pair resolution. In *Proceedings of the National Academy of Sciences of the United States of America.*
118. Robb, N. C., Cordes, T., Hwang, L. C., Gryte, K., Duchi, D., Craggs, T. D., et al. (2013). The transcription bubble of the RNA polymerase-promoter open complex exhibits conformational heterogeneity and millisecond-scale dynamics: Implications for transcription start-site selection. *Journal of Molecular Biology, 425,* 875–885.
119. Robb, N. C., Te Velthuis, A. J., Wieneke, R., Tampe, R., Cordes, T., Fodor, E., & Kapanidis, A. N. (2016). Single-molecule FRET reveals the pre-initiation and initiation conformations of influenza virus promoter RNA. *Nucleic Acids Research.*
120. Roberts, J. W., Shankar, S., & Filter, J. J. (2008). RNA polymerase elongation factors. *Annual Review of Microbiology, 62,* 211–233.
121. Robinson, A., & van Oijen, A. M. (2013). Bacterial replication, transcription and translation: Mechanistic insights from single-molecule biochemical studies. *Nature Reviews Microbiology, 11,* 303–315.
122. Ruff, E. F., Record, M. T., Jr., & Artsimovitch, I. (2015). Initial events in bacterial transcription initiation. *Biomolecules, 5,* 1035–1062.
123. Rutkauskas, M., Krivoy, A., Szczelkun, M. D., Rouillon, C., & Seidel, R. (2017). Single-molecule insight into target recognition by CRISPR-Cas complexes. *Methods in Enzymology, 582,* 239–273.
124. Saecker, R. M., Record, M. T., Jr., & Dehaseth, P. L. (2011). Mechanism of bacterial transcription initiation: RNA polymerase—Promoter binding, isomerization to initiation-competent open complexes, and initiation of RNA synthesis. *Journal of Molecular Biology, 412,* 754–771.
125. Santybayeva, Z., & Pedaci, F. (2017). Optical torque wrench design and calibration. *Methods in Molecular Biology, 1486,* 157–181.

126. Saunders, A., Core, L. J., & Lis, J. T. (2006). Breaking barriers to transcription elongation. *Nature Reviews. Molecular Cell Biology, 7,* 557–567.
127. Schafer, D. A., Gelles, J., Sheetz, M. P., & Landick, R. (1991). Transcription by single molecules of RNA polymerase observed by light microscopy. *Nature, 352,* 444–448.
128. Schulz, S., Gietl, A., Smollett, K., Tinnefeld, P., Werner, F., & Grohmann, D. (2016). TFE and Spt4/5 open and close the RNA polymerase clamp during the transcription cycle. *Proceedings of the National Academy of Sciences of the United States of America, 113,* E1816–E1825.
129. Schulz, S., Kramm, K., Werner, F., & Grohmann, D. (2015). Fluorescently labeled recombinant RNAP system to probe archaeal transcription initiation. *Methods, 86,* 10–18.
130. Selvin, P. R., Lougheed, T., Tonks Hoffman, M., Park, H., Balci, H., Blehm, B. H., & Toprak, E. (2007). Fluorescence imaging with one-nanometer accuracy (FIONA). *CSH Protocols 2007*, pdb top27.
131. Shaevitz, J. W., Abbondanzieri, E. A., Landick, R., & Block, S. M. (2003). Backtracking by single RNA polymerase molecules observed at near-base-pair resolution. *Nature, 426,* 684–687.
132. Smith, E. C. (2017). The not-so-infinite malleability of RNA viruses: Viral and cellular determinants of RNA virus mutation rates. *PLoS Pathogens, 13,* e1006254.
133. Stubbs, T. M., & Te Velthuis, A. J. (2014). The RNA-dependent RNA polymerase of the influenza A virus. *Future Virology, 9,* 863–876.
134. te Velthuis, A. J. (2014). Common and unique features of viral RNA-dependent polymerases. *Cellular and Molecular Life Sciences: CMLS, 71,* 4403–4420.
135. Te Velthuis, A. J., & Fodor, E. (2016). Influenza virus RNA polymerase: Insights into the mechanisms of viral RNA synthesis. *Nature Reviews Microbiology, 14,* 479–493.
136. Te Velthuis, A. J., Robb, N. C., Kapanidis, A. N., & Fodor, E. (2016). The role of the priming loop in influenza A virus RNA synthesis. *Nature Microbiology, 1,* 16029.
137. Tinoco, I., Jr., & Gonzalez, R. L., Jr. (2011). Biological mechanisms, one molecule at a time. *Genes & Development, 25,* 1205–1231.
138. Tomescu, A. I., Robb, N. C., Hengrung, N., Fodor, E., & Kapanidis, A. N. (2014). Single-molecule FRET reveals a corkscrew RNA structure for the polymerase-bound influenza virus promoter. *Proceedings of the National Academy of Sciences of the United States of America, 111,* E3335–E3342.
139. Tomko, E. J., Fishburn, J., Hahn, S., & Galburt, E. A. (2017). TFIIH generates a six-base-pair open complex during RNAP II transcription initiation and start-site scanning. *Nature Structural & Molecular Biology, 24,* 1139–1145.
140. van de Meent, J. W., Bronson, J. E., Wiggins, C. H., & Gonzalez, R. L., Jr. (2014). Empirical Bayes methods enable advanced population-level analyses of single-molecule FRET experiments. *Biophysical Journal, 106,* 1327–1337.
141. van Loenhout, M. T., de Grunt, M. V., & Dekker, C. (2012). Dynamics of DNA supercoils. *Science (New York, N.Y.), 338,* 94–97.
142. Vilfan, I. D., Lipfert, J., Koster, D. A., Lemay, S. G., & Dekker, N. H. (2009). Magnetic tweezers for single-molecule experiments. In *Handbook of single-molecule biophysics* (pp. 371–395).
143. Visscher, K., Schnitzer, M. J., & Block, S. M. (1999). Single kinesin molecules studied with a molecular force clamp. *Nature, 400,* 184–189.
144. Wang, F., Redding, S., Finkelstein, I. J., Gorman, J., Reichman, D. R., & Greene, E. C. (2013). The promoter-search mechanism of *Escherichia coli* RNA polymerase is dominated by three-dimensional diffusion. *Nature Structural & Molecular Biology, 20,* 174–181.
145. Wang, M. D., Schnitzer, M. J., Yin, H., Landick, R., Gelles, J., & Block, S. M. (1998). Force and velocity measured for single molecules of RNA polymerase. *Science (New York, N.Y.), 282,* 902–907.
146. Washburn, R. S., & Gottesman, M. E. (2015). Regulation of transcription elongation and termination. *Biomolecules, 5,* 1063–1078.
147. Werner, F., & Grohmann, D. (2011). Evolution of multisubunit RNA polymerases in the three domains of life. *Nature Reviews Microbiology, 9,* 85–98.

148. Werner, F., & Weinzierl, R. O. (2002). A recombinant RNA polymerase II-like enzyme capable of promoter-specific transcription. *Molecular Cell, 10,* 635–646.
149. Woodside, M. T., & Block, S. M. (2014). Reconstructing folding energy landscapes by single-molecule force spectroscopy. *Annual Review of Biophysics, 43,* 19–39.
150. Wuite, G. J., Davenport, R. J., Rappaport, A., & Bustamante, C. (2000). An integrated laser trap/flow control video microscope for the study of single biomolecules. *Biophysical Journal, 79,* 1155–1167.
151. Xie, S. N. (2001). Single-molecule approach to enzymology. *Single Molecules, 2,* 229–236.
152. Yin, H., Wang, M. D., Svoboda, K., Landick, R., Block, S. M., & Gelles, J. (1995). Transcription against an applied force. *Science (New York, N.Y.), 270,* 1653–1657.
153. Yu, L., Winkelman, J. T., Pukhrambam, C., Strick, T. R., Nickels, B. E., & Ebright, R. H. (2017). The mechanism of variability in transcription start site selection. *eLife, 6.*
154. Zamft, B., Bintu, L., Ishibashi, T., & Bustamante, C. (2012). Nascent RNA structure modulates the transcriptional dynamics of RNA polymerases. *Proceedings of the National Academy of Sciences of the United States of America, 109,* 8948–8953.
155. Zhou, J., Ha, K. S., La Porta, A., Landick, R., & Block, S. M. (2011). Applied force provides insight into transcriptional pausing and its modulation by transcription factor NusA. *Molecular Cell, 44,* 635–646.
156. Zong, J., Yao, X., Yin, J., Zhang, D., & Ma, H. (2009). Evolution of the RNA-dependent RNA polymerase (RdRP) genes: Duplications and possible losses before and after the divergence of major eukaryotic groups. *Gene, 447,* 29–39.

Chapter 7
Single-Molecule Optical Tweezers Studies of Translation

Xiaohui Qu

Abbreviations

AFM	Atomic force microscopy
bp	Base pair
CI	Confidence interval
mRNA	Messenger RNA
ms	Millisecond
nm	Nanometer
nt	Nucleotide
pN	picoNewton
SD	Shine-Dalgarno sequence
s.d.	Standard deviation
UTR	Untranslated region
WLC	Worm-like chain model

7.1 Introduction

Translation is the process of ribosome reading the codons on a messenger RNA (mRNA) and making a corresponding polypeptide. Translation is essential for the transduction of genetic information from DNA to protein. Besides being a reliable supply of new proteins, translation is also a key step for regulating protein expression levels [1]. New approaches and techniques continue to be developed for

X. Qu (✉)
Molecular Biology Program, Memorial Sloan Kettering Cancer Center, New York, NY 10065, USA
e-mail: qux@mskcc.org

C. Joo and D. Rueda (eds.), *Biophysics of RNA-Protein Interactions*,
Biological and Medical Physics, Biomedical Engineering,
https://doi.org/10.1007/978-1-4939-9726-8_7

a better understanding of translation, including the use of optical tweezers. It was first demonstrated 30 years ago that a focused laser beam can trap a micron-sized dielectric particle, and therefore, can be used as 'optical tweezers' to move such particles around [2]. Over the past several decades, optical tweezers have evolved into a sophisticated instrument that can achieve high-resolution manipulation and measurement on molecular interactions in three aspects simultaneously: distance at nanometer (nm) resolution, kinetics at millisecond (ms) resolution, and force at picoNewton (pN) resolution. This unique combination of high resolutions enables optical tweezers to complement bulk and other single-molecule techniques by providing a distinctive perspective for studying molecular interactions. Magnetic tweezers and atomic force microscopy (AFM) are the other two commonly used techniques for single-molecule mechanical measurements. In comparison with these other techniques, optical tweezers technique excels in spatial, temporal, and force resolutions, the flexibility in manipulating single molecules, and the ease of reagent exchange during experiments [3]. Consequently, translation has been almost exclusively studied by optical tweezers, particularly for high-resolution experiments. Specifically, single-molecule optical tweezers technique has been used to study mRNA structure disruption during translation initiation, the peptide chain elongation kinetics, and the interactions between ribosome and nascent polypeptide. In this chapter, we will first briefly introduce the single-molecule optical tweezers technique and then give a comprehensive overview of its application in translation.

7.2 The Single-Molecule Optical Tweezers Technique

The principles and technical aspects of optical tweezers technique have been extensively reviewed ([3–5] as examples). Therefore, the discussion here will be limited to the basic concepts that are necessary for understanding its application in translation studies. It is also noteworthy to point out that single-molecule optical tweezers techniques continue to evolve in complexity and capability beyond the ones that have been implemented in translation studies. Examples include attaching a bead to double-stranded DNA to track its rotation [6], the combination of fluorescence and optical tweezers [7], and using nanofabricated quartz cylinder for torque application and detection in an angular optical trap [8]. The readers are encouraged to refer to other literature to explore the full potential of optical tweezers for single-molecule biophysical studies.

How does the optical tweezers technique achieve the high-resolution measurement on single molecules? Figure 7.1a shows the experimental scheme used to unfold a structured RNA molecule. Each end of the RNA is extended with several hundred base pairs of RNA·DNA duplex and attached to a micron-sized dielectric bead. The RNA·DNA handle serves as a rigid spacer between the bead and the RNA sequence of interest. One bead is controlled by an optical trap, while the other bead is either held fixed in space or controlled by another optical trap. Having the second bead fixed or movable only matters for the technical detail of converting direct instrument

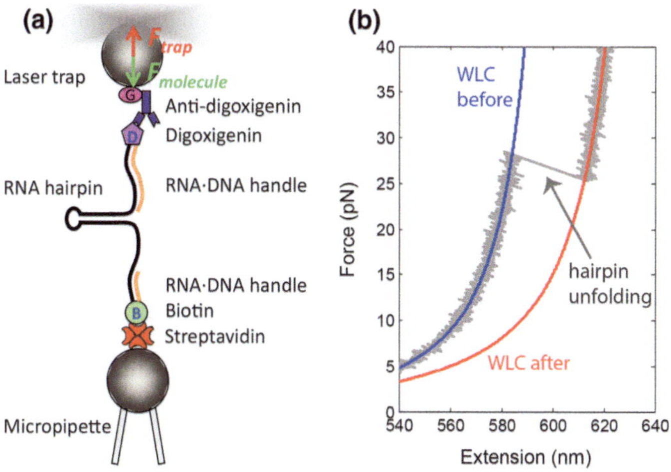

Fig. 7.1 **RNA hairpin unfolding by optical tweezers.** **a** The molecular arrangement. **b** An example force-extension curve (gray) resulting from pulling apart the two ends of a hairpin-forming RNA by optical tweezers. The RNA stretches elastically under force, except for the cooperative unfolding of the hairpin structure. The RNA elastic stretching both before and after the hairpin unfolding fit well to the WLC model (blue and red). The only parameter that differs between the two WLC fitting curves is the RNA contour length, reflecting the increase in the number of single-stranded RNA after hairpin unfolding

output to the experimental observables—force and extension. For clarity, we will use the case with the fixed second bead to explain the principle of measurements on optical tweezers. The optical trap exerts a force field on the trapped bead and the trap force (F_{trap}) scales linearly with the bead's deviation from the center of the trap (Δd_{bead}):

$$F_{\text{trap}} = k \cdot \Delta d_{\text{bead}} \tag{7.1}$$

Here, k is the stiffness of the trap, which depends on the specific instrument and can be characterized conveniently using a free bead. The direction of F_{trap} always points toward the center of the trap, therefore restraining a bead from straying away from the trap. But when the trapped bead is attached to a constrained molecule as illustrated in Fig. 7.1a, moving the trapped bead away from the fixed bead stretches the molecule, which generates a counteracting force (F_{molecule}) on the trapped bead. Consequently, the trapped bead settles at a position away from the trap center, where F_{molecule} and F_{trap} balance each other. Typically, the direct instrument outputs are Δd_{bead} and the moving distance of the trap (Δd_{trap}), from which the extension and force of the molecule can be calculated as $\Delta d_{\text{trap}} - \Delta d_{\text{bead}}$ and $k \cdot \Delta d_{\text{bead}}$, respectively.

Under the basic measurement principle explained above, several operational modes of optical tweezers [9] have been the most commonly used in translation studies. In the 'force ramp' mode, the optical trap moves back and forth between two set positions with a constant speed, resulting in the repeated stretching and relaxing of the molecule. Force ramp is typically used to obtain the unfolding/refolding curves, such as the one in Fig. 7.1b, to characterize the mechano-structural relationship for RNAs or proteins. In the 'constant force' mode, the instrument maintains the molecular force at a set value by moving the trapped bead to instantaneously compensate for any detected force change. This mode is used when the structural change in the tethered molecule between the two beads is expected at the set force value, such as due to enzyme unwinding of RNA structures or RNA and protein conformational fluctuations. In this mode, the distance that the trapped bead has to be moved to restore the set force value gives a direct measure of the tether length change. An example application of the constant force mode can be seen in Fig. 7.2 in Sect. 7.3.1.

The basic idea behind optical tweezers measurements is that applying an external force on a biological macromolecule can perturb its conformational states. How the conformation of a molecule changes under force is determined by its structural properties. For example, RNA molecules that differ in contour length, secondary or tertiary structures, percentage of G·C content or sequence arrangement in the structured region will all behave differently under mechanical perturbation. Therefore, each biological molecule has a characteristic mechanical signature, which is often presented in the form of force–extension relation in optical tweezers studies. Figure 7.1b shows the force–extension curve for a hairpin-forming RNA containing 30 G·C base pairs (bp) and a 4-nucleotide (nt) loop ('tetraloop'). In this example, the optical tweezers pull apart the two ends of the mRNA at a constant speed. The force increases continuously as the mRNA is stretched, up to approximately 28 pN. But then, the force drops instantaneously by about 2.8 pN, accompanied by a simultaneous 28-nm increase in RNA extension. Further stretching results again in continuous force increase.

Although the force–extension curve in Fig. 7.1b was obtained with a specific RNA, the two different types of conformational changes observed here are general for all nucleic acids and proteins under tension: (i) elastic stretching and (ii) disruption of structural motifs. Elastic stretching is well described by the empirical worm-like chain (WLC) model [11], which is a quantitative relation between the external force (F) and the molecular end-to-end distance (x):

$$\frac{F \cdot P}{k_B T} = \frac{1}{4}\left(1 - \frac{x}{L_0}\right)^{-2} - \frac{1}{4} + \frac{x}{L_0} \tag{7.2}$$

In this equation, L_0 is the contour length of the unstructured region of the molecule, P is the polymer's persistence length, k_B is the Boltzmann constant, and T is the absolute temperature. Qualitatively speaking, the extension of a WLC polymer increases nonlinearly with the external force between two extremes: $x/L_0 \to 0$ when $F = 0$, and $x/L_0 \to 1$ when $F \to +\infty$. Disruption of structural motifs by the external force often happens in a cooperative way and releases a number of unstructured residues at once. The instantaneous release of the additional unstructured residues gives rise to a more relaxed molecule between the two beads and, consequently, is accompanied by a simultaneous drop in force. The elastic stretching after structure unfolding can again be described by the WLC model with a larger L_0 to account for the lengthening of the unstructured region (Fig. 7.1b).

7.3 mRNA Structure Disruption in Translation Initiation

Initiation is the process of ribosome binding to the mRNA and locating the start codon. Initiation differs greatly between prokaryotes and eukaryotes. In prokaryotes, a strong initiation site is composed of a start codon and an upstream Shine-Dalgarno (SD) sequence, and ribosomal particles 30S and 50S bind directly to the start codon with the help of three initiation factors [12]. In eukaryotes [13], a strong initiation site is composed of a start codon flanked by the Kozak sequence, and the mRNAs are typically capped with the m^7G structure at the 5′ terminus and polyadenylated at the 3′ terminus. More than 30 eukaryotic initiation factors have been identified. Canonical cap-dependent initiation occurs in a multi-step fashion including 5′ cap recognition by eIF4F complex, small 40S ribosomal subunit binding close to the 5′ end, 40S scanning of the 5′ untranslated region (UTR) with a 5′ to 3′ directionality, 40S recognition of the start codon, and lastly the binding of the large 60S ribosomal subunit. So far, optical tweezers have not been applied to study the actual initiation process, but rather the mechanism of mRNA structure disruption during initiation for both prokaryotic (Sect. 7.3.1) and eukaryotic (Sect. 7.3.2) translations.

7.3.1 S1 Melting of mRNA Structures in Prokaryotic Initiation

It is known that secondary structures upstream of the initiation site help regulate prokaryotic translation [14–16], although nucleic acid helicases have not been identified to facilitate initiation. The small ribosomal protein S1 is required for in vivo translation of most natural mRNAs in *Escherichia coli* [17], particularly mRNAs with a highly structured 5′ region or lacking a strong SD sequence [18, 19]. Therefore, S1 was believed to be involved in the melting of mRNA structures around the initiation site. Qu et al. conducted a quantitative characterization of the S1 melting of RNA structures, using the optical tweezers scheme in Fig. 7.2a [10]. In this experiment, a hairpin-forming RNA containing 274 bp and a tetraloop is attached between two beads. When holding the RNA at a constant force above 18.9 pN but well below its mechanical unfolding force, the step-wise unwinding of the hairpin was observed in the presence of S1 proteins (Fig. 7.2b left). Although step-wise unwinding of structured RNA or DNA has been observed for several processive helicases using optical tweezers ([20, 21] as examples), the observation of S1 differed from the examples of processive helicases in several aspects: (i) unwinding happened in buffer, without

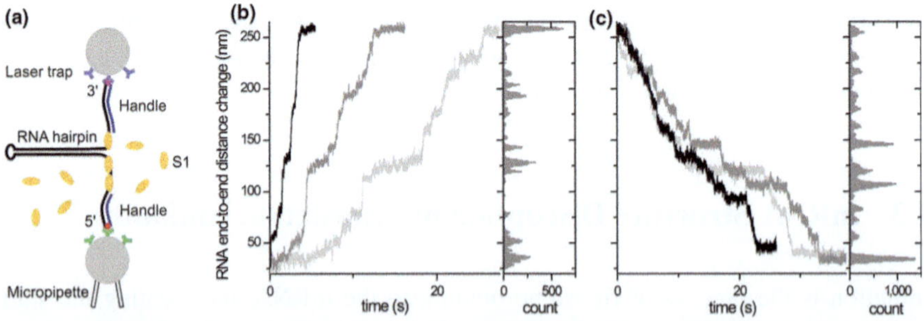

Fig. 7.2 Prokaryotic ribosomal protein S1 melting of RNA structures. **a** Schematic of the experimental setup. **b (left)** Typical unwinding trajectories for 30 nM (light gray), 100 nM (dark gray), and 300 nM (black) S1, when held at a constant force above 18.9 pN. **b (right)** Histogram of the number of data points in the dark gray trajectory along the RNA sequence. **c (left and right)** The same representation as in (**b**) for re-zipping events at force <17 pN. Both unwinding and re-zipping occur in a step-wise (pause-step-pause) fashion. The unwinding rate increases with S1 concentration, while the re-zipping rate is independent of S1 concentration. Reprinted from Ref. [10] with permission. Copyright 2012 National Academy of Sciences

any NTP as the energy source; (ii) the unwinding rate increased linearly with the S1 concentration; and (iii) when holding the unwound RNA at a constant force below 17 pN, step-wise re-zipping was observed with the same step size as the unwinding process (Fig. 7.2c left). These observations strongly suggest that S1 is a double-stranded RNA-melting protein instead of a processive helicase. More specifically, as the RNA junction between the single strands and the base-paired region undergoes very fast thermal fluctuations between open and closed states ('junction thermal breathing'), S1 can bind to the transiently released single strand at the hairpin junction, hence prohibiting the single strands from re-annealing. When the force is high enough to favor the open state in junction thermal breathing, multiple S1 proteins can bind consecutively, resulting in the step-wise unwinding of the RNA structure. Reversely, when the force is low enough to favor the closed state in junction thermal breathing, consecutive dissociation of S1 proteins occurs and gives rise to the step-wise re-zipping of the RNA structure.

The resolution of optical tweezers allows one to identify individual events of S1 binding or dissociation from the unwinding or re-zipping trajectories, respectively. As S1 binding or dissociation leads to instantaneous lengthening or shortening of the single strands, a fast increase or decrease of RNA end-to-end distance is observed ('stepping'). In the intervals between binding and dissociation events, the RNA end-to-end distance stays constant except for thermal fluctuations ('pausing'). Accordingly, the histograms of the number of data points along with the RNA sequence form multiple Gaussian peaks for both the unfolding and re-zipping trajectories (Fig. 7.2b, c right panels), with each peak representing a pause position. The distance between the adjacent Gaussian peaks is equivalent to the binding size (δ_0) of an individual S1. The average of δ_0 is 5 ± 1.4 (s.d.) nm for unwinding and 5 ± 1 (s.d.) nm for re-zipping, respectively. This binding size corresponds to 10 nt, consistent with a previous cryo-EM study [22]. Strikingly, it was observed that S1-bound single-stranded RNA has a force-independent length of 0.5 nm/nt. In stark contrast, it takes about 46.5 pN force to stretch protein-free single-stranded RNA to such an extent, based on the WLC model. Therefore, S1 appears to act as a rigid scaffold for RNA binding with little curvature in the binding surface, so that S1-bound RNAs stay in an extended form. Furthermore, analysis of the force dependence of the kinetics yielded important information on the substeps of unwinding or re-zipping that cannot be directly resolved from the trajectories. Specifically, the average value of the duration of the pause before an unwinding (τ_u) or re-zipping (τ_r) step depends exponentially on the size of the rate-limiting substep for the unwinding (δ_u) or re-zipping (δ_r) processes with the following relation:

$$\langle \tau_u \rangle \sim e^{\delta_u \cdot C(F)} \tag{7.3a}$$

$$\langle \tau_r \rangle \sim e^{\delta_r \cdot C(F)} \tag{7.3b}$$

where $C(F)$ is a force-dependent parameter that can be calculated from the WLC model for each force value. There are several possible scenarios. If a single S1 protein binds or dissociates in a single step, $\delta_0 = \delta_u = \delta_r$. If either binding or dissociation occurs in multiple substeps, $\delta_u < \delta_0$ or $\delta_r < \delta_0$ will be observed, respectively. The force dependence analysis yielded that $\delta_u = 5 \pm 1$ nt (s.d.) for unwinding and $\delta_u = 2.2 \pm 0.4$ nt (s.d.) for re-zipping, both smaller than the 10 nt S1 binding size. Therefore, both S1 binding and dissociation occur in multiple substeps and are each rate limited by a different substep. Furthermore, since $\delta_u + \delta_r < \delta_0$, there is at least one substep that does not rate limit either the unwinding or re-zipping process and hence is 'hidden' in the kinetic analysis. Taking into consideration that S1 has four RNA-binding domains (D3–D6) [23] and that D4 and D5 are tightly associated even in the absence of RNA [24], it was proposed that the unwinding process is rate limited by D4/D5 binding and re-zipping is rate limited by the dissociation of D3 or D6, or by both if their dissociation rates are similar.

7.3.2 *eIF4A Helicase Activity in Eukaryotic Initiation*

Eukaryotic initiation factor 4A (eIF4A) is a DEAD-box helicase [25] and is essential for mRNA secondary structure disruption during 40S binding and scanning [26]. Garcia-Garcia et al. studied eIF4A unwinding of RNA structures on optical tweezers, with a particular focus on the processivity of eIF4A function [27]. Specifically, the authors wanted to distinguish between two possible unwinding mechanisms: distributive (each eIF4A unwinds a small segment of RNA structure and multiple eIF4A molecules are required to unwind a long stretch of RNA base pairs) versus processive (a single eIF4A molecule can move along and unwind a long stretch of RNA base pairs). The experimental geometry is similar to Fig. 7.2a with a hairpin-forming

RNA attached between two beads. The unwinding activity was characterized for purified human eIF4A alone or in combination with its known accessory proteins eIF4G, eIF4B, and eIF4H [26]. One consideration in the experimental design concerns eIF4G. Wild-type eIF4G contains the auto-inhibitory eIF4E binding site and requires the cap-binding protein eIF4E to have full activity [28]. However, the experimental geometry does not provide an accessible RNA 5′ end for eIF4E to function properly. Therefore, this study used a truncated mutant of eIF4G (eIF4G$_{682–1105}$) that lacks the eIF4E-binding domain but retains the one for eIF4A [28–30]. Another consideration concerns the protein concentrations. As the central question is whether eIF4A can function as a processive helicase, limiting concentrations of eIF4A and accessory proteins were used so that the chance of multiple proteins binding to a single RNA molecule was very low in the experimental time frame. Constant force mode was used to track RNA structural change in the presence of different protein combinations. It was found that eIF4A alone typically gave rise to a single unwinding step of 11 ± 2 bp. The additional introduction of any single accessory protein did not change the unwinding step size, but slightly increased the unwinding activity by allowing two or three consecutive unwinding steps. Interestingly, when eIF4B or eIF4H was introduced together with eIF4G$_{682–1105}$, they synergistically enhanced the processivity of eIF4A and permitted complete unwinding of the entire 72-bp reporter hairpin, while also increasing the average unwinding speed by about threefold.

DEAD-box helicases are typically considered to be nonprocessive [31]. This study provides intriguing evidence that eIF4A gains processivity with the help of eIF4G$_{682–1105}$/eIF4B or eIF4G$_{682–1105}$/eIF4H. The experimental strategy should be compatible with more biophysical characterizations of eIF4A helicase activity. Of particular interest is the measurement of the force dependence of the unwinding rate. Using the quantitative analysis similar to Eq. (7.3a) above and the Betterton model discussed in Sect. 7.4.2 (b), the force dependence analysis can determine the size of the rate-limiting substep δ_u and the destabilization energy ΔG_d (i.e., how much the helicase weakens the hairpin junction to allow efficient unwinding). Such quantification can allow comparative studies to see whether the accessory proteins change δ_u, ΔG_d, or both, which may yield important insights on how accessory proteins help eIF4A to gain processivity.

7.4 The Decoding Process

Decoding is the process of ribosome reading the codons on mRNA and synthesizing the corresponding polypeptide. Decoding includes a few major steps: aminoacyl-tRNA selection and binding, peptidyl transfer, and ribosome translocation to the next codon [32–34]. These steps repeat until the ribosome reaches the stop codon. The mechanism of decoding is generally believed to be well preserved in all organisms. Up to date, all optical tweezers studies of the decoding process were carried out with prokaryotic ribosomes using either of the two geometries illustrated in Fig. 7.3. In the ‘tug-of-war’ geometry (Fig. 7.3a), one bead is attached to 30S and the other bead

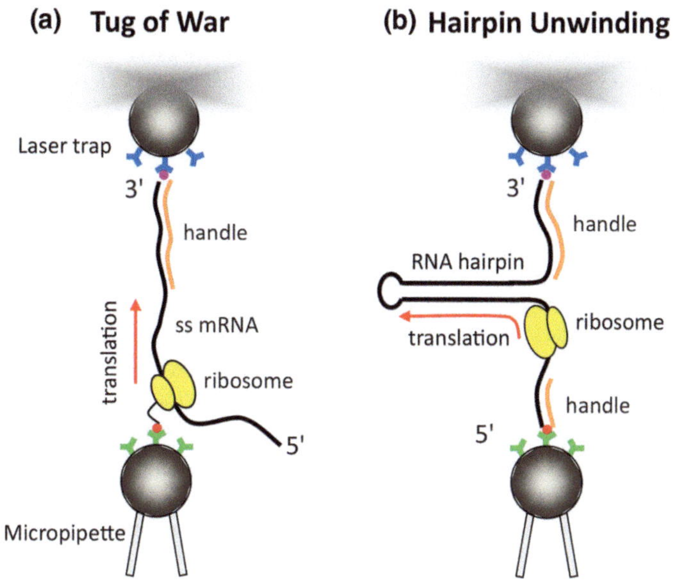

Fig. 7.3 Two commonly used geometries in optical tweezers studies of the decoding process. **a** The tug-of-war geometry. **b** The hairpin unwinding geometry

is attached to either 5′ or 3′ end of the mRNA. This geometry can directly track the ribosome movement on mRNA (Sect. 7.4.3) and also apply force to perturb ribosome movement (Sect. 7.4.3) or to disrupt the mRNA/ribosome complex (Sect. 7.4.1). In the 'hairpin unwinding' geometry (Fig. 7.3b), a hairpin-forming mRNA is attached between two beads and the RNA structural change resulting from ribosome helicase activity is used as a reporter of ribosome movement on mRNA during decoding (Sect. 7.4.2).

7.4.1 Mechanical Stability of the mRNA/Ribosome Complex

The ribosome has extensive interactions with mRNA and tRNAs [32–34]. mRNA wraps around the 30S small ribosomal subunit in a U-shaped channel. The center of the mRNA channel locates at the 30S decoding center, where the tRNAs base pair with mRNA codons. The 50S large ribosomal subunit catalyzes the formation of the peptide bonds. To achieve high translation fidelity, the ribosome is expected to maintain steady interactions with the codons during tRNA selection and peptidyl transfer and to weaken these interactions during translocation.

Using the tug-of-war geometry, Uemura et al. studied how the SD sequence and tRNA identities in the ribosomal A- (aminoacyl-) and P- (peptidyl-) sites modulate the mechanical stability of mRNA/ribosome interactions [35]. Specifically, the ribosome is assembled on a 57-nucleotide mRNA in the presence of different tRNA species, and the mRNA 5′ end and 30S are each attached to a bead. When pulling apart the two beads at a constant speed, an increasingly higher external force applies on the ribosome/tRNA/mRNA complex, until the ribosome/tRNA dissociates from the mRNA and breaks the molecular linkage between the two beads. The rupture force is a measure of the mechanical stability of the ribosome/tRNA/mRNA complex. For the mRNA containing a natural SD sequence, the complexes between the mRNA and (i) 70S alone, (ii) 70S and a P-site non-acylated initiator $tRNA^{fMet}$, and (iii) 70S, a P-site non-acylated $tRNA^{fMet}$ and an A-site Phe-$tRNA^{Phe}$ had a rupture force distribution centered at 10.6, 15.2, and 26.5 pN, respectively. Therefore, both A- and P-site tRNA bindings stabilize the ribosome/mRNA complex. The SD sequence can basepair with the 3′ end of the 16S rRNA on the 30S subunit and increases the ribosome binding affinity to mRNA [36, 37]. When the SD sequence was mutated to weaken its interaction with 30S, the rupture force reduced by about 10 pN for all three mRNA/tRNA/ribosome complexes mentioned above, but the force difference between the complexes was not affected. Therefore, the SD sequence and tRNA binding work additively to increase the ribosome/mRNA binding affinity. Interestingly, the post-peptidyl transfer state mimic that has a P-site $tRNA^{fMet}$ and an A-site peptidyl-tRNA analogue N-acetyl-Phe-$tRNA^{Phe}$ shows a SD-independent rupture force of approximately 12 pN, significantly smaller than the complex with a P-site $tRNA^{fMet}$ and an A-site Phe-$tRNA^{Phe}$. Furthermore, ribosome/mRNA complexes with P-site fMet-$tRNA^{fMet}$ and A-site Phe-$tRNA^{Phe}$, which allows efficient ribosome-catalyzed peptide bond formation, also give rise to a similarly reduced rupture force. Therefore, peptide bond formation weakens the ribosome/mRNA interactions, including the SD interactions, which should facilitate the subsequent ribosome translocation on mRNA.

Note that in an earlier study of the mechanical stability of mRNA/ribosome interactions, Vanzi et al. [38] also used the tug-of-war geometry and measured the rupture force for complexes between a long poly(U) mRNA and (i) 70S alone or (ii) 70S and a P-site N-acetyl-Phe-$tRNA^{Phe}$. In both cases, the rupture force showed a multimodal distribution with peaks situated at approximately 1.5, 12, and 19 pN. However, the relative population of the peaks shifted toward the higher force peaks in the presence of the P-site N-acetyl-Phe-$tRNA^{Phe}$, which is consistent with the scenario of P-site tRNA binding stabilizing mRNA/ribosome interactions. Besides the difference in mRNA sequence and length, this study differed from the study by Uemura et al. [35] in the strategy of ribosome attachment. Here, ribosomes were covalently linked to surface via free thiol groups on the ribosome surface, rendering a random attachment point. The heterogeneity in ribosome attachment suggested heterogeneity in the geometry of how mRNA/ribosome complexes were stretched under the applied optical tweezers force, which consequently could affect the magnitude of the rupture force. This difference in the experimental setup is likely one of the major factors that gave rise to clearly different rupture force values between the two studies.

7.4.2 Decoding Kinetics

a. *The first single-molecule optical tweezers assay of the decoding kinetics*

Wen et al. developed the first single-molecule assay to follow the real-time single ribosome translation dynamics at the single codon resolution, using the hairpin unwinding geometry [39]. Approximately, 30 nucleotides of single-stranded RNA sit in the mRNA channel on the 30S subunit, and the distance from the P-site codon to the entry site is 13 ± 2 (s.d.) nt [40–42]. The mRNA entry site, formed by S3, S4, S5 ribosomal proteins, allows only single-stranded RNA to enter. Therefore, in the hairpin unwinding geometry, the translocation of the ribosome from one codon to the next in the decoding center is accompanied by the unwinding of the hairpin structure by three base pairs at the mRNA entry site. As the accompanying hairpin unwinding releases twice the single-stranded length that the ribosome translocates, this geometry allows a better resolution for tracking individual ribosome translocation steps than the tug-of-war geometry. Wen et al. demonstrated that individual ribosome translocation steps were clearly visible from the lengthening of the single-strand region of the mRNA. Analysis of the step size and the dwell time for individual steps yielded that 70S translocates on mRNA at precisely three bases per step with an average peptide elongation rate of 0.45 ± 0.17 codons per second. Furthermore, long ribosome pauses were frequently observed for one reporter mRNA, one-third of which occurred just downstream from an internal SD-like AGGAGG sequence. With synonymous mutation of A$\underline{\text{G}}$ GA$\underline{\text{G}}$ G to A$\underline{\text{A}}$ GA$\underline{\text{A}}$ G, the ribosomal pauses at this position disappeared. This observation indicates that SD interactions can stall translating ribosomes, similarly as during the initiation process.

b. *The helicase activity of translating ribosomes*

Folded structures in the coding region of an mRNA represent a kinetic barrier for the peptide elongation process and are exploited in diverse strategies for regulating the decoding process ([43–45] as examples). The strand separation activity is inherent to the ribosome, requiring no exogenous helicases [41]. Qu et al. studied the helicase activity of translating 70S, using the hairpin unwinding geometry and two hairpin-forming mRNAs, $\text{hpVal}_{\text{GC50}}$ and $\text{hpVal}_{\text{GC100}}$ [42]. The hairpin region of each mRNA is composed of ten valine (Val) codons, followed by four codons with ~50 or 100% G·C content (Fig. 7.4a). Due to the 13 ± 2 (s.d.) nt distance from the first nucleotide in the P-site to the mRNA entry site [40–42], translation of the 7th to the 10th Val codons on both mRNAs is accompanied by unwinding of the four codons subsequent to the Val codons, whose G·C content differs between the two mRNAs. As the same Val codons are being translated in the decoding center, any difference in the translation kinetics between the two mRNAs comes from the

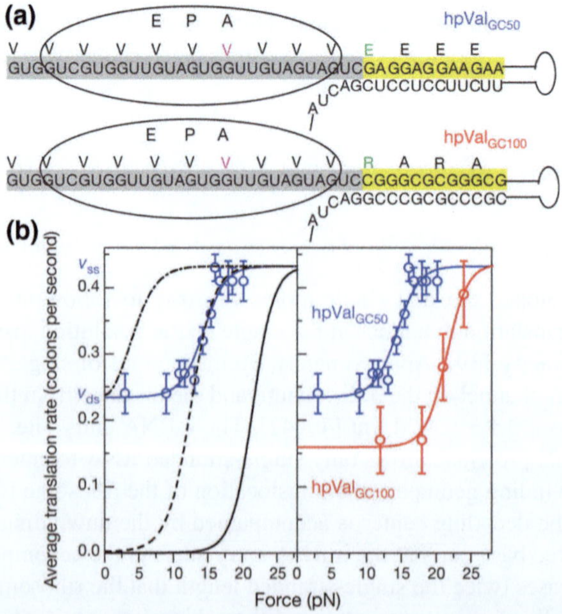

Fig. 7.4 Helicase activity of a translating prokaryotic ribosome studied using the hairpin unwinding geometry. a The hairpin region sequence of the two mRNAs used in this study. Each mRNA has ten valine (Val) codons (highlighted in gray) followed by four other codons (highlighted in yellow) with either 50% (hpVal$_{GC50}$) or 100% (hpVal$_{GC100}$) G·C content. Translation of the 7th (magenta) to 10th valine codons in the A-site is coupled with unwinding of the four codons with varying G·C content at the mRNA entry site. **b (left)** The translation rate dependence on force for hpVal$_{GC50}$ mRNA (blue circles) can only be partially fit to the Betterton model (black lines): the case of a totally passive helicase (solid), the best fit to the force-dependent region and the high-force plateau (dashed), and the best fit to the two plateaus (dot-dash). The blue solid line represents the best fit to the modified unwinding model (Scheme 7.1 and Eq. 7.4a, b). **b (right)** The translation rate dependence on force for hpVal$_{GC50}$ (blue circles) and hpVal$_{GC100}$(red circles) mRNAs and the best fit to the modified unwinding model (blue and red lines). Adapted from Ref. [42] with permission. Copyright 2011 Macmillan Publishers Limited

ribosomal helicase activity. It was observed that the average translation rate for both mRNAs shows a sigmoid dependence on force (Fig. 7.4b right panel). The high-force and low-force plateaus represent the translation rate without external force on single-stranded or double-stranded mRNAs, respectively. Except for the high-force plateau, the hpVal$_{GC100}$ mRNA has a slower translation rate at all other forces than the hpVal$_{GC50}$ mRNA, indicating that the more stable hairpin poses a stronger barrier for ribosome translocation.

The Betterton model [46, 47] has been instrumental in the quantitative analysis of the helicase activity of several nucleic acid helicases [47–49] and HIV-1 reverse transcriptase on DNA templates [50]. The Betterton model applied to a translation reaction can be illustrated by the following scheme, *excluding* the pathway indicated by the dashed arrow:

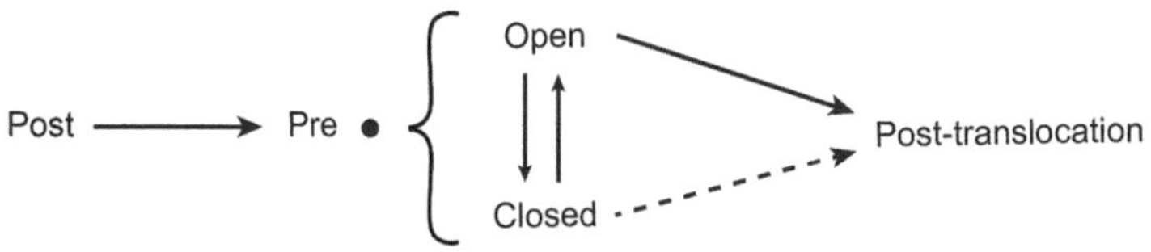

Scheme 7.1 The proposed kinetic scheme for the helicase activity of translating prokaryotic ribosomes

The first arrow from 'post' to 'pre' represents all biochemical steps in a translation cycle other than the translocation step. When the ribosome attempts to translocate at the 'Pre' state, it can encounter either an open or closed hairpin junction at the mRNA entry site, owing to the junction thermal fluctuations. The Betterton model postulates that (i) a helicase only translocates through an open junction and that (ii) an active helicase can bias thermal fluctuations of the junction toward the open conformation by lowering the free energy difference between the open and closed states with the amount ΔG_d. For a totally active helicase, the destabilization energy ΔG_d is much greater than the base pair free energy (ΔG_{bp}), so that the junction is always open and no longer hinders translocation. A passive helicase ($\Delta G_d = 0$) solely relies on junction opening by thermal fluctuations to translocate. In general, a helicase will show an unwinding activity between the two extremes. However, as shown in Fig. 7.4b left panel for the hpVal$_{GC50}$ mRNA, the Betterton model can only fit some features of the force dependence of the translation kinetics (black solid, dashed, and dot-dash lines) but lacks an overall agreement with the data. Clearly, additional interactions need to be incorporated into the model for the mRNA unwinding kinetics by a translating ribosome. Noting that considerable translation rates were observed for both mRNAs at the low-force plateau wherein the RNA hairpin junction thermal breathing predominantly favors the closed state, the ribosome appears to have an active mechanism that can directly break open a closed junction to translocate. Therefore, an additional pathway between the 'closed' and 'post-translocation' state was added to the Betterton model to describe the helicase activity of translating ribosomes (dashed arrow in Scheme 7.1).

In this modified scheme, $v(F)$, the overall translation rate under force F is given by:

$$v(F) = v_{ss} \cdot f_{open}(F) + v_{ds} \cdot \left(1 - f_{open}(F)\right) \tag{7.4a}$$

where $f_{open}(F)$ is the probability that the junction is open at force F, $v_{ds} = v(f_{open} = 0)$ is the rate of ribosome translation through an always closed junction (the low-force plateau), and $v_{ss} = v(f_{open} = 1)$ is the rate of ribosome translation through an always open junction (the high-force plateau). Specifically, $f_{open}(F)$ depends on ΔG_{bp}, ΔG_d, and the effect of force applied to the ends of the hairpin, ΔG_F, with the following relation:

$$f_{open}(F) = 1/\left(1 + \exp\left[(\Delta G_{bp} + \Delta G_F - \Delta G_d)/k_B T\right]\right) \tag{7.4b}$$

Here, ΔG_{bp} is known given the mRNA sequence, and ΔG_F is calculated using the WLC model. The parameters to be determined by fitting to the experimental results are ΔG_d, v_{ss}, and v_{ds}. The 50 and 100% G·C hairpin unwinding were fit independently (Fig. 7.4b right panel) and yielded the same values of $\Delta G_d = 0.9$ kcal/mole per base pair, and $v_{ss} = 0.43$ or 0.44 codon/s. The best fit values of v_{ds} are 0.23 and 0.16 codon/s for the unwinding of 50 and 100% G·C-containing hairpins, respectively.

This study provides strong mechano-kinetic evidence that the ribosome uses two active mechanisms to promote junction unwinding: open-state stabilization (the role traditionally described for active helicases in the Betterton model, characterized by ΔG_d) and mechanical unwinding (a new active mechanism in which the ribosome translocates by applying force to break open the closed junction, characterized by v_{ds}). For open-state stabilization, no known nucleic acid helicase motifs are found in ribosomal proteins. However, a mutational study implicated several positively charged residues on ribosomal proteins S3 and S4 at the mRNA entry site in ribosome helicase activity. It was proposed that these residues preferentially interact with phosphate groups on the single-stranded mRNA backbone [41]. Consistent with this hypothesis, the identical values of ΔG_d determined for the unwinding of the 50 and 100% G·C-containing hairpins suggest that the open-state stabilization mechanism in ribosomes has no significant base preference. The mechanical unwinding mechanism is so far a unique characteristic of translating ribosomes. In comparison with other enzymes whose helicase activity has been characterized by similar mechano-kinetic approaches, the ribosome has a unique property that several large-scale inter- and intra-subunit conformational changes are necessary to promote translocation [51–56]. Such large motions have great potential to generate a force that pulls on the tRNA/mRNA complex to promote unwinding at the mRNA entry site.

c. *Ribosome translocation dynamics during programmed frameshifting*

Normal translation is highly accurate with an error rate of less than 0.1% [57]. However, ribosomes can be programmed to frameshift, i.e., changing to either −1 or +1 reading frame during decoding [58]. Yan et al. [59] used a combination of mass spectrometry and optical tweezers to investigate frameshift-programming mRNAs derived from the *E. coli* dnaX gene. This mRNA promotes −1 frameshift with 80% efficiency in vivo [58]. Three elements in this mRNA's sequence have been identified to be essential to promote frameshifting [58]: the slippery sequence AAAAAAG, a flanking internal SD sequence located 10 nt upstream, and a flanking 11 bp hairpin located 6 nt downstream (Fig. 7.5a). Ribosomes are thought to backshift by 1 nt on the mRNA slippery sequence [58], because such slippage involves minimal base-pairing difference between the lysine codons, AAA and AAG, and the UUU anticodon used in *E. coli* [60]. However, the mass spectrometry characterization in this study showed unexpectedly that ribosomes slip by −1, −4, or +2 nt at various codon positions around the slippery sequence region, producing a collection of products that terminated at the −1 frame stop codon. To further elucidate the underlying molecular mechanism, Yan et al. used the hairpin unwinding geometry to track the real-time ribosome translocation kinetics on frameshifting mRNAs (Fig. 7.5a). Interestingly, ~90% of the trajectories exhibit distinct fluctuations in mRNA extension specifically

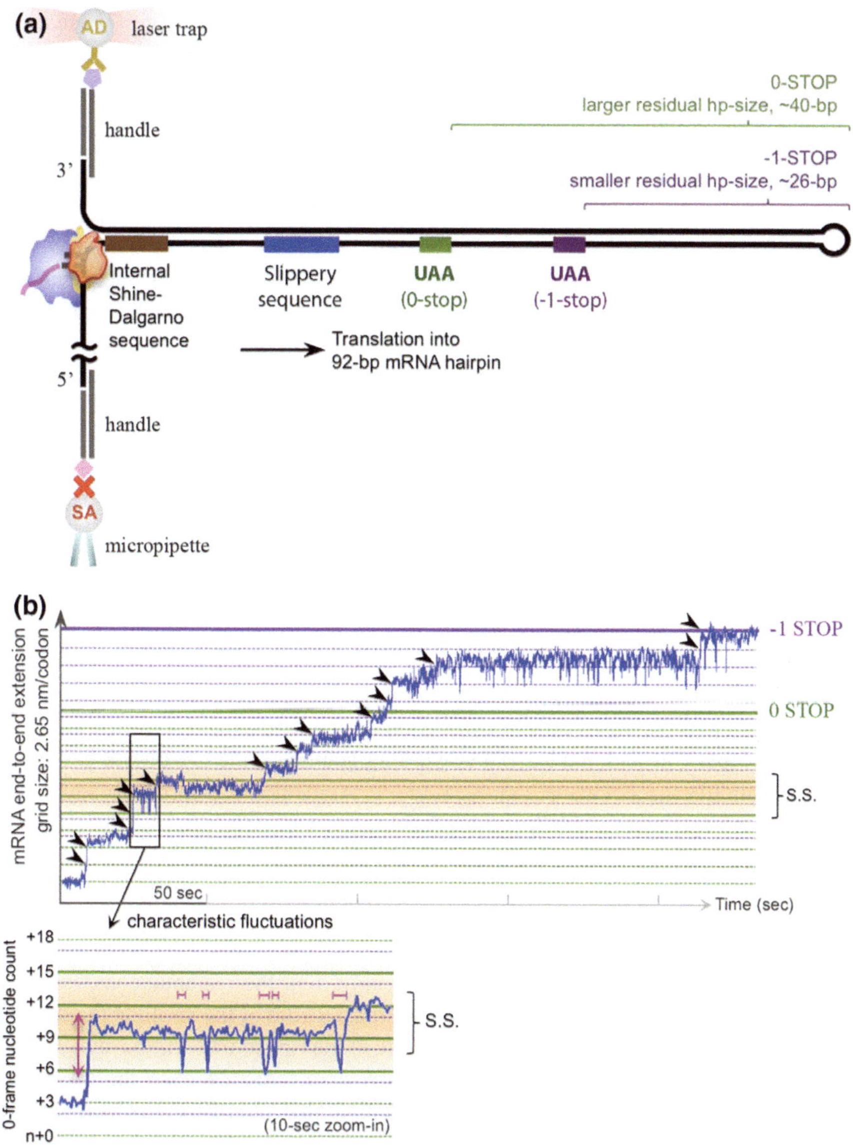

Fig. 7.5 Probing ribosome translocation dynamics during programmed frameshifting. **a** Experimental setup using the hairpin unwinding geometry. The hairpin region of the reporter mRNA contains all three features required to promote efficient −1 frameshifting: the SD sequence, the slippery sequence, and a stable downstream hairpin (served by the remaining hairpin after the slippery sequence). Ribosomes without or with −1 frameshifting will terminate at the downstream 0- or −1-stop, respectively, and leave a residual hairpin with different sizes. **b** An example translation trajectory for −1 frameshifting. The ribosome translocates in a step-wise fashion until it reaches the −1 stop codon. The ≥1 codon translocation fluctuations (black-squared section on the blue trace; expanded underneath) are commonly observed around the slippery sequence (orange-shaded area). The figure contains parts of Figs. 3 and 5A from Ref. [59] with modifications. Adapted with permission from Ref. [59]. Copyright 2015 Elsevier Inc.

around the slippery sequence region (orange-shaded area in Fig. 7.5b). These large displacement fluctuations between the ribosome and the mRNA around the slippery sequence indicate that multiple ribosome translocation attempts occur at this region and that large slipping sizes such as −4 nt are indeed attainable. Frequency analysis of the translation kinetics showed that fluctuations with characteristic frequencies of 2, 30, 85, and 180 Hz take place exclusively in the slippery sequence region, as compared to elsewhere in the trajectory. These timescales are similar to those reported for the 30S conformational dynamics during regular translation, particularly the head forward rotation at 80 Hz and reverse rotation at ~4–5 Hz [61]. It is likely that the fluctuations captured at the slippery sequence region in the tweezers data reflect the conformational excursions of the 30S head during multiple ribosome forward translocation attempts. Altogether, these findings suggest a dynamic frameshifting scheme via alternative reading frame sampling, which is accessed upon multiple ribosome translocation attempts.

7.4.3 Molecular Motor Property of the Translating Ribosome

Applying force on a processive enzyme directly to perturb its movement has been an important biophysical approach to characterize a molecular motor [62]. Liu et al. used the tug-of-war geometry to measure the effect of an opposing force on the movement of a translating ribosome by attaching the 3′ end of the mRNA and 30S each to a bead [63]. In comparison with the hairpin unwinding geometry, the tug-of-war geometry lacks an amplifying mechanism of ribosome movement and also has to operate at much lower forces to avoid stalling the ribosome. Accordingly, this assay has a much lower signal-to-noise ratio in extension measurement, making it difficult to resolve individual ribosome translocation steps. Therefore, this study utilized a threshold method to calculate the 'pause-free' velocity of ribosome movement. Specifically, the raw 1 kHz tether extension data at a constant force is filtered down to 1 Hz to calculate the instantaneous rate of tether length shortening, i.e., the instantaneous velocity. Regions with instantaneous velocity lower than 2.5× fold of the baseline fluctuation were considered to lack translation activity due to the ribosome temporarily stalling. These regions with paused translation activity were removed from the trajectory and the rest of the trajectory was used to calculate the 'pause-free' velocity of the ribosome.

It was found that the pause-free velocity decreased exponentially with the opposing force. The force at which the velocity approaches zero (the stall force) represents the maximum force that can be intrinsically generated by the motor in a cycle and was found to be 13 ± 2 pN for the ribosome. This finding provides direct evidence that the translating ribosome can generate a significant amount of force, corroborating the finding of the mechanical unwinding mechanism in the study of the ribosome helicase activity [42]. Furthermore, in this geometry, force affects the physical translocation process directly and should not perturb the biochemical reactions. Hence, the force dependence of the pause-free velocity ('v') can be fitted to the following expression to dissect how force affects translocation:

$$v(F) = v_0 \exp\left(-\frac{F \cdot \tilde{x}}{k_B T}\right) \tag{7.5}$$

where $\tilde{x}$ is the typical distance over which the force acts and v_0 is the zero-force translocation velocity. The fit yields $v_0 = 2.9$ codons/s [95% confidence interval (CI): 1.8, 4.0 codons/s] and $\tilde{x} = 1.4$ nm (95% CI: 0.9, 1.8 nm). The faster v_0 in this study, when compared to the v_{ss} determined in the ribosome helicase study [42], may result from the bias of the pause-free velocity analysis toward a faster rate and/or the different codons being translated in the two experiments. Analogous to the analysis of S1 unwinding of RNA structures (Eq. 7.3a, b), the magnitude of $\tilde{x}$ relative to the total distance of a single codon translocation has implications for whether translocation occurs as a single step or successive smaller substeps. Crystal structures show that the distance between A- and P-site mRNA codons is 1.48 nm [64], indistinguishable from $\tilde{x}$. Therefore, codon translocation is performed by the ribosome in a single step.

7.5 Interactions Between Nascent Polypeptide and Ribosome

Small proteins and single domains can fold into their native structures within microseconds in vitro [65]. Given the maximum peptide elongation rate of ~20 amino acids per second in *E. coli* [66], the nascent peptide chain has sufficient time to begin to fold while still being elongated. Kaiser et al. developed an optical tweezers assay to study the effect of the ribosome on nascent polypeptide folding [67]. The assay starts with running an in vitro translation reaction programmed with an mRNA missing a stop codon, so that the ribosome/nascent polypeptide will be stably stalled at the 3' end of the mRNA template. The stalled ribosome/polypeptide/mRNA complex is then studied on optical tweezers with the N-terminal of the nascent polypeptide and 50S each attached to a bead. In this geometry, the applied force selectively perturbs the stability of ribosome-bound nascent polypeptides and does not disrupt the structural integrity of the ribosome. The model protein used in this study is a cysteine-free version of T4 lysozyme [68], whose native fold requires interactions between the N- and C-terminal sequences. To generate a nascent polypeptide that has the entire T4 lysozyme sequence emerging from the narrow ribosomal exit tunnel, the translation reaction was programmed with an mRNA that codes for the protein followed by an unstructured C-terminal extension of 41 amino acids [69]. Interestingly, free and ribosome-bound full-length T4 lysozyme showed the same behavior in unfolding, but the folding rate of the ribosome-bound protein is more than two orders of magnitude slower than the free protein. When increasing the C-terminal extension length by 19 amino acids, which provides ~2.1 nm of additional separation from the ribosomal surface at 3.6 pN force based on the WLC model, approximately a 20-fold increase in the folding rate was observed relative to the case with the shorter extension. Furthermore, increasing the potassium chloride concentration from 150 to 500 mM for more effective screening of electrostatic interactions increased the folding rate of the ribosome-bound protein, but not the free protein. These observations clearly demonstrated that the ribosome surface can affect nascent polypeptide folding and the effect is mediated at least in part by electrostatic interactions. Given the diversity in protein structure and folding properties and the complex chemical compositions of the ribosome surface, we await the results of similar studies on other proteins for a comprehensive understanding of the impact of ribosome on nascent protein folding.

Furthermore, interactions of specific nascent chain sequences [70, 71] with the ribosome exit tunnel [72] can result in reduced rates of elongation. The bacterial SecM protein represents an example of a stalling sequence that interacts with the ribosome exit tunnel and allosterically represses the peptidyl transferase activity of the ribosome [72–75]. Release of stalling in vivo requires interactions between nascent SecM and the translocon machinery [76, 77]. It has been suggested that mechanical force exerted by the translocon relieves elongation arrest and leads to translation restart [78]. Goldman et al. adapted the above geometry to investigate the effect of force on the release of SecM-stalled ribosome–nascent chains [79]. Because tracking real-time translation kinetics has not yet been achieved in this geometry, the experimental design took advantage of the unique response of SecM-arrested ribosomes to the antibiotic puromycin. Puromycin binds to the empty ribosomal A-site and is incorporated into the nascent polypeptide, leading to its release from the ribosome [80]. SecM-arrested ribosomes contain a prolyl-tRNA^{pro} stably bound in the A-site and, therefore, are refractory to treatment with puromycin. However, the A-site becomes accessible to puromycin after arrest release, proline incorporation, and translocation [81]. Therefore, in the presence of puromycin and the translocation promoting factor EF-G, rupture of the tether due to puromycin binding and consequent polypeptide release can be used to track the timing of arrest release. The rate of stalling rescue, measured under constant force in the range of 10–30 pN, increased with the external force. The force dependence analysis of the stalling rescue rate yielded a distance to the transition state of 0.4 nm (95% CI: 0.1, 0.8 nm) and a zero-force rupture rate of 3×10^{-4} s^{-1} (95% CI: 0.5×10^{-4}, 20×10^{-4} s^{-1}). This rate is in agreement with biochemical ensemble experiments, in which no force was applied. Over the above force range, the release of SecM-mediated arrest is accelerated by more than an order of magnitude, supporting the hypothesis that SecM arrest is relieved by the mechanical force generated by the translocon.

7.6 Concluding Remarks

The past decade witnessed the successful application of the optical tweezers technique to study various aspects of the translation process, spanning from initiation, decoding, to nascent polypeptide and ribosome interactions. These successes capitalized on the much longer history of the development of high-resolution optical tweezers technique and the quantitative analysis framework for biophysical studies of a broad range of molecular interactions. However, it is important to note that the endeavor to establish these translation assays often took several years of laborious work, particularly for observing real-time translation dynamics. Both the intrinsic low throughput of optical tweezers, i.e., only one molecule is measured at a time, and the requirement of attaching the molecules of interest between micron-sized beads, make it particularly demanding to find an optimized condition wherein the ribosome remains highly active on optical tweezers. Nonetheless, these past achievements demonstrated the strong potential of optical tweezers techniques for multi-faceted studies of translation. Many exciting new developments can be expected in coming years, such as the application of combined fluorescence and optical tweezers measurements to translation, the ability to track real-time translation dynamics for eukaryotic systems, and the ability to monitor cofactor-ribosome interactions.

Acknowledgements X. Qu was supported by the National Institute of General Medical Sciences of the National Institutes of Health under Award Number R01GM121847 and Memorial Sloan Kettering Cancer Center (MSKCC) Functional Genomics Initiative. The content is solely the responsibility of the author and does not necessarily represent the official views of the National Institutes of Health.

References

1. Hershey, J. W. B., Sonenberg, N., & Mathews, M. (2012). *Protein synthesis and translational control: A subject collection from Cold Spring Harbor perspectives in biology* (vii, 352 pp.). Cold Spring Harbor perspectives in biology. Cold Spring Harbor, N.Y.: Cold Spring Harbor Laboratory Press.
2. Ashkin, A., et al. (1986). Observation of a single-beam gradient force optical trap for dielectric particles. *Optics Letters, 11*(5), 288–290.
3. Neuman, K. C., & Nagy, A. (2008). Single-molecule force spectroscopy: Optical tweezers, magnetic tweezers and atomic force microscopy. *Nature Methods, 5*(6), 491–505.
4. Svoboda, K., & Block, S. M. (1994). Biological applications of optical forces. *Annual Review of Biophysics and Biomolecular Structure, 23,* 247–285.
5. Moffitt, J. R., et al. (2008). Recent advances in optical tweezers. *Annual Review of Biochemistry, 77,* 205–228.
6. Liu, S. X., et al. (2014). A viral packaging motor varies its DNA rotation and step size to preserve subunit coordination as the capsid fills. *Cell, 157*(3), 702–713.
7. Comstock, M. J., Ha, T., & Chemla, Y. R. (2011). Ultrahigh-resolution optical trap with single-fluorophore sensitivity. *Nature Methods, 8*(4), 335–U82.
8. Deufel, C., et al. (2007). Nanofabricated quartz cylinders for angular trapping: DNA supercoiling torque detection. *Nature Methods, 4*(3), 223–225.
9. Li, P. T., et al. (2006). Probing the mechanical folding kinetics of TAR RNA by hopping, force-jump, and force-ramp methods. *Biophysical Journal, 90*(1), 250–260.
10. Qu, X. H., et al. (2012). Ribosomal protein S1 unwinds double-stranded RNA in multiple steps. *Proceedings of the National Academy of Sciences of the United States of America, 109*(36), 14458–14463.
11. Tinoco, I., & Bustamante, C. (2002). The effect of force on thermodynamics and kinetics of single molecule reactions. *Biophysical Chemistry, 101,* 513–533.
12. Laursen, B. S., et al. (2005). Initiation of protein synthesis in bacteria. *Microbiology and Molecular Biology Reviews, 69*(1), 101–123.

13. Hinnebusch, A. G. (2014). The scanning mechanism of eukaryotic translation initiation. *Annual Review of Biochemistry, 83,* 779–812.
14. Marzi, S., et al. (2007). Structured mRNAs regulate translation initiation by binding to the platform of the ribosome. *Cell, 130*(6), 1019–1031.
15. Studer, S. M., & Joseph, S. (2006). Unfolding of mRNA secondary structure by the bacterial translation initiation complex. *Molecular Cell, 22*(1), 105–115.
16. Kozak, M. (2005). Regulation of translation via mRNA structure in prokaryotes and eukaryotes. *Gene, 361,* 13–37.
17. Sorensen, M. A., Fricke, J., & Pedersen, S. (1998). Ribosomal protein S1 is required for translation of most, if not all, natural mRNAs in *Escherichia coli* in vivo. *Journal of Molecular Biology, 280*(4), 561–569.
18. Farwell, M. A., Roberts, M. W., & Rabinowitz, J. C. (1992). The effect of ribosomal protein S1 from *Escherichia coli* and *Micrococcus luteus* on protein synthesis in vitro by *E. coli* and *Bacillus subtilis*. *Molecular Microbiology, 6*(22), 3375–3383.
19. Vandieijen, G., Vanknippenberg, P. H., & Vanduin, J. (1976). Specific role of ribosomal protein S1 in recognition of native phage RNA. *European Journal of Biochemistry, 64*(2), 511–518.
20. Dumont, S., et al. (2006). RNA translocation and unwinding mechanism of HCV NS3 helicase and its coordination by ATP. *Nature, 439*(7072), 105–108.
21. Spies, M. (2014). Two steps forward, one step back: Determining XPD helicase mechanism by single-molecule fluorescence and high-resolution optical tweezers. *DNA Repair (Amst), 20,* 58–70.
22. Sengupta, J., Agrawal, R. K., & Frank, J. (2001). Visualization of protein S1 within the 30S ribosomal subunit and its interaction with messenger RNA. *Proceedings of the National Academy of Sciences of the United States of America, 98*(21), 11991–11996.
23. Subramanian, A. R. (1983). Structure and functions of ribosomal protein S1. *Progress in Nucleic Acid Research and Molecular Biology, 28,* 101–142.
24. Aliprandi, P., et al. (2008). S1 ribosomal protein functions in translation initiation and ribonuclease RegB activation are mediated by similar RNA-protein interactions. *Journal of Biological Chemistry, 283*(19), 13289–13301.
25. Fairman-Williams, M. E., Guenther, U.-P., & Jankowsky, E. (2010). SF1 and SF2 helicases: Family matters. *Current Opinion in Structural Biology, 20*(3), 313–324.
26. Parsyan, A., et al. (2011). mRNA helicases: The tacticians of translational control. *Nature Reviews Molecular Cell Biology, 12*(4), 235–245.
27. Garcia-Garcia, C., et al. (2015). Factor-dependent processivity in human eIF4A DEAD-box helicase. *Science, 348*(6242), 1486–1488.
28. Feoktistova, K., et al. (2013). Human eIF4E promotes mRNA restructuring by stimulating eIF4A helicase activity. *Proceedings of the National Academy of Sciences of the United States of America, 110*(33), 13339–13344.
29. De Gregorio, E., Preiss, T., & Hentze, M. W. (1999). Translation driven by an eIF4G core domain in vivo. *EMBO Journal, 18*(17), 4865–4874.
30. Korneeva, N. L., et al. (2005). Interaction between the NH2-terminal domain of eIF4A and the central domain of eIF4G modulates RNA-stimulated ATPase activity. *Journal of Biological Chemistry, 280*(3), 1872–1881.
31. Pyle, A. M. (2008). Translocation and unwinding mechanisms of RNA and DNA helicases. *Annual Review of Biophysics*, 317–336.
32. Noller, H. F. (1984). Structure of ribosomal RNA. *Annual Review of Biochemistry, 53,* 119–162.
33. Wintermeyer, W., et al. (2004). Mechanisms of elongation on the ribosome: Dynamics of a macromolecular machine. *Biochemical Society Transactions, 32,* 733–737.
34. Green, R., & Noller, H. F. (1997). Ribosomes and translation. *Annual Review of Biochemistry, 66,* 679–716.
35. Uemura, S., et al. (2007). Peptide bond formation destabilizes Shine-Dalgarno interaction on the ribosome. *Nature, 446*(7134), 454–457.
36. Shine, J., & Dalgarno, L. (1974). The 3′-terminal sequence of *Escherichia coli* 16s ribosomal RNA complementarity to nonsense triplets and ribosome binding sites. *Proceedings of the National Academy of Sciences of the United States of America, 71*(4), 1342–1346.

37. Calogero, R. A., et al. (1988). Selection of the messenger-RNA translation initiation region by *Escherichia coli* ribosomes. *Proceedings of the National Academy of Sciences of the United States of America, 85*(17), 6427–6431.
38. Vanzi, F., et al. (2005). Mechanical studies of single ribosome/mRNA complexes. *Biophysical Journal, 89*(3), 1909–1919.
39. Wen, J.-D., et al. (2008). Following translation by single ribosomes one codon at a time. *Nature, 452*(7187), 598–603.
40. Yusupova, G. Z., et al. (2001). The path of messenger RNA through the ribosome. *Cell, 106*(2), 233–241.
41. Takyar, S., Hickerson, R. P., & Noller, H. F. (2005). MRNA helicase activity of the ribosome. *Cell, 120*(1), 49–58.
42. Qu, X., et al. (2011). The ribosome uses two active mechanisms to unwind messenger RNA during translation. *Nature, 475*(7354), 118–121.
43. Tsuchihashi, Z. (1991). Translational frameshifting in the *Escherichia coli* dnaX gene in vitro. *Nucleic Acids Research, 19*(9), 2457–2462.
44. Nackley, A. G., et al. (2006). Human catechol-O-methyltransferase haplotypes modulate protein expression by altering mRNA secondary structure. *Science, 314*(5807), 1930–1933.
45. Watts, J. M., et al. (2009). Architecture and secondary structure of an entire HIV-1 RNA genome. *Nature, 460*(7256), 711–U87.
46. Betterton, M. D., & Julicher, F. (2005). Opening of nucleic-acid double strands by helicases: Active versus passive opening. *Physical Review E, 71*(1).
47. Johnson, D. S., et al. (2007). Single-molecule studies reveal dynamics of DNA unwinding by the ring-shaped T7 helicase. *Cell, 129*(7), 1299–1309.
48. Lionnet, T., et al. (2007). Real-time observation of bacteriophage T4 gp41 helicase reveals an unwinding mechanism. *Proceedings of the National Academy of Sciences of the United States of America, 104*(50), 19790–19795.
49. Manosas, M., et al. (2010). Active and passive mechanisms of helicases. *Nucleic Acids Research, 38*(16), 5518–5526.
50. Kim, S., Schroeder, C. M., & Xie, X. S. (2010). Single-molecule study of DNA polymerization activity of HIV-1 reverse transcriptase on DNA templates. *Journal of Molecular Biology, 395*(5), 995–1006.
51. Fischer, N., et al. (2010). Ribosome dynamics and tRNA movement by time-resolved electron cryomicroscopy. *Nature, 466*(7304), 329–333.
52. Moazed, D., & Noller, H. F. (1989). Intermediate states in the movement of transfer RNA in the ribosome. *Nature, 342*(6246), 142–148.
53. Frank, J., & Agrawal, R. K. (2000). A ratchet-like inter-subunit reorganization of the ribosome during translocation. *Nature, 406*(6793), 318–322.
54. Schuwirth, B. S., et al. (2005). Structures of the bacterial ribosome at 3.5 Å resolution. *Science, 310*(5749), 827–834.
55. Valle, M., et al. (2003). Locking and unlocking of ribosomal motions. *Cell, 114*(1), 123–134.
56. Peske, F., et al. (2000). Conformationally restricted elongation factor G retains GTPase activity but is inactive in translocation on the ribosome. *Molecular Cell, 6*(2), 501–505.
57. Drummond, D. A., & Wilke, C. O. (2009). The evolutionary consequences of erroneous protein synthesis. *Nature Reviews Genetics, 10*(10), 715–724.
58. Farabaugh, P. J. (1996). Programmed translational frameshifting. *Microbiological Reviews, 60*(1), 103–&.
59. Yan, S. N., et al. (2015). Ribosome excursions during mRNA translocation mediate broad branching of frameshift pathways. *Cell, 160*(5), 870–881.
60. Tsuchihashi, Z., & Brown, P. O. (1992). Sequence requirements for efficient translational frameshifting in the Escherichia coli dnaX gene and the role of an unstable interaction between transfer RNA(Lys) and an AAG lysine codon. *Genes & Development, 6*(3), 511–519.
61. Guo, Z., & Noller, H. F. (2012). Rotation of the head of the 30S ribosomal subunit during mRNA translocation. *Proceedings of the National Academy of Sciences of the United States of America, 109*(50), 20391–20394.

62. Bustamante, C., et al. (2004). Mechanical processes in biochemistry. *Annual Review of Biochemistry, 73,* 705–748.
63. Liu, T. T., et al. (2014). Direct measurement of the mechanical work during translocation by the ribosome. *Elife, 3*.
64. Jenner, L. B., et al. (2010). Structural aspects of messenger RNA reading frame maintenance by the ribosome. *Nature Structural & Molecular Biology, 17*(5), 555–U48.
65. Kubelka, J., et al. (2006). Sub-microsecond protein folding. *Journal of Molecular Biology, 359*(3), 546–553.
66. Liang, S. T., et al. (2000). mRNA composition and control of bacterial gene expression. *Journal of Bacteriology, 182*(11), 3037–3044.
67. Kaiser, C. M., et al. (2011). The ribosome modulates nascent protein folding. *Science, 334*(6063), 1723–1727.
68. Matsumura, M., & Matthews, B. W. (1989). Control of enzyme activity by an engineered disulfide bond. *Science, 243*(4892), 792–794.
69. Voss, N. R., et al. (2006). The geometry of the ribosomal polypeptide exit tunnel. *Journal of Molecular Biology, 360*(4), 893–906.
70. Ito, K., & Chiba, S. (2013). Arrest peptides: Cis-acting modulators of translation. *Annual Review of Biochemistry, 82,* 171–202.
71. Deutsch, C. (2014). Tunnel vision: Insights from biochemical and biophysical studies. In K. Ito (Ed.), *Regulatory nascent polypeptides* (pp. 61–86). Tokyo: Springer Japan.
72. Wilson, D. N., & Beckmann, R. (2011). The ribosomal tunnel as a functional environment for nascent polypeptide folding and translational stalling. *Current Opinion in Structural Biology, 21*(2), 274–282.
73. Tsai, A., et al. (2014). The dynamics of SecM-induced translational stalling. *Cell Reports, 7*(5), 1521–1533.
74. Nakatogawa, H., & Ito, K. (2002). The ribosomal exit tunnel functions as a discriminating gate. *Cell, 108*(5), 629–636.
75. Gumbart, J., et al. (2012). Mechanisms of SecM-mediated stalling in the ribosome. *Biophysical Journal, 103*(2), 331–341.
76. Yap, M.-N., & Bernstein, H. D. (2011). The translational regulatory function of SecM requires the precise timing of membrane targeting. *Molecular Microbiology, 81*(2), 540–553.
77. Nakamori, K., Chiba, S., & Ito, K. (2014). Identification of a SecM segment required for export-coupled release from elongation arrest. *FEBS Letters, 588*(17), 3098–3103.
78. Butkus, M. E., Prundeanu, L. B., & Oliver, D. B. (2003). Translocon "Pulling" of nascent SecM controls the duration of its translational pause and secretion-responsive secA regulation. *Journal of Bacteriology, 185*(22), 6719–6722.
79. Goldman, D. H., et al. (2015). Mechanical force releases nascent chain-mediated ribosome arrest in vitro and in vivo. *Science, 348*(6233), 457–460.
80. Nathans, D. (1964). Puromycin inhibition of protein synthesis: incorporation of puromycin into peptide chains. *Proceedings of the National Academy of Sciences of the United States of America, 51,* 585–592.
81. Muto, H., Nakatogawa, H., & Ito, K. (2006). Genetically encoded but nonpolypeptide prolyl-tRNA functions in the A site for SecM-mediated ribosomal stall. *Molecular Cell, 22*(4), 545–552.

Part III
RNA-Guided Protein Machineries

Chapter 8
Biophysical and Biochemical Approaches in the Analysis of Argonaute–MicroRNA Complexes

Sujin Kim and Yoosik Kim

8.1 Introduction

One of the key posttranscriptional gene regulatory mechanisms in eukaryotes is mediated by small regulatory RNAs such as microRNAs (miRNAs). miRNAs are ~22 nucleotides (nt) long, small noncoding RNAs that induce translational repression and degradation of mRNAs that are complementary to seed sequences of the miRNA (reviewed in [1, 2]). A summary of miRNA biogenesis process is presented in Fig. 8.1. Briefly, miRNA biogenesis begins with the transcription of the miRNA gene by RNA polymerase II [3–6]. A cluster of miRNAs is transcribed together as a long polycistronic transcript known as primary miRNA (pri-miRNA), which folds back on itself to form multiple hairpin structures in a single transcript (Fig. 8.1). These hairpins undergo endonucleolytic cleavage by RNase III-type enzyme Drosha in a complex with DiGeorge syndrome critical region gene 8 (DGCR8) [7–11]. The complex, known as the microprocessor, recognizes the junction between the hairpin structure and the single-stranded RNA and cleaves the RNA ~11 bases away from the junction [9, 12, 13]. More recently, structural and biochemical investigations have identified the molar composition of the microprocessor (one molecule of Drosha and two molecules of DGCR8), Drosha-binding motif in the basal segment of the pri-miRNA, as well as DGCR8-binding motif in the hairpin region of the RNA [13–15].

Microprocessor cleaves pri-miRNAs and releases ~65–70 nt long stem-loop structured RNAs known as precursor miRNAs (pre-miRNAs). Pre-miRNAs are then exported to the cytoplasm by Exportins including Exportin-5 where they are recognized by another RNase III-type enzyme Dicer [16]. Dicer recognizes both the phosphate group at the 5′ end and the 2 nt overhang structure of the pre-miRNA and cleaves the RNA ~22 nt from the ends [17–19]. The resulting miRNA duplex is then

S. Kim · Y. Kim (✉)
Department of Chemical and Biomolecular Engineering, KAIST Institute for Health Science and Technology, KAIST, Daejeon 34141, South Korea
e-mail: ysyoosik@kaist.ac.kr

C. Joo and D. Rueda (eds.), *Biophysics of RNA-Protein Interactions*,
Biological and Medical Physics, Biomedical Engineering,
https://doi.org/10.1007/978-1-4939-9726-8_8

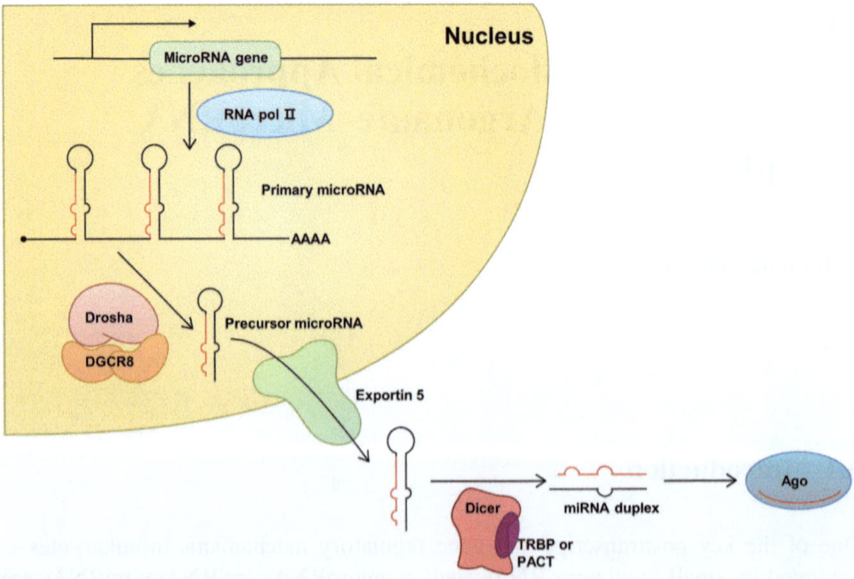

Fig. 8.1 A schematic depicting biogenesis of miRNA from transcription by RNA polymerase II to Ago loading in the cytosol

loaded onto Argonaute (Ago) family of proteins which discards one of the strands (known as the passenger strand) and retains the other strand (known as the guide strand). Ago–miRNA complex constitutes the core of the RNA-induced silencing complex (RISC) and uses the miRNA seed sequences as the guide to search for target mRNAs to induce posttranscriptional gene silencing [20, 21].

Numerous studies analyzed the Ago–miRNA and RISC–mRNA interactions using biochemical and biophysical single-molecule approaches. Their experimental findings were further complemented by the structural knowledge of Ago and RISC. Together, these studies have significantly advanced our understanding of the mechanism of gene regulation mediated by miRNAs. In this chapter, we present these studies.

8.2 Functional Domains of Ago

The overall structure of Ago family of proteins is a bilobate architecture that consists of four distinct domains: the *N*-terminal, PAZ, MID, and Piwi domains (Fig. 8.2) [22, 23]. Biological functions of these domains are summarized in Table 8.1. The *N*-terminal region forms one lobe with the PAZ domain. The function of the *N*-terminal region is unclear, but it may assist in the release of the target mRNA by disrupting its base pairing with the miRNA [24]. The PAZ domain can be subdi-

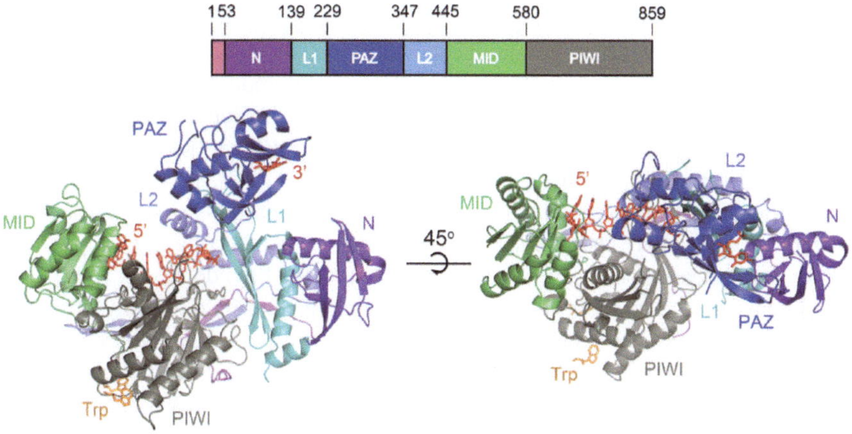

Fig. 8.2 A schematic of different functional domains (top) and the ternary structure of human Ago2 (bottom). The figure is adapted from [22] with American Association for the Advancement of Science, Copyright 2012

Table 8.1 Summary of Ago domains and their functions

Ago domain	Function	References
N-terminal	• May assist the release of the target mRNA	[24]
PAZ	• Anchors 3′ end of the miRNA	[31, 34–36]
	• Provides steric hindrance to prevent extended miRNA-target interaction	[31, 32]
MID	• Induces translational repression by binding to the cap of the mRNA	[47, 48]
Piwi	• Mediates target cleavage for hAgo2	[21, 37]
	• Recognizes target mRNA	[28, 42, 45]

vided into two subdomains separated by threonine 667; one domain consists mostly of aromatic residues, while the other subdomain folds into a structure similar to oligonucleotide/oligosaccharide binding (or OB-fold) structure that is capable of binding to single-stranded nucleic acids [22, 25–27]. The possibility of the interaction between the PAZ domain and the single-stranded nucleic acids is confirmed via crystallographic studies and biochemical experiments where the PAZ domain binds to single-stranded RNAs, although with low affinity [28–30].

The PAZ domain can interact and anchor the 3′ end of the miRNA [31]. The anchoring incurs steric hindrance and prevents the interaction between the last few nucleotides of the miRNA with its target mRNA. This reduces the degree of interaction between the miRNA and the target mRNA, facilitating the target release and allowing RISC to act as a multi-turnover complex [31, 32]. Furthermore, anchoring of the 3′ end of the miRNA is important for the loading of miRNA duplex onto Ago. Dicer cleavage product (miRNA duplex) contains two nucleotide 3′ overhangs which is a common characteristic of RNase type-III enzyme products [33]. This recognition of the 3′ overhang allows Ago to distinguish miRNA duplex from other small RNAs such as degradation by-products or small duplex RNAs that are derived from non-related pathways [34–36].

The human genome encodes four paralogs of Ago proteins (hAgo1–4). While all four proteins share the characteristic domains of the Ago family, only hAgo2 shows target cleavage activity, which is mediated by the Piwi domain [21, 37]. This domain has an RNaseH-like fold and is responsible for the endonucleolytic activity of the protein. RNaseH is an endonuclease that recognizes DNA–RNA hybrid and cleaves RNA using DNA as the template. The catalytic activity of RNaseH requires a conserved Asp-Asp-Glu/Asp motif in the catalytic center and two divalent metal ions [38]. The Piwi domain of cleavage competent Agos including hAgo2 has a very similar motif (Asp-Asp-Asp/Glu/His/Lys) [23]. Mutagenesis of this region resulted in the loss of the catalytic activity [23]. In addition, these Agos require divalent metal ions to induce RNA cleavage [21, 39, 40]. Moreover, the products of Agos and RNaseH both show 3′-OH and 5′-phosphate groups, suggesting that the two proteins induce RNA cleavage in a similar manner [21, 40, 41].

Unlike hAgo2, other three paralogs of human Agos (hAgo1, hAgo3, and hAgo4) do not show slicing activity. Examination of their Piwi domains reveals that the RNaseH-like motif in hAgo1 and hAgo4 does not match the consensus sequence and hence accounts for their inability to cleave target mRNAs. Human Ago3 shows Asp-Asp-His consensus sequence which matches the one from hAgo2, yet studies reported that hAgo3 does not show RNA cleavage activity [42]. Therefore, simple RNaseH fold structure may not account for the action mechanism of Agos.

One possible explanation is the difference in the target cleavage efficiency. In *Drosophila,* two Agos (Ago1 and Ago2) have the identical consensus motif, but Ago1 shows much higher cleavage efficiency than that of Ago2 due to faster target release kinetics [43]. Applying similar logic to the human Agos, hAgo3 may have much slower dissociation kinetics with the target compared to that of hAgo2, which can make hAgo3 effectively a single turnover enzyme and show much lower cleavage efficiency. Through a series of biochemical experiments using recombinant hAgo2 and hAgo3, Park et al. showed that hAgo3 loaded with miR-20a can cleave target mRNAs [44]. However, when incubated with other miRNAs such as let-7a, miR-19b, or miR-16, recombinant hAgo3 failed to induce target cleavage [44]. The authors attributed this phenomenon to the differences in the miRNA-target interaction channel between hAgo2 and hAgo3, indicating that hAgo3 has more strict substrate requirement in addition to simple sequence complementarity in order to induce target cleavage [44]. As Ago protein structure plays a key role in the miRNA-target interaction as well as during target dissociation from RISC (see below for details), the difference in action mechanism of hAgo2 and hAgo3 may arise from the differences in their interaction with the target rather than in the RNaseH-like motif in their Piwi domains. Together, these evidences call for more detailed investigation of miRNA-target interaction in combination with information on Ago protein structures to better understand the action mechanism of miRNAs.

For both cleavage competent and incompetent Agos, the Piwi domain plays an essential role in substrate recognition. While the 3′ end of the miRNA is recognized by the PAZ domain, the 5′-phosphate of the miRNA is anchored at the interface between the Piwi and the MID domains [42, 45]. Biochemical studies further showed that a divalent cation binds to this interface and interacts with the 5′-phosphate of the miRNA [28]. Furthermore, the preferential nucleotide is shown to be uridine although the effect of the nucleotide identity and the efficiency of Ago loading in cell need further investigation [46].

The main silencing effect by RISC is mediated not by target cleavage, but mostly through translational repression and subsequent RNA degradation via deadenylation and decapping [42]. The MID domain holds the key to explain the latter function of Agos as hAgo2 contains an MC motif which shows high homology to the cap-binding motif of the eukaryotic initiation factor 4E (eIF4E) [47]. Biochemical studies further showed that the MID domain can bind to the cap of the mRNA, and this interaction is required for efficient translational repression [48]. As hAgo1, hAgo3, and hAgo4 are not capable of inducing target cleavage, the cap-binding ability provided by the MID domain may be critical for these Agos to induce repressive effects. However, the MID domains of these cleavage incompetent Agos do not show the motif homologous to that of eIF4E and the exact mechanism of the cap-binding ability of these Agos remains to be investigated. Perhaps, this process may be mediated by Ago-interacting proteins that are components of RISC [49–51]. In addition, it remains unclear how the MID domain and Ago induce deadenylation and decapping of the target mRNA. Recently, it has been shown that targeting by miRNAs induces uridylation of the mRNA at the end of its poly(A) tail which facilitates RNA turnover [52]. Considering that mRNA turnover is responsible for most of the gene silencing effect by RISC, the role of MID domain and its cap-binding ability to induce deadenylation and decapping needs further investigation in the future.

8.3 Assembly of Ago–MiRNA Complex

During the posttranscriptional regulation by miRNAs, miRNAs use their seed sequences to guide the RISC to find the target mRNAs while Ago functions as the effector protein of the complex [53–55]. The interaction between miRNAs and Agos begins with the assembly of the RISC. Since the mechanism of RISC assembly has been key aspects in understanding miRNA-mediated gene silencing, it was under extensive investigation over the past few decades. RISC assembly begins with the loading of duplex Dicer cleavage product onto Ago protein. This process is most thoroughly studied using *Drosophila* Ago2, but in this chapter, we will focus on the mammalian system (Fig. 8.3).

One of the key questions in RISC assembly is how Dicer releases its cleavage product and delivers it to Ago. Numerous studies have suggested that Dicer together

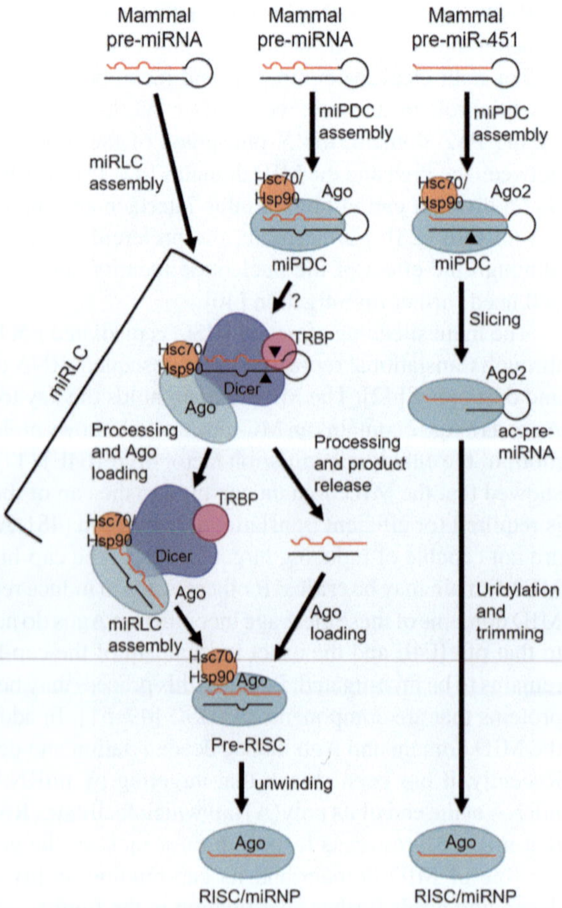

Fig. 8.3 RISC assembly in mammals. Pre-miRNAs are first loaded onto miRNA RISC loading complex (miRLC). However, the transfer mechanism of Dicer cleavage products to Ago and the function of the direct interaction between pre-miRNAs and Ago remain to be investigated. The figure is modified from [66] with Elsevier, Copyright 2012

with Ago and TAR RNA-binding protein (TRBP) forms miRNA RISC loading complex (miRLC) which plays a key role in loading of the miRNA duplex (Fig. 8.3). In this conventional model, Dicer product is oriented by Dicer–TRBP heterodimer and is handed over to Ago in the miRLC [56–58]. This is further supported by the EM data where Ago is bound to miRNA duplex in complex with Dicer and TRBP [59].

However, accumulating in vitro biochemical evidences suggests that Dicer and miRLC may not be required for miRNA loading [60–63]. Using Dicer knockout embryonic stem cells, it has been shown that the loading of small RNA duplexes to Ago can occur in the absence of Dicer [64]. In addition, Ago can directly bind to pre-miRNAs and form a complex called miRNA deposit complex (miPDC; Fig. 8.3). For specific miRNAs such as miR-451 whose pre-miRNA is too short for Dicer processing, miPDC formation is responsible for mature miRNA biogenesis [65]. In other cases, miPDC can incorporate Dicer and TRBP to form miRLC and deliver pre-miRNA to Dicer for RNA processing [66].

Interestingly, it has been shown that dissociation of Ago from miRLC to form RISC with mature miRNA requires catalytically active Dicer [57]. This supports the conventional model where Ago is present as a complex with Dicer, and Dicer physically hands over its cleavage product, the miRNA duplex, to Ago [57]. Furthermore, the loading of miRNA duplex may trigger conformational change on Ago such that it can now dissociate from miRLC to form mature RISC [60, 67]. However, since small duplex RNAs can be loaded on to Ago without Dicer, it still remains unclear how the miRNA RISC loading occurs. In the end, there exist two models: Dicer product is released to the bulk solution and then loaded onto Ago in the vicinity, and Dicer is physically handing over the cleaved product to the Ago protein [66]. In either of the two cases, the interaction between Ago and Dicer and the formation of miRLC are required in order to minimize the searching process of Ago to find duplex miRNAs [58, 68]. Further investigation is required to elucidate the in-depth mechanism of Ago loading during miRNA biogenesis in mammals.

8.4 Target Recognition by Minimal RISC

Once the RISC assembly has been completed, Ago uses the guide strand of the miRNA embedded within the complex to search for its target mRNAs to induce posttranscriptional regulation [69, 70]. During the process, Ago searches for mRNAs whose 3′ UTR sequences are complementary to the miRNAs' seed sequences, 2–7 or 2–8 nt from the 5′ end of the miRNA [42]. The importance of the miRNA seed sequences has been demonstrated by numerous high-throughput sequencing studies. They showed that when expression of a given miRNA is perturbed, levels of mRNAs that contain sequences complementary to seed sequence of the miRNA are significantly affected [71–74]. Detailed examination of the interaction between miRNA and its target mRNA revealed that the two RNAs hybridize in a stepwise process that is accompanied by structural changes in Ago [75].

miRNA loading and target mRNA recognition by RISC can be subdivided into five steps. The first step is the loading of the miRNA onto Ago and its effect on the accessibility of the individual nucleotides. The 5′-monophosphate and the first base of the miRNA are anchored at the interface between the MID and the Piwi domains, and the 3′ end of the miRNA is recognized by the PAZ domain of Ago [25, 28, 29, 31, 45, 76]. This anchoring of miRNA guide strand makes the first nucleotide inaccessible and opens up the seed region to the media [28, 45]. Therefore, the seed region becomes the first nucleotides that can interact with mRNAs and play an essential role in target selection by the miRNA.

Although the seed region is crucial for target recognition, not all seven bases hybridize with their complementary bases on the target mRNA simultaneously. The latter part of the seed region is inaccessible to the media due to steric hindrance imposed by the Ago protein (Fig. 8.4) [75]. Initially, only the second–fifth positions of the guide RNA are exposed and are able to interact with the target mRNAs [75]. These sequences are known as the sub-seed sequences and are responsible for the weak

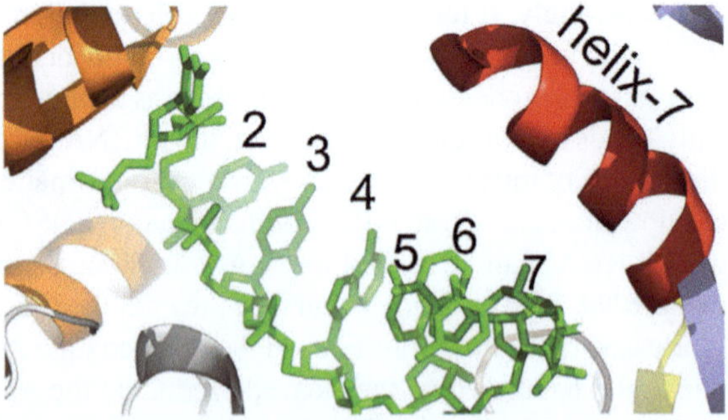

Fig. 8.4 In RISC, the access of the latter half of the seed region (5th–7th nt) of the miRNA is blocked by the helix-7 motif of Ago. The figure is adapted from [77] with Elsevier, Copyright 2017

recognition of the target by RISC. When RISC finds an mRNA with complementary match to the sub-seed sequences, the hybridization between miRNA and mRNA is stabilized by Ago protein which assists the two to form into an A-form helix [75, 77]. This pre-organization of miRNAs to base pair only the sub-seed region greatly accelerates the target finding speed by increasing the on-rate as much as 250-folds [78].

The initial interaction between the sub-seed sequences and the mRNAs has been supported by a number of single-molecule studies. For example, Salomon et al. designed an experiment where they introduced a series of di-nucleotide changes on mRNAs and measured the dissociation rate of the RISC–mRNA complex [78]. They found that mismatches in the first two sequences of the seed region had the greatest effects on the target binding rates compared to mismatches in other regions [78]. Interestingly, mismatches in the last two sequences of the seed had significantly weaker effects on target binding rate [78]. Similarly, Chandradoss et al. compared the binding rate of mRNA with full seed complementarity and one with partial seed complementarity. They found that the first three nucleotides of the seed region are critical for the target recognition and the latter part of the seed region did not have significant effects on the binding rate between the RISC and the target mRNA [79]. Of note, while the structural study subclassified 2–5 nt as sub-seed sequences, these follow-up single-molecule studies showed that only the 2–4 nt may act as the "mini-seed" region [75, 78, 79]. These studies support the notion that RISC uses the first three or maybe four nucleotides of the seed region for the initial target search.

The hypothesis of the existence and the significance of the sub-seed match are further supported by structural analysis of Ago protein in complex with guide miRNA. Without interaction with the target, only the second–fifth positions rather than the entire seed region are exposed to the media [22, 45, 80, 81]. This is because Ago protein induces structural constraint and makes the guide kink away from the A-form helix, in particular at the position 7. With this conformation, mRNA will not be able to hybridize and form duplex RNA beyond the fifth position of the miRNA (Fig. 8.4) [77].

This structural constraint at the position 7 is relieved by the initial interaction between the mRNA and the sub-seed sequences of the guide miRNA. The hybridization triggers structural change on Ago such that it undergoes 4 Å displacement at

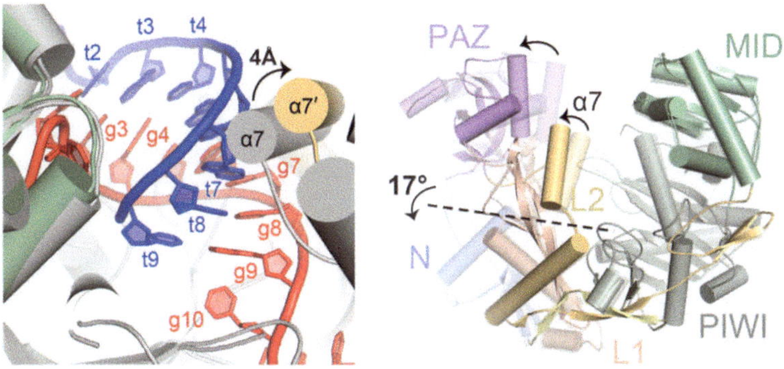

Fig. 8.5 Base pairing in the sub-seed region induces conformational change such that the helix-7 is shifted by 4 Å which allows further base pairing with the target. The figure is adapted from [75] with American Association for the Advancement of Science, Copyright 2014

the region around the position 7, allowing the seventh nucleotide to adapt A-helix configuration (Fig. 8.5) [77]. This allows the sixth–eighth positions of the guide RNA to base pair with the target mRNA [77].

The conformational change of Ago and the subsequent extension of hybridization between the miRNA and the target mRNA are supported by a number of single-molecule studies. They showed that there is a sharp increase in the binding affinity when the number of seed matches is increased from six to seven [78, 79]. This result is consistent with the idea that the sub-seed interaction induces changes in Ago structure such that the sixth–eighth position of the guide RNA has become accessible to hybridize with the target mRNA [82]. Without such change, these positions of the seed region will not be able to hybridize with the mRNA and thus complementarity in these bases will not affect the binding affinity with the target.

The combination of structural, biochemical, and biophysical single-molecule experiments provides a powerful approach in understanding RISC–mRNA interaction. Together, these studies converge on the idea that the target recognition by RISC is a multi-step process (Fig. 8.6). First, RISC anchors the two ends of the miRNA to orient and stabilize the miRNA into proper A-helix conformation. Second, the sub-seed bases, in particular positions 2–4, provide the initial searching platform that mediates the interaction with the target mRNA. Lastly, the hybridization between the mRNA and the sub-seed bases of the miRNA induces conformational change on Ago to allow further seed match for extended hybridization between the two RNAs. This increased complementarity significantly lengthens the residence time of Ago bound on the target mRNA, which may be necessary for sufficient gene silencing [78, 79].

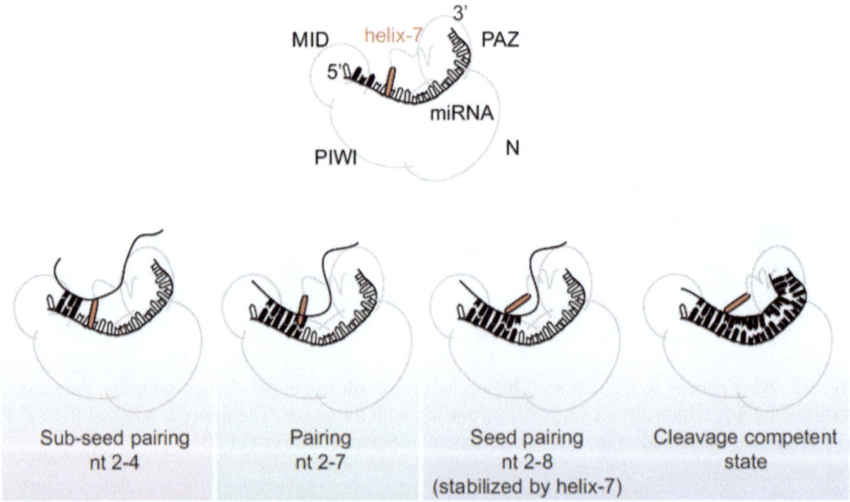

Fig. 8.6 Model summarizing the conformational changes of Ago during miRNA-target interaction. The figure is adapted from [77] with Elsevier, Copyright 2017

8.5 Implications of the Sub-seed Region: 1-D Target Search

One of the unresolved questions in RISC-target interaction is how the miRNA embedded in RISC can effectively find its target mRNAs in a complex media like inside the cell. The sequential target recognition process by RISC suggests one clue: By using the sub-seed sequences, Ago may find an mRNA with the partial seed match first and then slide along the RNA to search for better binding sites, i.e., sites with extended seed match sequences if such sites do exist. By doing so, the target search in the three-dimensional space has effectively become one-dimensional sliding problem [79]. In fact, this kind of one-dimensional search algorithm is employed by transcription factors in search for their optimal binding sites on the DNA. Previous studies on the mechanism of the recognition of the *lac* operon by LacI repressor in *Escherichia coli* (*E. coli*) showed that the protein first finds the DNA in the three-dimensional space and slides along the DNA to find the optimal binding site located within the *lac* operon [83–85]. Therefore, the three-dimensional diffusion has essentially become one-dimensional sliding which significantly facilitates the target search process.

Using the combination of three- and one-dimensional search mechanisms can be advantageous in multiple ways. First, once the RISC finds an mRNA with sub-seed match, it can undergo fast lateral diffusion along the RNA to search for the optimal binding site with extended seed match (Fig. 8.7) [79]. This lateral searching process may require hopping and sliding along the mRNA rather than trying to find the optimal site through searching in three-dimensional space of the cytosol [79]. The facilitated search mechanism may allow RISC to act as a multi-turnover type of regulator as it can quickly move from one target to another. In addition, the ability to

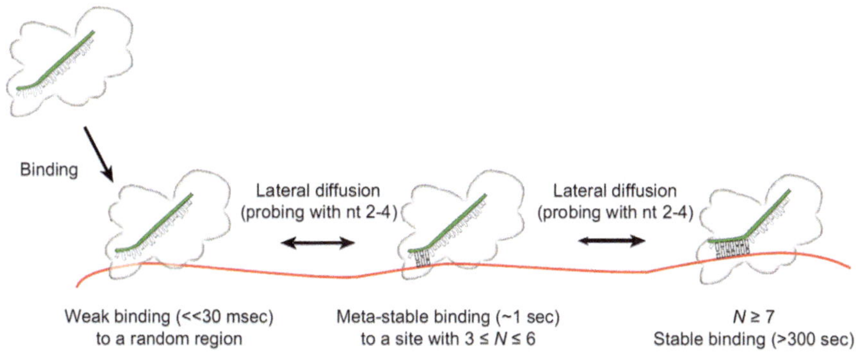

Fig. 8.7 Model summarizing dynamic target search by RISC using sub-seed match and lateral diffusion along the target mRNA. The figure is adapted from [79] with Elsevier, Copyright 2015

search in three-dimensional space is necessary as RISC, unlike transcription factors, targets mRNAs. As the sub-seed region only contains three or four nucleotides, it is likely that mRNAs with the sub-seed match may not have the sequences that are complementary to the rest of the seed region. In this case, the RISC should detach from the mRNA and diffuse through the cytosol, searching for an unprobed new target mRNA. Of note, the partial seed match assists this step as it has lower binding affinities to the mRNA compared to that of the full seed match [78, 79]. The search for the new target strictly depends on the diffusion in the three-dimensional space. Once the RISC finds another mRNA with sub-seed match, it will again undergo one-dimensional sliding and hopping along the RNA to find the optimal binding site (Fig. 8.7). Therefore, the sub-seed match allows the RISC to scan through many different mRNAs to find the true target, and the optimal target search will depend on the proper distribution of three-dimensional search and one-dimensional scan mode of the RISC [79].

The structure of Ago and its conformational changes during RISC-target recognition play a role in optimizing the balance between three-dimensional search and one-dimensional scan processes. Previous investigation of the one-dimensional scanning of transcription factor argued that it is not possible for the transcription factor to have both fast searching and stable binding [86–88]. This is because fast searching requires weak interaction between the protein and the DNA and consequently transcription factor with fast searching speed is likely to miss many of its true binding sites. Similarly, stable protein–DNA interaction implies slow dissociation which results in slow lateral diffusion as the protein gets trapped at nontarget sites. This results in speed–stability paradox in DNA scanning where fast and specific target search is difficult to achieve simultaneously [87, 89]. However, this problem can be resolved if the protein can adopt multiple configurations each with different DNA binding affinities. During the initial search mode, the protein may show weak interaction with the DNA and only when it recognizes sequences similar to its binding

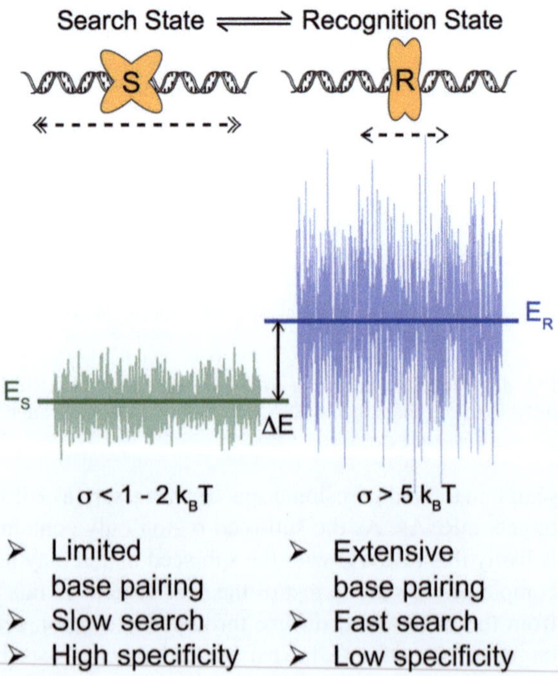

Fig. 8.8 Speed–stability paradox of DNA–protein interaction. In order to overcome the speed–stability paradox, transcription factors often use two state DNA interactions. The search mode is characterized by weak DNA–protein interaction which allows fast search with relatively smooth energy landscape. The recognition mode shows increased interaction with the DNA which results in decreased speed with increased specificity. The recognition mode also shows a large energy variation. The figure is modified from [89] with Elsevier, Copyright 2016

site, then the protein may change its configuration and slowly scan the vicinity to find the optimal binding site [87, 89] (Fig. 8.8).

The structure of Ago provides an ideal example for the configuration changes required for optimal target search process. When miRNA is loaded onto Ago, the protein arranges the miRNA such that only the mini-seed bases can adapt A-form helical structure [22]. In other words, the kink at the position 7 imposed by Ago prevents the target interaction beyond the fifth position of the guide RNA, restricting the interaction and lowering binding affinity between miRNA and mRNA pairs [75]. Therefore, during the scan mode, Ago effectively limits miRNA–mRNA interactions to reduce the binding affinity such that it can quickly scan through the 3′ UTR to find the sequences that match the sub-seed region of the miRNA [79].

In addition, once the sub-seed match region is found, the hybridization between miRNA and mRNA induces the conformational change in Ago such that it now allows full seed interaction [79]. This change in Ago configuration allows base pairing beyond the sub-seed region and can significantly increase the RISC–mRNA-binding affinity. As a result, the scanning process has slowed down sufficiently to find the full or nearly full seed match sites on the mRNA [79]. Therefore, the steric hindrance imposed by Ago and conformational change of Ago by miRNA–mRNA interaction provide the necessary conditions for the optimal target search by RISC as suggested by Slutsky and Mirny: facilitated diffusion and one-dimensional target search along the mRNA and slowed diffusion near the target site in order to converge on the site with full seed match nucleotides [86–89]. Overall, structural understanding of Ago and the change induced by sub-seed base pairing strongly suggest that the target search mechanism by the RISC is a multi-step process with at least one scan mode and one recognition mode.

8.6 Toward Target Cleavage

Gene silencing by RISC is mediated through at least three mechanisms: (1) RISC components interact with the cap-binding proteins and suppress translation at the initiation step; (2) RISC induces RNA decay by triggering deadenylation and decapping of the target mRNA; (3) RISC directly cleaves mRNA at the binding site. The direct cleavage requires hAgo2, the only cleavage competent Ago in the human genome [21, 37]. While the miRNA-mediated gene silencing mostly occurs through the first two mechanisms, small interfering RNAs (siRNAs) can also induce target cleavage when loaded onto hAgo2. The difference between the two pathways (miRNA vs. siRNA) lies in the extent of the target complementarity. Target cleavage can occur when the 10th and 11th nucleotides of the guide siRNA/miRNA pair with the mRNA [90–92]. However, while siRNAs are designed to have extended base pairing with the target, most of miRNAs do not base pair at these positions, resulting in only the siRNA being able to induce target cleavage.

Consistent with this idea, bioinformatics studies have shown that most of the targets of miRNAs in human do not show complementarity beyond the seed sequences and thus are not cleaved by hAgo2 [71–74]. Moreover, numerous studies have shown that seed sequences of the miRNA are the key determinant of target selection and the rest of the sequences do not affect RISC-target interactions [32, 72, 82, 93–96]. Yet, structural and single-molecule studies have shown that seed pairing triggers an additional conformational change in Ago, which provides a key understanding of the mechanism and efficiency of gene silencing by RISC [75, 81, 97, 98].

First, the sub-seed match relieves the kink at the position 7 and allows extended seed match with the mRNA [77]. However, the pairing beyond the eighth position is still restricted and requires the widening of the channel between the PAZ and the *N*-terminal domains [75]. Recently, Jo et al. observed that many of the targets of the miRNA are not cleaved despite the perfect complementarities [99, 100]. One possibility is that Ago imposes structural hindrance such that the 10th and 11th positions of the miRNA cannot base pair with the corresponding complementary nucleotides on the mRNA. This result suggests that the identity of miRNA may be important in predicting its target cleavage capability. Therefore, the simple identity and complementarity are not enough to predict the target cleavage and further investigation on the conformational change on Ago due to miRNA–mRNA interaction is required [77].

Although the seed sequence match may not induce conformational change in Ago to allow target cleavage, it does rearrange the protein such that the 13th–16th

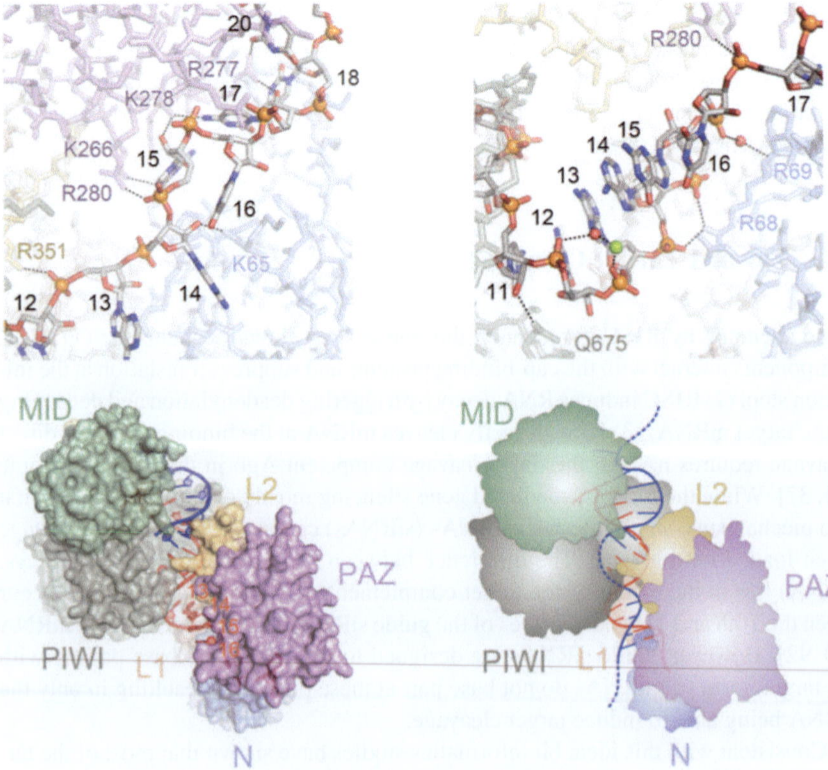

Fig. 8.9 Seed match with the target mRNA triggers conformational change of Ago such that the supplementary region of the guide miRNA arranges into A-form helical structure and may base pair with the target mRNA. The figure is adapted from [75] with American Association for the Advancement of Science, Copyright 2014

nucleotides of the miRNA (also known as the supplementary region) are now configured into an A-helical form and may base pair with the target RNA (Fig. 8.9) [75]. The extended target complementarity in this region of the miRNA further enhances the binding affinity and increases the residence time of Ago on the target mRNA [75, 78, 79].

The sequences beyond the 16th position of the miRNA cannot interact with mRNA due to structural constraint imposed by Ago protein [77]. Similar to the first sequence of the miRNA, the 3′ end of the miRNA is anchored at the PAZ domain [22]. Ago does not release the 3′ end of the miRNA even after the pairing at the supplementary region and prevents the access of the sequences beyond the 16th position [31, 45, 101, 102]. This tight association between the PAZ domain and the 3′ end of the miRNA is necessary to ensure efficient target release (Fig. 8.10). It has been shown that RNA duplex longer than 12 base pairs has half-life of approximately one year, indicating that the two will form an extremely stable complex and are not likely to dissociate [103]. However, miRNA loaded onto Ago dissociates with the target

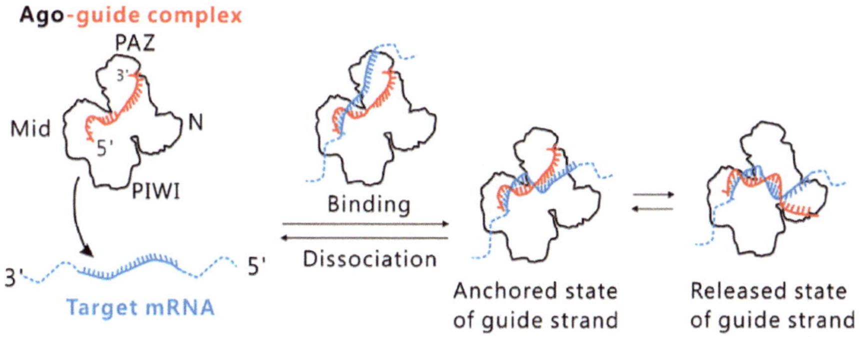

Fig. 8.10 During miRNA-target interaction, the 3′ end of the miRNA is anchored at the PAZ domain and is prohibited from base pairing with the target, which may play a role in efficient target release. The figure is adapted from [31] with American Chemical Society, Copyright 2013

mRNA quickly and allows the protein to act as a multi-turnover enzyme [100]. This reversible interaction between Ago and its target might be possible because the 3′ end of the miRNA is anchored at the PAZ domain, which lowers the binding affinity between the Ago and the target mRNA [31, 45, 67, 101, 102, 104].

Lastly, evidences suggest that Ago may directly interact with mRNAs and contribute to the target recognition process [79]. The first sequence of the miRNA is anchored and cannot interact with the target [21, 90, 105, 106]. However, when the first sequence of the miRNA is uridine, it may interact with adenine nucleotide of the mRNA and anchor the mRNA onto the MID domain of Ago [107]. This provides an additional sequence pairing between miRNA and mRNA and can increase the efficiency of gene silencing effect [108]. Moreover, this interaction may account for the phenomenon where uridine is the preferred sequence at the 5′ end of the guide strand [21, 90, 105, 106]. Interestingly, using a single-molecule approach, Schirle et al. showed that the adenine anchoring in Ago does not influence the initial target recognition process, but does increase the residence time of Ago on the mRNA [108]. Therefore, RISC can still search for its target using the sub-seed sequences, and the base pairing at the first position only affects the gene silencing efficiency.

8.7 Concluding Remarks

In this chapter, we present structural, biochemical, and biophysical single-molecule studies on the Ago–miRNA-target interactions. As a key posttranscriptional regulatory molecule, miRNA has received increasing attention over the recent decades [109]. In particular, miRNAs are recognized as key potential biomarkers for diagnosis of human disease and prognosis during clinical treatments as well as indicators of cellular status [109]. The dynamic changes in miRNA expressions are associated with a variety of human diseases including heart disease, neurological diseases, immune function disorders, and age-related diseases. Furthermore, dramatic changes in miRNA profiles have been observed during disease progression, drug treatment, and differentiation of stem cells.

Despite its significance, our understanding of miRNA-mediated gene silencing is limited due to lack of knowledge of the in-depth mechanism of RISC assembly and RISC-target interactions. Numerous structural studies have significantly expanded our understanding of how miRNAs are loaded onto Ago. Loading of the miRNA to the MID and PAZ domains of Ago positions the RNA such that it can efficiently search for targets. In addition, miRNA–Ago complex constantly undergoes changes in its configuration as it interacts with the target mRNA to effectively find and associate with the true targets. Lastly, anchoring of the 3′ end of the miRNA to the PAZ domain ensures the efficient target release, which may be important for the effective gene silencing by RISC.

These structural studies inspired the development of new models of target search and regulation that are further confirmed by single-molecule experiments. Particularly, these studies investigated the binding dynamics of miRNA-target interactions. Like transcription factors, RISC also utilizes one-dimensional scanning in order to quickly search for the optimal binding site on an mRNA. This mode of quick scanning is possible as Ago protein induces steric hindrance and prevents the complementary pairing beyond the sub-seed region. More importantly, RISC also can readily dissociate from the mRNA that lacks full seed match sequences and undergo diffusion in the three-dimensional space in the cytosol to find a new potential target. Ago structure also plays a key role in this switch between the two target search modes. First, complementary pairing at the sub-seed region triggers structural change such that the helix-7 motif can no longer block the miRNA–mRNA interaction beyond the sub-seed region. Furthermore, seed pairing also can switch the protein to the recognition mode and allow base pairing even at the supplementary region. Therefore, Ago subdivides miRNA into multiple functional domains and changes its configuration

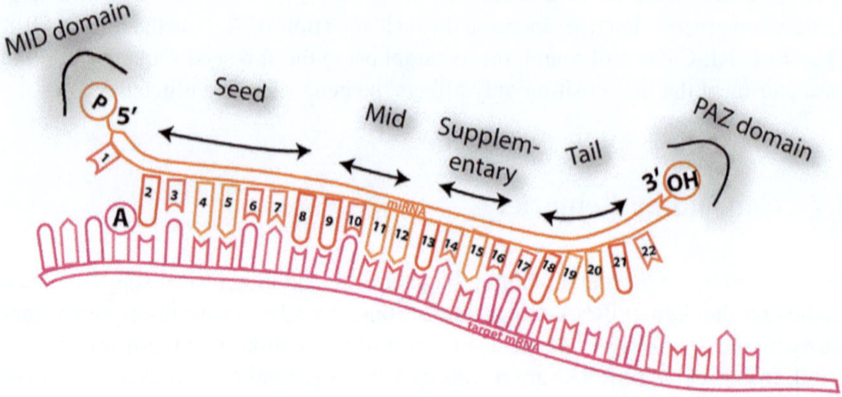

Fig. 8.11 Schematic of miRNA-target interaction depicting subdivision of the miRNA by Ago. The figure is adapted from [77] with Elsevier, Copyright 2017

to determine which regions to expose to the media in order to optimize its target search and release processes (Fig. 8.11).

The one-dimensional scanning mechanism as well as the dynamic conformational change induced by target interaction is also observed in various other systems. As discussed above, LacI repressor scans through the DNA to find the optimal binding site on the *lac* operon. In addition, the interaction between RecA and DNA is restricted to 7–8 nt due to the steric hindrance imposed by RecA protein [110]. Qi and colleagues termed these sequences as "microhomology motif" which serves as an initial platform for target recognition by RecA [111]. Such restriction may allow RecA to quickly scan the DNA to find the optimal binding site. Lastly, CRISPR/Cas protein also utilizes multi-step target recognition mechanism. It first scans the DNA to find the PAM motif and subsequently scans the vicinity to find the optimal binding site. Furthermore, binding to the extended complementary target induces conformational change to bring the nuclease domain to the target DNA [112–114]. These series of target recognition steps may account for the remarkable efficiency and specificity of the CRISPR system.

Given the generality of the target search mechanism, we expect that one-dimensional scanning as well as the conformational change induced by the target interaction is an important regulatory strategy in RNA/DNA binding proteins. Quantitative and single-molecule approaches in combination with the structural crystallographic information will provide valuable insights into a comprehensive understanding of RNA–protein interactions.

Acknowledgements This work was supported by Samsung Research Funding & Incubation Center of Samsung Electronics under Project Number SRFC-MA1702-08.

References

1. Bartel, D. P. (2004). MicroRNAs: Genomics, biogenesis, mechanism, and function. *Cell, 116,* 281–297.
2. Ha, M., & Kim, V. N. (2014). Regulation of microRNA biogenesis. *Nature Reviews Molecular Cell Biology, 15,* 509–524.
3. Cai, X., Hagedorn, C. H., & Cullen, B. R. (2004). Human microRNAs are processed from capped, polyadenylated transcripts that can also function as mRNAs. *RNA, 10,* 1957–1966.
4. Lee, Y., Jeon, K., Lee, J. T., Kim, S., & Kim, V. N. (2002). MicroRNA maturation: Stepwise processing and subcellular localization. *The EMBO Journal, 21,* 4663–4670.
5. Lee, Y., et al. (2004). MicroRNA genes are transcribed by RNA polymerase II. *The EMBO Journal, 23,* 4051–4060.
6. Ozsolak, F., et al. (2008). Chromatin structure analyses identify miRNA promoters. *Genes & Development, 22,* 3172–3183.
7. Denli, A. M., Tops, B. B., Plasterk, R. H., Ketting, R. F., & Hannon, G. J. (2004). Processing of primary microRNAs by the microprocessor complex. *Nature, 432,* 231–235.
8. Gregory, R. I., et al. (2004). The Microprocessor complex mediates the genesis of microRNAs. *Nature, 432,* 235–240.
9. Han, J., et al. (2004). The Drosha-DGCR9 complex in primary microRNA processing. *Genes & Development, 18,* 3016–3027.

10. Landthaler, M., Yalcin, A., & Tuschl, T. (2004). The human DiGeorge syndrome critical region gene 8 and Its *D. melanogaster* homolog are required for miRNA biogenesis. *Current Biology, 14,* 2162–2167.
11. Lee, Y., et al. (2003). The nuclear RNase III Drosha initiates microRNA processing. *Nature, 425,* 415–419.
12. Kim, B., Jeong, K., & Kim, V. N. (2017). Genome-wide mapping of DROSHA cleavage sites on primary microRNAs and noncanonical substrates. *Molecular Cell, 66,* 258–269. e255.
13. Nguyen, T. A., et al. (2015). Functional anatomy of the human microprocessor. *Cell, 161,* 1374–1387.
14. Kwon, S. C., et al. (2016). Structure of human DROSHA. *Cell, 164,* 81–90.
15. Auyeung, V. C., Ulitsky, I., McGeary, S. E., & Bartel, D. P. (2013). Beyond secondary structure: Primary-sequence determinants license pri-miRNA hairpins for processing. *Cell, 152,* 844–858.
16. Yi, R., Qin, Y., Macara, I. G., & Cullen, B. R. (2003). Exportin-5 mediates the nuclear export of pre-microRNAs and short hairpin RNAs. *Genes & Development, 17,* 3011–3016.
17. MacRae, I. J., Zhou, K., & Doudna, J. A. (2007). Structural determinants of RNA recognition and cleavage by Dicer. *Nature Structural & Molecular Biology, 14,* 934–940.
18. Park, J. E., et al. (2011). Dicer recognizes the 5′ end of RNA for efficient and accurate processing. *Nature, 475,* 201–205.
19. Macrae, I. J., et al. (2006). Structural basis for double-stranded RNA processing by Dicer. *Science, 311,* 195–198.
20. Hutvagner, G., & Simard, M. J. (2008). Argonaute proteins: Key players in RNA silencing. *Nature Reviews Molecular Cell Biology, 9,* 22–32.
21. Liu, J., et al. (2004). Arogonaute2 Is the catalytic engine of mammalian RNAi. *Science, 305,* 1437–1441.
22. Schirle, N. T., & MacRae, I. J. (2012). The crystal structure of human Argonaute2. *Science, 336,* 1037–1040.
23. Song, J. J., Smith, S. K., Hannon, G. J., & Joshua-Tor, L. (2004). Crystal structure of Argonaute and its implications for RISC slicer activity. *Science, 305,* 1434–1437.
24. Gan, H. H., & Gunsalus, K. C. (2015). Assembly and analysis of eukaryotic Argonaute-RNA complexes in microRNA-target recognition. *Nucleic Acids Research, 43,* 9613–9625.
25. Lingel, A., Simon, B., Izaurralde, E., & Sattler, M. (2003). Structure and nucleic-acid binding of the *Drosophila* Argonaute 2 PAZ domain. *Nature, 426,* 465–469.
26. Lingel, A., Simon, B., Izaurralde, E., & Sattler, M. (2004). Nucleic acid 3′-end recognition by the Argonaute2 PAZ domain. *Nature Structural & Molecular Biology, 11,* 576–577.
27. Ma, J. B., Ye, K., & Patel, D. J. (2004). Structural basis for overhang-specific small interfering RNA recognition by the PAZ domain. *Nature, 429,* 318–322.
28. Ma, J. B., et al. (2005). Structural basis for 5′-end-specific recognition of guide RNA by the A. fulgidus Piwi protein. *Nature, 434,* 666–670.
29. Parker, J. S., Roe, S. M., & Barford, D. (2004). Crystal structure of a PIWI protein suggests mechanisms for siRNA recognition and slicer activity. *The EMBO Journal, 23,* 4727–4737.
30. Yuan, Y. R., et al. (2005). Crystal structure of A. aeolicus argonaute, a site-specific DNA-guided endoribonuclease, provides insights into RISC-mediated mRNA cleavage. *Molecular Cell, 19,* 405–419.
31. Jung, S. R., et al. (2013). Dynamic anchoring of the 3′-end of the guide strand controls the target dissociation of Argonaute-guide complex. *The Journal of the American Chemical Society, 135,* 16865–16871.
32. Haley, B., & Zamore, P. D. (2004). Kinetic analysis of the RNAi enzyme complex. *Nature Structural & Molecular Biology, 11,* 599–606.
33. Zamore, P. D. (2001). Thirty-three years later, a glimpse at the ribonuclease III active site. *Molecular Cell, 8,* 1158–1160.
34. Elbashir, S. M. (2001). RNA interference is mediated by 21- and 22-nucleotide RNAs. *Genes & Development, 15,* 188–200.

35. Elbashir, S. M., Martinez, J., Patkaniowska, A., Lendeckel, W., & Tuschl, T. (2001). Functional anatomy of siRNAs for mediating efficient RNAi in *Drosophila* melanogaster embryo lysate. *The EMBO Journal, 20,* 6877–6888.
36. Nykanen, A., Haley, B., & Zamore, P. D. (2001). ATP requirements and small interfering RNA structure in the RNA interference pathway. *Cell, 107,* 309–321.
37. Meister, G., et al. (2004). Human Argonaute2 mediates RNA cleavage targeted by miRNAs and siRNAs. *Molecular Cell, 15,* 185–197.
38. Nowotny, M., Gaidamakov, S. A., Crouch, R. J., & Yang, W. (2005). Crystal structures of RNase H bound to an RNA/DNA hybrid: Substrate specificity and metal-dependent catalysis. *Cell, 121,* 1005–1016.
39. Rivas, F. V., et al. (2005). Purified Argonaute2 and an siRNA form recombinant human RISC. *Nature Structural & Molecular Biology, 12,* 340–349.
40. Schwarz, D. S., Tomari, Y., & Zamore, P. D. (2004). The RNA-induced silencing complex is a Mg2+ -dependent endonuclease. *Current Biology, 14,* 787–791.
41. Martinez, J., & Tuschl, T. (2004). RISC is a 5′ phosphomonoester-producing RNA endonuclease. *Genes & Development, 18,* 975–980.
42. Jinek, M., & Doudna, J. A. (2009). A three-dimensional view of the molecular machinery of RNA interference. *Nature, 457,* 405–412.
43. Forstemann, K., Horwich, M. D., Wee, L., Tomari, Y., & Zamore, P. D. (2007). *Drosophila* microRNAs are sorted into functionally distinct argonaute complexes after production by dicer-1. *Cell, 130,* 287–297.
44. Park, M. S., et al. (2017). Human Argonaute3 has slicer activity. *Nucleic Acids Research, 45,* 11867–11877.
45. Wang, Y., Sheng, G., Juranek, S., Tuschl, T., & Patel, D. J. (2008). Structure of the guide-strand-containing argonaute silencing complex. *Nature, 456,* 209–213.
46. Kim, V. N. (2008). Sorting out small RNAs. *Cell, 133,* 25–26.
47. Kiriakidou, M., et al. (2007). An mRNA m7G cap binding-like motif within human Ago2 represses translation. *Cell, 129,* 1141–1151.
48. Frank, F., et al. (2011). Structural analysis of 5′-mRNA-cap interactions with the human AGO2 MID domain. *EMBO Reports, 12,* 415–420.
49. Behm-Ansmant, I., et al. (2006). mRNA degradation by miRNAs and GW182 requires both CCR49:NOT deadenylase and DCP1:DCP2 decapping complexes. *Genes & Development, 20,* 1885–1898.
50. Braun, J. E., Huntzinger, E., Fauser, M., & Izaurralde, E. (2011). GW182 proteins directly recruit cytoplasmic deadenylase complexes to miRNA targets. *Molecular Cell, 44,* 120–133.
51. Fabian, M. R., et al. (2011). miRNA-mediated deadenylation is orchestrated by GW182 through two conserved motifs that interact with CCR51-NOT. *Nature Structural & Molecular Biology, 18,* 1211–1217.
52. Lim, J., et al. (2014). Uridylation by TUT4 and TUT7 marks mRNA for degradation. *Cell, 159,* 1365–1376.
53. Doench, J. G., & Sharp, P. A. (2004). Specificity of microRNA target selection in translational repression. *Genes & Development, 18,* 504–511.
54. Lewis, B. P., Shih, I. H., Jones-Rhoades, M. W., Bartel, D. P., & Burge, C. B. (2003). Prediction of mammalian microRNA targets. *Cell, 115,* 787–798.
55. Stark, A., Brennecke, J., Russell, R. B., & Cohen, S. M. (2003). Identification of *Drosophila* microRNA targets. *PLOS Biology, 1,* E60.
56. Gregory, R. I., Chendrimada, T. P., Cooch, N., & Shiekhattar, R. (2005). Human RISC couples microRNA biogenesis and posttranscriptional gene silencing. *Cell, 123,* 631–640.
57. MacRae, I. J., Ma, E., Zhou, M., Robinson, C. V., & Doudna, J. A. (2008). In vitro reconstitution of the human RISC-loading complex. *Proceedings of the National Academy of Sciences of the United States of America, 105,* 512–517.
58. Maniataki, E., & Mourelatos, Z. (2005). A human, ATP-independent, RISC assembly machine fueled by pre-miRNA. *Genes & Development, 19,* 2979–2990.

59. Wang, H. W., et al. (2009). Structural insights into RNA processing by the human RISC-loading complex. *Nature Structural & Molecular Biology, 16,* 1148–1153.
60. Kanellopoulou, C., et al. (2005). Dicer-deficient mouse embryonic stem cells are defective in differentiation and centromeric silencing. *Genes & Development, 19,* 489–501.
61. Martinez, J., Patkaniowska, A., Urlaub, H., Luhrmann, R., & Tuschl, T. (2002). Single-stranded antisense siRNAs guide target RNA cleavage in RNAi. *Cell, 110,* 563–574.
62. Murchison, E. P., Partridge, J. F., Tam, O. H., Cheloufi, S., & Hannon, G. J. (2005). Characterization of Dicer-deficient murine embryonic stem cells. *Proceedings of the National Academy of Sciences of the United States of America, 102,* 12135–12140.
63. Ye, X., et al. (2011). Structure of C3PO and mechanism of human RISC activation. *Nature Structural & Molecular Biology, 18,* 650–657.
64. Betancur, J. G., & Tomari, Y. (2012). Dicer is dispensable for asymmetric RISC loading in mammals. *RNA, 18,* 24–30.
65. Cheloufi, S., Dos Santos, C. O., Chong, M. M., & Hannon, G. J. (2010). A dicer-independent miRNA biogenesis pathway that requires Ago catalysis. *Nature, 465,* 584–589.
66. Kim, Y., & Kim, V. N. (2012). MicroRNA factory: RISC assembly from precursor microRNAs. *Molecular Cell, 46,* 384–386.
67. Kawamata, T., & Tomari, Y. (2010). Making RISC. *Trends in Biochemical Sciences, 35,* 368–376.
68. Tomari, Y., Matranga, C., Haley, B., Martinez, N., & Zamore, P. D. (2004). A protein sensor for siRNA asymmetry. *Science, 306,* 1377–1380.
69. Eulalio, A., Huntzinger, E., & Izaurralde, E. (2008). Getting to the root of miRNA-mediated gene silencing. *Cell, 132,* 9–14.
70. Filipowicz, W., Bhattacharyya, S. N., & Sonenberg, N. (2008). Mechanisms of post-transcriptional regulation by microRNAs: Are the answers in sight? *Nature Reviews Genetics, 9,* 102–114.
71. Baek, D., et al. (2008). The impact of microRNAs on protein output. *Nature, 455,* 64–71.
72. Brennecke, J., Stark, A., Russell, R. B., & Cohen, S. M. (2005). Principles of microRNA-target recognition. *PLOS Biology, 3,* e85.
73. Selbach, M., et al. (2008). Widespread changes in protein synthesis induced by microRNAs. *Nature, 455,* 58–63.
74. Kim, D., et al. (2016). General rules for functional microRNA targeting. *Nature Genetics, 48,* 1517–1526.
75. Schirle, N. T., Sheu-Gruttadauria, J., & MacRae, I. J. (2014). Structural basis for microRNA targeting. *Science, 346,* 608–613.
76. Song, J. J., et al. (2003). The crystal structure of the Argonaute2 PAZ domain reveals an RNA binding motif in RNAi effector complexes. *Nature Structural Biology, 10,* 1026–1032.
77. Klein, M., Chandradoss, S. D., Depken, M., & Joo, C. (2017). Why Argonaute is needed to make microRNA target search fast and reliable. *Seminars in Cell and Developmental Biology, 65,* 20–28.
78. Salomon, W. E., Jolly, S. M., Moore, M. J., Zamore, P. D., & Serebrov, V. (2015). Single-molecule imaging reveals that Argonaute reshapes the binding properties of its nucleic acid guides. *Cell, 162,* 84–95.
79. Chandradoss, S. D., Schirle, N. T., Szczepaniak, M., MacRae, I. J., & Joo, C. (2015). A dynamic search process underlies microRNA targeting. *Cell, 162,* 96–107.
80. Elkayam, E., et al. (2012). The structure of human argonaute-2 in complex with miR-20a. *Cell, 150,* 100–110.
81. Nakanishi, K., Weinberg, D. E., Bartel, D. P., & Patel, D. J. (2012). Structure of yeast Argonaute with guide RNA. *Nature, 486,* 368–374.
82. Lai, E. C. (2002). Micro RNAs are complementary to $3'$ UTR sequence motifs that mediate negative post-transcriptional regulation. *Nature Genetics, 30,* 363–364.
83. Berg, O. G., Winter, R. B., & von Hippel, P. H. (1981). Diffusion-driven mechanisms of protein translocation on nucleic acids. *Biochemistry, 20,* 6929–6948.

84. Riggs, A. D., Bourgeois, S., & Cohn, M. (1970). The lac represspr-operator interaction III. Kinetic studies. *The Journal of Molecular Biology, 53,* 401–417.
85. von Hippel, P. H., & Berg, O. G. (1989). Facilitated target location in biological systems. *The Journal of Biological Chemistry, 264,* 675–678.
86. Mirny, L., et al. (2009). How a protein searches for its site on DNA: The mechanism of facilitated diffusion. *Journal of Physics A: Mathematical and Theoretical, 42*
87. Slutsky, M., & Mirny, L. A. (2004). Kinetics of protein-DNA interaction: Facilitated target location in sequence-dependent potential. *The Biophysical Journal, 87,* 4021–4035.
88. Gerland, U., Moroz, J. D., & Hwa, T. (2002). Physical constraints and functional characterists of transcription factor-DNA interaction. *Proceedings of the National Academy of Sciences of the United States of America, 99,* 12015–12020.
89. Kong, M., & Van Houten, B. (2017). Rad4 recognition-at-a-distance: Physical basis of conformation-specific anomalous diffusion of DNA repair proteins. *Progress in Biophysics & Molecular Biology, 127,* 93–104.
90. Chiu, Y.-L., & Rana, T. M. (2002). RNAi in human cells. *Molecular Cell, 10,* 549–561.
91. Doench, J. G., Petersen, C. P., & Sharp, P. A. (2003). siRNAs can function as miRNAs. *Genes & Development, 17,* 438–442.
92. Hutvagner, G., & Zamore, P. D. (2002). A microRNA in a multiple-turnover RNAi enzyme complex. *Science, 297,* 2056–2060.
93. Bartel, D. P. (2009). MicroRNAs: Target recognition and regulatory functions. *Cell, 136,* 215–233.
94. Krek, A., et al. (2005). Combinatorial microRNA target predictions. *Nature Genetics, 37,* 495–500.
95. Lewis, B. P., Burge, C. B., & Bartel, D. P. (2005). Conserved seed pairing, often flanked by adenosines, indicates that thousands of human genes are microRNA targets. *Cell, 120,* 15–20.
96. Lim, L. P., et al. (2005). Microarray analysis shows that some microRNAs downregulate large numbers of target mRNAs. *Nature, 433,* 769–773.
97. Faehnle, C. R., Elkayam, E., Haase, A. D., Hannon, G. J., & Joshua-Tor, L. (2013). The making of a slicer: Activation of human Argonaute-1. *Cell Reports, 3,* 1901–1909.
98. Nakanishi, K., et al. (2013). Eukaryote-specific insertion elements control human ARGONAUTE slicer activity. *Cell Reports, 3,* 1893–1900.
99. Jo, M. H., et al. (2015). Human Argonaute 2 has diverse reaction pathways on target RNAs. *Molecular Cell, 59,* 117–124.
100. Wee, L. M., Flores-Jasso, C. F., Salomon, W. E., & Zamore, P. D. (2012). Argonaute divides its RNA guide into domains with distinct functions and RNA-binding properties. *Cell, 151,* 1055–1067.
101. Sasaki, H. M., & Tomari, Y. (2012). The true core of RNA silencing revealed. *Nature Structural & Molecular Biology, 19,* 657–660.
102. Zander, A., Holzmeister, P., Klose, D., Tinnefeld, P., & Grohmann, D. (2014). Single-molecule FRET supports the two-state model of Argonaute action. *RNA Biology, 11,* 45–56.
103. Herschlag, D. (1991). Implications of ribozyme kinetics for targeting the cleavage of specific RNA molecules in vivo: More isn't always better. *Proceedings of the National Academy of Sciences of the United States of America, 88,* 6921–6925.
104. Deerberg, A., Willkomm, S., & Restle, T. (2013). Minimal mechanistic model of siRNA-dependent target RNA slicing by recombinant human Argonaute 2 protein. *Proceedings of the National Academy of Sciences of the United States of America, 110,* 17850–17855.
105. Bofill-De Ros, X., & Gu, S. (2016). Guidelines for the optimal design of miRNA-based shRNAs. *Methods, 103,* 157–166.
106. Seitz, H., Tushir, J. S., & Zamore, P. D. (2011). A 5′-uridine amplifies miRNA/miRNA* asymmetry in *Drosophila* by promoting RNA-induced silencing complex formation. *Silence, 2,* 4.
107. Mi, S., et al. (2008). Sorting of small RNAs into Arabidopsis argonaute complexes is directed by the 5′ terminal nucleotide. *Cell, 133,* 116–127.

108. Schirle, N. T., Sheu-Gruttadauria, J., Chandradoss, S. D., Joo, C., & MacRae, I. J. (2015). Water-mediated recognition of t1-adenosine anchors Argonaute2 to microRNA targets. *Elife, 4*
109. Casey, M. C., Kerin, M. J., Brown, J. A., & Sweeney, K. J. (2015). Evolution of a research field-a micro (RNA) example. *PeerJ, 3,* e829.
110. Ragunathan, K., Liu, C., & Ha, T. (2012). RecA filament sliding on DNA facilitates homology search. *Elife, 1,* e00067.
111. Qi, Z., et al. (2015). DNA sequence alignment by microhomology sampling during homologous recombination. *Cell, 160,* 856–869.
112. Shvets, A. A., & Kolomeisky, A. B. (2017). Mechanism of genome interrogation: How CRISPR RNA-guided Cas9 proteins locate specific targets on DNA. *The Biophysical Journal, 113,* 1416–1424.
113. Sternberg, S. H., LaFrance, B., Kaplan, M., & Doudna, J. A. (2015). Conformational control of DNA target cleavage by CRISPR-Cas9. *Nature, 527,* 110–113.
114. Westra, E. R., et al. (2013). Type I-E CRISPR-cas systems discriminate target from non-target DNA through base pairing-independent PAM recognition. *PLOS Genetics, 9,* e1003742.

Chapter 9
Biophysics of RNA-Guided CRISPR Immunity

Luuk Loeff and Chirlmin Joo

9.1 Introduction

9.1.1 CRISPR Immunity

RNA molecules play essential roles in living organisms. RNA is commonly known as a transmitter of genetic information that is stored in DNA. However, in the last two decades, it has become clear that the function of RNA lies far beyond an information carrier. Non-coding RNAs are involved in many different cellular processes, such as translation (tRNA), protein synthesis (rRNA) and gene regulation (RNAi, see Chap. 8, Kim and Kim). More recently, it was found that non-coding RNA encoded by clustered regularly interspaced short palindromic repeats (CRISPR) loci and CRISPR-associated proteins (Cas) provide rapid and robust immunity against invading bacteriophages in prokaryotes (Fig. 9.1) [1–5]. Because the RNA-guided CRISPR effector complexes of this immune system are programmable and highly specific, the effector complexes have been repurposed (e.g. Cas9 in Class II systems) as a tool for genome engineering applications [6, 7] in a broad spectrum of organisms [6, 8–10]. This discovery led to a "CRISPR craze" that fast-tracked the characterization of new CRISPR-Cas systems [11].

CRISPR immunity is divided into three distinct stages. First, when the host encounters invasive mobile genetic elements [2], Cas proteins integrate small fragments of foreign DNA into the host CRISPR locus (Fig. 9.1). This process is commonly referred to as adaptation and results in the formation of genetic memory against the invading mobile genetic elements (Fig. 9.1) [3, 4]. Subsequent transcription and processing of the CRISPR-array (called crRNA biogenesis) produce small

L. Loeff · C. Joo (✉)
Kavli Institute of Nanoscience and Department of Bionanoscience, Delft University of Technology, Delft, The Netherlands
e-mail: c.joo@tudelft.nl

C. Joo and D. Rueda (eds.), *Biophysics of RNA-Protein Interactions*,
Biological and Medical Physics, Biomedical Engineering,
https://doi.org/10.1007/978-1-4939-9726-8_9

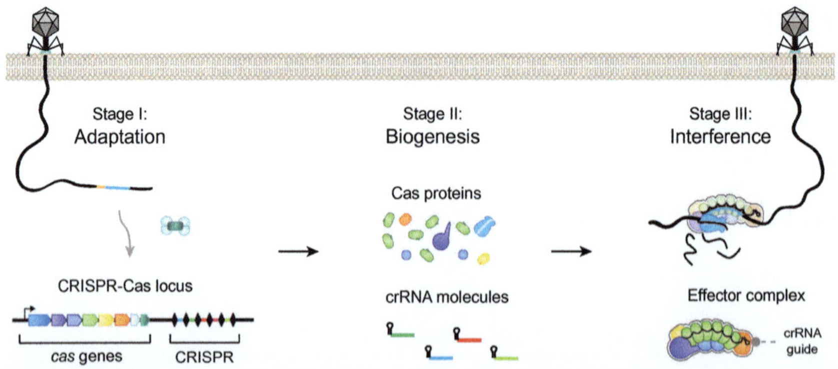

Fig. 9.1 Schematic overview of CRISPR-Cas adaptive immunity in prokaryotes. CRISPR immunity is conveyed in three distinct stages. During the adaptation stage, small fragments of invading DNA are incorporated into the CRISPR locus. The second stage of CRISPR immunity is crRNA biogenesis, in which the CRISPR locus is transcribed and processed into small guide RNA molecules. The last stage is CRISPR immunity is interference, where the invading DNA is located and destroyed by the CRISPR-associated proteins

non-coding CRISPR RNA (crRNA) molecules that form an effector complex with a single-Cas protein (Class II systems) or multiple Cas proteins (Class I systems) (Figs. 9.1 and 9.2) [12, 13]. In the last stage of immunity, called CRISPR interference, the effector complexes locate target sites (protospacers) that are complementary to their crRNA to trigger the destruction of invading DNA and/or RNA molecules (Fig. 9.1).

The constant evolutionary arms' race between prokaryotes and their invaders has resulted in an extreme diversity of CRISPR systems [12–14]. To date, CRISPR-Cas systems are classified using a two-step classification system that consists of two classes, six types and 21 subtypes (Fig. 9.2) [12–15]. First, CRISPR-Cas systems are divided into two broad classes, namely Class I and Class II [13, 14]. Class I systems are characterized by the presence of multi-subunit effector complexes (e.g. Cascade), whereas Class II systems encode for single-protein effector complexes (e.g. Cas9) (Fig. 9.2) [13, 14]. These classes are further divided into types (Class I into types I, III and IV; Class II into types II, V and VI) and subtypes based on the presence of signature Cas proteins and the mechanisms of crRNA processing, target recognition and destruction (Fig. 9.2) [13]. Despite this wealth in diversity, CRISPR-Cas systems share a common architecture: an array of alternating repeat and spacer sequences and a set of Cas proteins that convey immunization and immunity (Fig. 9.1).

To date, most of the knowledge on how the CRISPR immune system functions originates from ensemble-averaged measurements. While these assays provide valuable information on the collective behaviour of the population, they mask the molecular dynamics of individual molecules. Single-molecule biophysics (Fig. 9.3) has emerged as a powerful tool to visualise the molecular dynamics of single proteins with high spatial and temporal resolution [16–28]. Most of the single-molecule stud-

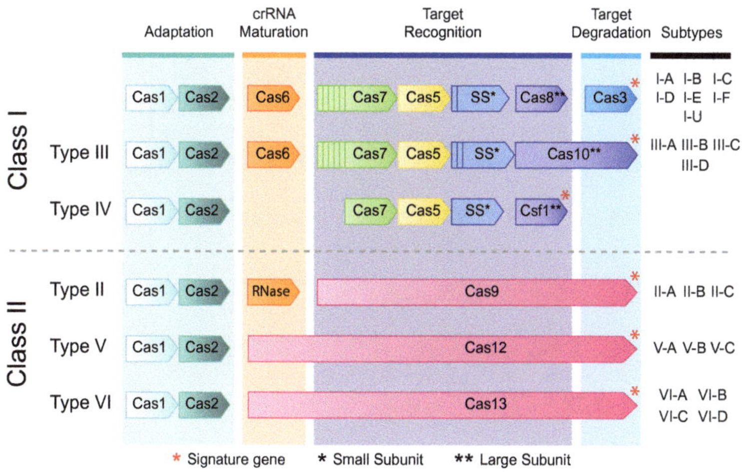

Fig. 9.2 Classification of CRISPR-Cas systems. CRISPR-Cas systems can be classified using a two-step classification system. First, CRISPR-Cas systems are divided into two broad classes based on the presence of multi-subunit or single-protein effector complexes. The systems are then further divided into types and subtypes based on the presence of signature genes. As a result, CRISPR-Cas systems are divided into two classes, six types and 21 subtypes. Red asterisk (*) indicates the signature gene for the specific type. Black asterisk (*) indicates the small subunit (e.g. Cse2 of Cascade). Double asterisk (**) indicates the large subunit (e.g. Cse1 of Cascade)

ies on the CRISPR-Cas immune system have focused on how effector complexes, such as Cascade (Class I, type I) and Cas9 (Class II, type II), locate and destroy their viral targets. In this chapter, we will review the progress that has been made on understanding CRISPR-Cas immunity, through the use of these single-molecule techniques.

9.1.2 Single-Molecule Techniques

The biophysical aspects of CRISPR-Cas systems have been probed using a wide variety of single-molecule techniques, including fluorescence spectroscopy-based and force spectroscopy-based methods (Fig. 9.3). To date, the majority of single-molecule fluorescence assays to study CRISPR have used total internal reflection fluorescence microscopy (TIRFM) (Fig. 9.3a) [29–38]. TIRFM is based on the principle that when the incidence angle of an excitation beam is set to a critical angle relative to the sample (e.g. glass slide or coverslip), the excitation beam is totally internally reflected (Fig. 9.3a). This generates an electromagnetic field, called an evanescent wave, at the interface of the glass slide and the solvent. This evanescent wave decays exponentially and thereby illuminates only the molecules that are close to the surface (~100 nm) (Fig. 9.3c, d). TIRFM achieves a higher signal-to-noise ratio compared to conventional wide-field microscopy, allowing for the visualization of single fluorophores with a millisecond time resolution.

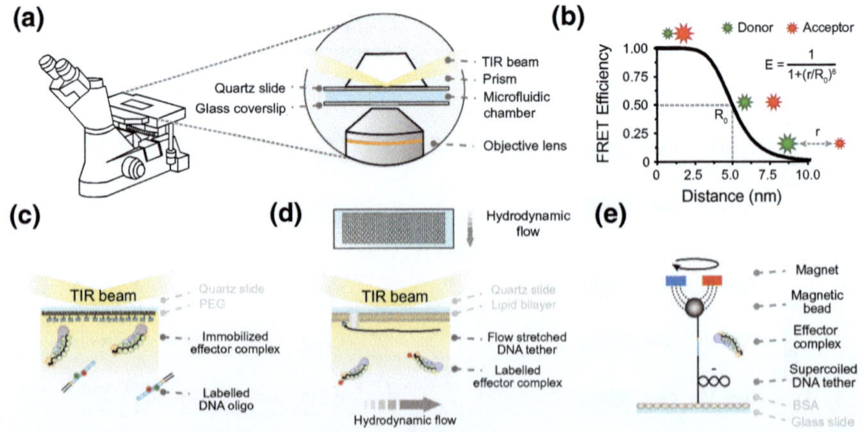

Fig. 9.3 Schematic overview of various single-molecule techniques **a** Schematic of total internal reflection microscopy. Total internal reflection microscopy is a light microscopy technique, which relies on total internal reflection of the excitation light to generate an evanescent wave at the interface of the glass slide and the solvent. The intensity of this evanescent wave decays exponentially and thereby selectively excites fluorophores that are immediately adjacent to the glass slide or coverslip. Thereby, TIRFM achieves much higher signal-to-noise ratios than conventional light microscopy, which is essential for harvesting a finite number of photons from single fluorophore. **b** FRET efficiency versus the distance in nanometres. The FRET efficiency changes with the 6th power of the distance between donor and acceptor, resulting in a steep decline when the distance increases. The Forster radius (R_0) corresponds to the distance at which 50% of the energy is transferred from the donor to the acceptor. **c** Schematic of a single-molecule fluorescence experiment. Typically, glass slides are passivated using polyethylene glycol (PEG) molecules to prevent non-specific binding. A fraction of these PEG molecules is biotinylated, allowing for the immobilization of biotinylated molecules (nucleic acids or protein). In this scheme, the interaction with fluorescently labelled DNA results in a sudden increase in the fluorescence signal. Image adapted from [30]. **d** Schematic of a single-molecule DNA curtain experiment. DNA curtains consist of an array of highly arranged DNA molecules that are tethered to the surface of a glass slide (top). These DNA molecules can be stretched by applying hydrodynamic flow, facilitating tethering of both sides of the DNA. The stretched form of the DNA allows one to observe interactions of fluorescently labelled proteins on several kilo-bases of DNA (bottom). **e** Schematic of a magnetic tweezers experiment. In magnetic tweezer experiments, the DNA is tethered with one end to a magnetic bead, whereas the other end is tethered to the surface of a glass slide. Precise modulation of a magnet (e.g. pulling or rotating) allows one to manipulate the surface-tethered DNA molecule

To elucidate the molecular details of CRISPR-Cas systems with single-molecule fluorescence, mainly two techniques have been used—single-molecule FRET and DNA curtains. Single-molecule Försterresonance energy transfer (smFRET) is based on non-radiative energy transfer between two fluorophores (termed donor and acceptor) (Fig. 9.3b). This energy transfer between the donor and acceptor takes place when the distance between the fluorophores is around 1–10 nm (Fig. 9.3b, c). Thereby, FRET can be used to probe the dynamics of proteins that would otherwise be masked by the physical diffraction limit of light microscopy.

To probe the long-range interactions of CRISPR–effector complexes using single-molecule fluorescence, DNA curtain assays have been used. A DNA curtain consists of an array of long DNA molecules that are tethered on either end to a glass slide in a highly-organized manner through the use of microfabrication (Fig. 9.3d, top). Double tethering is achieved applying hydrodynamic flow that evenly stretches the DNA molecules. This experimental set-up allows real-time visualization of protein–DNA interactions over several kilo-bases of DNA (Fig. 9.3d, bottom). While DNA curtains are used to probe long-range interactions that cannot be probed by smFRET, the resolution of DNA curtains is restricted by the diffraction limit and cannot be used to probe length scales smaller than ~250 nm. Thereby, smFRET and DNA curtain assays are considered complementary approaches.

Magnetic tweezers allow for precise micromanipulation and force measurement at the molecular level. These properties make magnetic tweezers a powerful tool to investigate protein-induced changes to DNA, such as target binding of RNA-guided CRISPR-Cas effector complexes (Fig. 9.3e) [39, 40]. In magnetic tweezer experiments, one end of a DNA substrate is tethered to the surface of a glass slide and the other end to a magnetic bead (Fig. 9.3e). Through the use of a magnetic field, the bead is precisely manipulated and the force and torque are applied on the DNA. For example, by pulling the magnetic field away from the surface, a stretching force can be exerted on the DNA (Fig. 9.3e). When the magnetic field is rotated at low forces, positive or negative supercoils are introduced into the DNA that decrease the length of the DNA and thereby lower the position of the bead. In this torsionally constrained configuration, magnetic tweezers are highly sensitive to changes in the length of DNA, enabling the detection of minute changes to the DNA (e.g. separation of two DNA strands over a few base pairs).

9.2 Target Search

Given the applications of CRISPR in genome engineering, there is a great interest in understanding how CRISPR effector complexes find target sites that are complementarity to their RNA guide. The RNA-guided effector complexes fulfil the daunting task of locating and identifying a 20–30 base pair protospacer (Fig. 9.4a) among the vast amount of DNA in the cell. Related biological systems, such as the RNAi-associated Argonaute protein (Chap. 8), use short-lived interactions with their targets to enable fast target search [41–44]. The transient nature of the interactions makes it difficult to investigate the mechanisms using conventional biochemical techniques. The advent of single-molecule fluorescence has allowed for visualization of these interactions with high spatio-temporal resolution.

To locate target sites, CRISPR effector complexes probe for a short sequence motif called protospacer adjacent motif (PAM), typically 2–6 base pairs, that is located immediately upstream of the protospacer (Fig. 9.4a) [45, 46]. The PAM sequence allows the immune system to distinguish self (CRISPR-array) from non-self (invading mobile genetic elements) (Fig. 9.1). However, PAM sequences are highly abundant in the genomes of prokaryotes. For example, the *Escherichia coli*

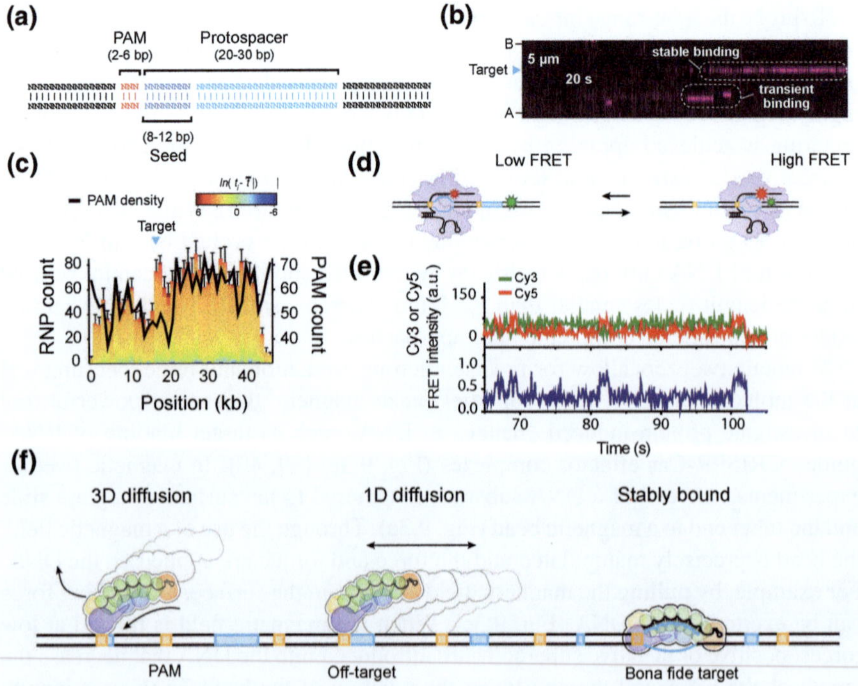

Fig. 9.4 Target search mechanism of CRISPR-Cas effector complexes. **a** Schematic representation of the sequence elements required for R-loop formation by CRISPR-Cas effector complexes. The protospacer adjacent motif (PAM, 2–6 nt) is highlighted in orange. The protospacer (20–30 bp) is highlighted in blue and purple. The seed is highlighted in blue (8–12 bp). **b** Representative kymograph displaying stable binding at the target site (blue arrow) and transient sampling of off-target sites. Figure adapted from [29]. **c** Distribution of transient binding events of CRISPR effector complexes along the DNA curtain DNA. The counts represent the number of binding events within bins of 1 kb of DNA. The black line represents the PAM counts along the DNA curtain. Target site is highlighted with a blue arrow. Figure adapted from [31]. **d** Schematic of a tandem-target assay to observe lateral diffusion. In this assay, the crRNA guide is labelled with an acceptor dye (red), whereas the target is labelled with a donor dye (green). The donor is placed such that when the CRISPR effector complex (purple) binds to the partial target site on the right, high FRET is observed, whereas binding to the left target site results in low FRET. **e** Representative time trajectory of donor (green) and acceptor (red) fluorescence and corresponding FRET values (blue), displaying lateral diffusion by the Cas9 effector complex. Figure adapted from [55]. **f** Distinct target search mechanisms of CRISPR effector complexes. Target search by CRISPR effector complexes is dominated by random 3D diffusion. Upon collision with the DNA, the effector complexes will use facilitated 1D diffusion to probe the DNA for potential target sites. When the effector complex locates a complementary target, it will form a R-loop, resulting in a stable interaction with the DNA

Cascade complex (hereafter called Cascade, Class I, type I, Fig. 9.2) requires a tri-nucleotide PAM (5′-CTT-3′) which is found every ~30 base pair on the *E. coli* genome [47]. The SpCas9 protein (hereafter called Cas9, Class II, type II, Fig. 9.2) from *Streptococcus pyogenesis* requires a dinucleotide PAM (5′-GG-3′), which on average is found every eighth base pair on the genome of *S. pyogenesis* [48]. If Cascade and Cas9 would probe complementarity over the full crRNA for every PAM sequence it encounters, the effector complexes would spend a considerable amount of time on the host genome before it finds an invading protospacer. Yet, both the Cascade and Cas9 effector complex manage to find their targets within the cell in an efficient matter [48, 49].

DNA curtain technology (Fig. 9.3d) has allowed direct visualization of the intermediates that lead to target-binding, for the effector complexes of *E. coli* (Cascade) and *S. pyogenesis* (Cas9) [29, 31, 36]. This single-molecule fluorescence approach revealed that both Cascade and Cas9 stably bind to *bona fide* target sites, whereas the effector complexes only transiently sample off-target sites on the 48-kb λ phage DNA (Fig. 9.4b) [29, 31]. Interestingly, these transient binding events are not uniformly distributed along the DNA, but instead correlate with the PAM density (Fig. 9.4c). As a result, more frequent binding of the CRISPR effector complexes was observed at sections of DNA with a high PAM density.

Dwell-time analysis of the binding events of Cas9 and Cascade at off-targets revealed two characteristic binding times [29, 31], suggesting that two kinetic intermediates exist on the pathway towards binding of a *bona fide* target [29, 31]. In agreement with this observation, a single-molecule FRET study on the kinetics of Cas9 displayed two distinct FRET states with characteristically different lifetimes [32]. Generally, non-specific interactions with the negatively charged backbone of the DNA are influenced by the presence of cations [50, 51]. However, the kinetic intermediates of the CRISPR effector complexes were almost unaffected by the salt concentration [29, 31]. Therefore, the two observed intermediate states observed in the DNA curtain assays may reflect protein-specific interaction with the PAM and the subsequent interrogation of the adjacent protospacer (Fig. 9.4a).

In contrast to the DNA curtain studies described above [29, 31], a recent DNA curtain study on a Cascade complex from *Thermobifida fusca* (TfuCascade) revealed that Cascade can non-specifically bind to the DNA and laterally scan the substrate via one-dimensional (1D) diffusion [36]. This 1D diffusion is facilitated by a conserved patch of positively charged residues on the PAM recognizing subunit (Cse1) of the Cascade complex [36, 52–54]. Although to a lesser extent, these charged residues are also present on the Cse1 subunit of *E.coli* Cascade [36]. Given that *T.fusca* is a thermophilic organism, it may have evolved additional charges in its PAM scanning subunit to compensate for its weakened interaction with DNA at elevated temperatures. Therefore, it is plausible that other CRISPR effector complexes from mesophilic organisms (e.g. *E. coli* Cascade and *S. pyogenesis* Cas9) might diffuse 1D over shorter distances within the diffraction limit of light microscopy (~250 nm) (Fig. 9.4f).

Single-molecule FRET was used to investigate if the Cas9 effector complex is capable of short-range lateral diffusion between potential targets [55]. Previous studies on the related RNA-guided Argonaute protein (Chap. 8, Kim and Kim) have shown that this protein uses lateral diffusion, which causes a synergistic effect between closely spaced target sites [41]. In this tandem-target assay, two partial target sites were placed at different distances from each other. Binding to one target would yield high FRET, whereas binding to the other target site would yield low FRET (Fig. 9.4d, e). These experiments showed that Cas9 laterally diffuses between potential targets with a range of approximately 20 bp [55]. Taken together, these results suggest that CRISPR effector complexes employ a combination of both 3D and 1D diffusion to locate their targets in an efficient manner (Fig. 9.4f).

9.3 crRNA-DNA Duplex Formation

Once a CRISPR effector complex encounters a PAM sequence, it locally melts the DNA [52, 56] and initiates R-loop formation. During R-loop formation, the RNA guide hybridizes with the complementary strand of the protospacer (target strand), while the non-complementary strand of the protospacer is displaced (non-target strand, Fig. 9.4a) [52, 57–59]. If the effector complexes would probe for complementarity over the entire RNA guide at every PAM it encounters, the effector complex would spend a substantial amount of time on off-targets. Given the high PAM density in prokaryotic genomes [47, 48], such mechanism would severely affect the capability of the effector complex to efficiently locate a complementary target.

High-throughput plasmid loss assays and biochemical experiments have shown that the CRISPR effector complexes tolerate mismatches in the PAM-distal end of the protospacer, whereas mismatches in the first 8–12 nucleotides at the PAM-proximal end of the protospacer abolish stable binding (Fig. 9.4a) [60–63]. The PAM-proximal end of the protospacer has therefore been suggested to "seed" the R-loop formation [64]. Hence, the first 8–12 nucleotides of the protospacer are commonly referred to as the seed sequence (Fig. 9.4a). This suggests that R-loops are formed in a directional manner, providing a mechanism to reject off-targets as soon as a mismatch is encountered. In line with this hypothesis, kinetic modelling of directional R-loop formation has shown that seed mutations are more likely to be rejected than PAM-distal mutations, due to unfavourable energetics (Fig. 9.5a) [65].

To provide experimental evidence for the directionality in R-loop formation, magnetic tweezers have been used to visualize *Streptococcus thermophilus* Cas9 and Cascade-dependent R-loop formation [39, 40]. In both systems, the PAM sequence dictates the frequency at which R-loops are formed, whereas the stability remains unaffected upon mutation of the PAM [39]. While StCas9 showed a strict regime in its PAM tolerance, St-Cascade is more flexible in PAM recognition, tolerating several mutated forms of the PAM sequence [39, 54, 61]. These results highlight the subtle differences among CRISPR–effector complexes, with a central role for the PAM sequence in the initiation of R-loop formation. Although PAM facilitates the initiation of the R-loop, PAM seems to be dispensable for the downstream process of crRNA–protospacer hybridization [30, 40].

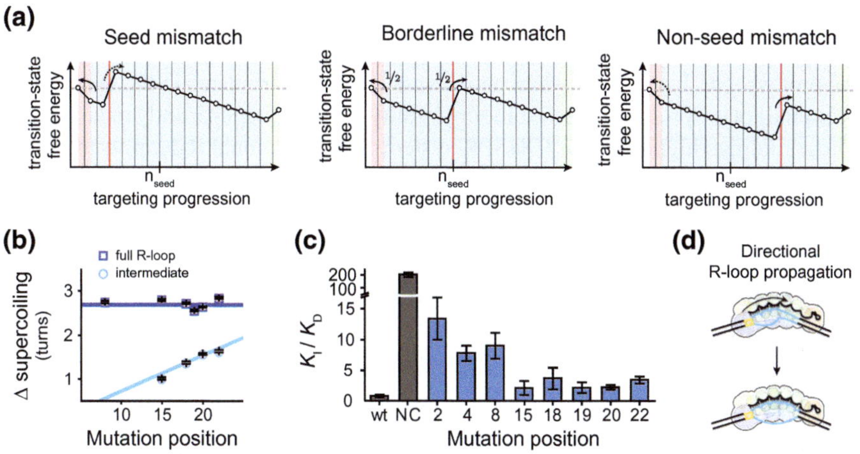

Fig. 9.5 R-loop formation by CRISPR-Cas effector complexes. **a** Transition landscapes displaying the influence of mismatches on R-loop formation. When a mismatch is placed in the seed region (left), the energy barrier for dissociation is lower than the energy barrier for R-loop propagation and thereby is likely to dissociate. When the mismatch is placed on the border of the seed region (middle), the energy barriers for R-loop propagation and dissociation are equally high. In contrast, when the mismatch is placed outside of the seed region (right), the energy barrier for dissociation is higher than R-loop propagation and R-loop propagation is lower than dissociation and thereby it is more likely to form a R-loop. **b** Mean supercoiling changes that are associated with the formation of a full R-loop (purple square) and R-loop intermediates (light blue circle). The solid line indicates the expected change in supercoiling when a R-loop is formed in a directional manner, starting from the PAM-proximal end of the protospacer. Figure adapted from [40]. **c** Quantification of a competitive EMSA assay. The K_I/K_D indicates the ratio between the dissociation constants of Cascade binding to a mutant competitor (x-axis) and a fully complementary protospacer (WT). NC indicates a protospacer that is not complementary to the RNA guide. Figure adapted from [40]. **d** Schematic model for directional R-loop propagation

To probe the directionality in R-loop formation, point mutations were introduced at either the PAM-proximal or PAM-distal end of the protospacer. Introduction of a single-mismatch revealed that the R-loop formation of the St-Cascade complex would stall, resulting in a partial R-loop (Fig. 9.5b) [40]. The stability of this R-loop is highly dependent on the position of the mismatch. Mismatches that are located in the PAM-proximal end of the protospacer lead R-loop intermediates that are less stable, compared to mutations that are located within the PAM-distal end (Fig. 9.5c) [40]. Upon stalling, the partial R-loop collapses and as a consequence the CRISPR effector complex dissociates. These data suggest that R-loop formation is initiated at the PAM-proximal end of the protospacer and subsequently propagates downstream towards the PAM-distal end of the protospacer (Fig. 9.5d).

Additional evidence for directional R-loop formation came from single-molecule FRET studies on the Cas9 effector complex [30, 32]. To probe the interaction between Cas9 and the DNA, the donor and acceptor dye were placed on the DNA and guide RNA, giving rise to high FRET when binding a complementary target site [32]. In line with the magnetic tweezer data [40], PAM-distal mutations minimally affect the lifetime of the bound Cas9, tolerating up to 11 mutations at the distal end of the protospacer. In contrast, when two mismatches are introduced in the seed (PAM-proximal) of the protospacer, the binding affinity of Cas9 is significantly reduced [32]. These findings highlight that CRISPR systems use directional R-loop formation to speed up their target search mechanism (Fig. 9.5d).

9.4 Conformational Dynamics of CRISPR Effector Complexes

To achieve high fidelity in target recognition, proteins generally undergo ordered target recognition that is accompanied by conformational changes [66]. For example, in the RNA-guided Argonaute protein (Chap. 8, Kim and Kim), the seed sequence is pre-ordered for initial target recognition. Upon pairing of the seed, the protein undergoes a conformational change that subsequently facilitates interactions with the downstream nucleotides [41, 42, 67, 68]. Similarly, structural studies on CRISPR effector complexes have shown that these complexes undergo significant conformational changes upon R-loop formation [33, 57, 58, 69–73]. Furthermore, genome-wide high-throughput sequencing studies on Cas9 binding and cleavage revealed that binding of Cas9 is far more promiscuous than cleavage [74, 75]. This discrepancy between binding and cleavage can be explained by the conformational dynamics of these programmable RNA-guided protein complexes.

9.4.1 CRISPR-Cas9 Uses a Conformational Checkpoint to License Target Degradation

To probe the conformational dynamics of the *S. pyogenesis* Cas9 protein, Cas9 was engineered for site-specific labelling by introducing cysteine residues into distinct domains of a cysteine-free variant of Cas9 (Fig. 9.6a) [33, 73]. The initial experiments on this labelled variant of Cas9 were performed in bulk and showed that the activation of the HNH nuclease domain is highly dependent on full R-loop formation. Moreover, the HNH domain allosterically activates the RuvC domain, suggesting that HNH dictates the double-stranded DNA cleavage activity of Cas9 [73].

Since bulk experiments average out population dynamics and thereby mask the underlying molecular dynamics, single-molecule FRET (Fig. 9.3c) was employed to further explore the conformational changes of Cas9 [33]. These experiments showed that the HNH domain visits an intermediate FRET state before activating its nuclease domain (Fig. 9.6a, b) [33]. The abundance of this intermediate FRET state is highly dependent on the presence of mismatches on the protospacer, suggesting Cas9 uses a conformational checkpoint through which the HNH domain must pass to trigger

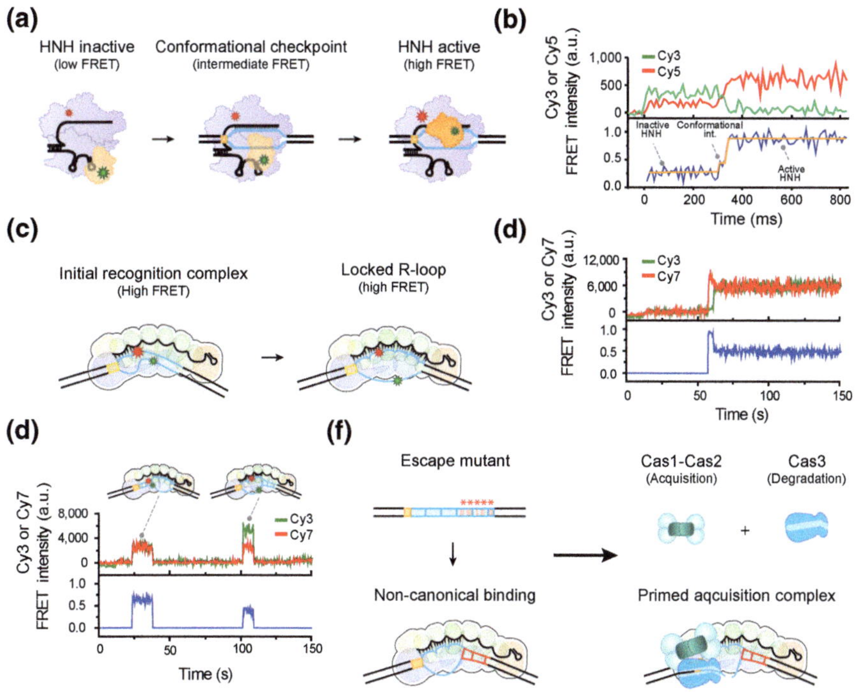

Fig. 9.6 Conformational dynamics of CRISPR effector complexes. **a** Schematic of the conformational dynamics of the Cas9 effector complex that were observed upon binding of a DNA substrate. In this assay, the Cas9 effector complex is labelled such that the donor (green star) and acceptor (red star) would report distinct conformational states of the protein. **b** Representative time trajectory of donor (green) and acceptor (red) fluorescence and corresponding FRET values (blue), displaying distinct conformational states of the Cas9 protein. **c** Schematic of the conformational dynamics of the Cascade complex that were observed upon binding of a canonical DNA substrate. In this assay, the DNA is labelled such that the donor (green star) and acceptor (red star) would be able to indirectly report on distinct conformational states of the Cascade complex. **d** Representative time trajectory of donor (green) and acceptor (red) fluorescence and corresponding FRET values (blue), displaying distinct conformational states of the Cascade complex upon encountering a bona fide target. **e** Representative time trajectory of donor (green) and acceptor (red) fluorescence and corresponding FRET values (blue), displaying the non-canonical binding mode of the Cascade complex. **f** Schematic of the formation of a primed acquisition complex. When Cascade binds a mutated target, it requires the Cas1/Cas2 integration complex to recruit the trans-acting Cas3 protein with both helicase and nuclease activities

degradation of the DNA [33]. Moreover, these experiments show how binding of DNA substrates is decoupled from cleavage activity, explaining why DNA binding of Cas9 is far more promiscuous than DNA cleavage [74, 75].

9.4.2 Cascade Uses Ordered Recognition to Achieve High Fidelity

While the conformational dynamics of Cas9 were monitored by directly tracking the movement of specific domains, the conformational dynamics of the *E.coli* Cascade complex were monitored by placing FRET probes onto the DNA (Fig. 9.3b) [30]. In this set-up, the observed FRET states reflect changes to the DNA that were induced by distinct conformations of the protein complex. This analysis revealed that Cascade forms an initial recognition complex where it hybridizes its seed sequence with the target DNA and thereby bends the DNA, resulting in a high FRET state (Fig. 9.6c, d) [30]. After the formation of this initial recognition complex, the R-loop propagates towards the PAM-distal end of the protospacer, resulting in a transition from the high FRET state to a low FRET state (Fig. 9.6c, d) [30]. Recent crystal and cryo-EM structures have captured snapshots of each of these states, providing a high-resolution map of all the conformational rearrangements that take place within the Cascade complex [52, 58, 70, 71, 76].

High-resolution structures of the Cascade complex display a global conformational change of Cascade upon R-loop formation, including rearrangements of the Cas6, Cse2 and Cse1 subunit of the complex [52, 57, 58, 69–71]. This global rearrangement of Cascade "lock" the R-loop resulting in a stable protein–DNA complex [39]. Such locking mechanism has not been observed for Cas9 effector complexes, and therefore, less torque is needed to dissociate Cas9 than Cascade in magnetic tweezer experiments [39]. The requirement for locking of the R-loop might be attributed to the fact that these systems have to recruit a trans-acting protein named Cas3 that is responsible for the subsequent target degradation. The recruitment of Cas3 is conveyed through a conformational change within the Cse1 subunit, the PAM recognizing subunit of the Cascade complex, which licenses the DNA for degradation [72].

9.4.3 Cascade Exhibits a Distinct Conformation to Flag Mutated Targets

In a co-evolutionary arms' race, the CRISPR immune system is constantly challenged by mutated phages that escape immunity. In response to escape mutants, some CRISPR systems (e.g. type I CRISPR-Cas systems) can initiate a response called primed spacer acquisition. During primed spacer acquisition, the immune system rapidly acquires new spacers and thereby restores immunity against mutants that escape immunity [77, 78]. While it was known that this response requires the CRISPR machinery, a mechanistic basis for this response remained elusive.

To obtain a mechanistic understanding of the recognition of mutated targets by the Cascade effector complex, the single-molecule FRET assay described above was used. The FRET probes, which were placed onto the DNA, could capture an additional binding mode of the Cascade complex. This short-lived non-canonical binding mode allows the complex to recognize mutated targets in a PAM- and seed-independent manner (Fig. 9.6e) [30]. In agreement with this observation, a Cascade complex from *T. fusca* was shown to bind partial complementary targets when DNA curtain assays were used [36]. Moreover, DNA curtain data showed that this binding mode promotes the formation of a primed acquisition complex, which consists of the Cascade complex (effector complex), the CRISPR-associated Cas3 protein (degradation) and Cas1-Cas2 (acquisition) proteins (Fig. 9.6f) [31, 36].

To explore if the non-canonical binding mode triggers a distinct conformation of the Cascade complex, two subunits of the Cascade complex were site-specifically labelled with a donor and acceptor dye. Bulk FRET measurements revealed that the PAM recognizing subunit (Cse1) adopts a distinct conformation when the Cascade binds a mutated target [72]. The equilibrium of this conformational change within Cse1 depends on the nature (e.g. PAM or seed) and number of the mutations and thereby provides a mechanism for the distinct functionalities of the complex [72]. It would be of interest to further understand the underlying the dynamics within the Cse1 subunit using single-molecule approaches.

9.5 CRISPR-Mediated DNA Degradation

To clear the viral infection, the CRISPR immune system degrades the invading nucleic acids, using its CRISPR-associated proteins [4]. Once the CRISPR effector complexes have formed an R-loop, the DNA is licensed for degradation. How this degradation takes places is dependent on the class and type of the CRISPR system. For all the Class II CRISPR systems (including Cas9), degradation of invading nucleic acids is conveyed by the effector complex itself. In contrast, the *E. coli* Cascade complex (Class I) relies on the trans-acting Cas3 protein with both nuclease and helicase activities.

9.5.1 Type I Systems

For target degradation, type I systems rely on the Cascade complex for targeting and the trans-acting protein called Cas3 for degradation [3]. Cas3 is a multi-domain protein that has both helicase and nuclease activities [79–83]. Once a full R-loop is formed, the Cas3 protein is recruited to the Cse1 subunit of the Cascade complex (Fig. 9.7a) [84]. Subsequently, Cas3 nicks the exposed non-target strand of the DNA, ~11 nucleotides downstream of the PAM sequence (Fig. 9.7a) [81, 82]. This initial nick generates a single-stranded overhang that facilitates loading of the helicase domain [35], which is followed by DNA unwinding in a 3′–5′ direction on the non-target strand with intermitted DNA cleavage [79–82].

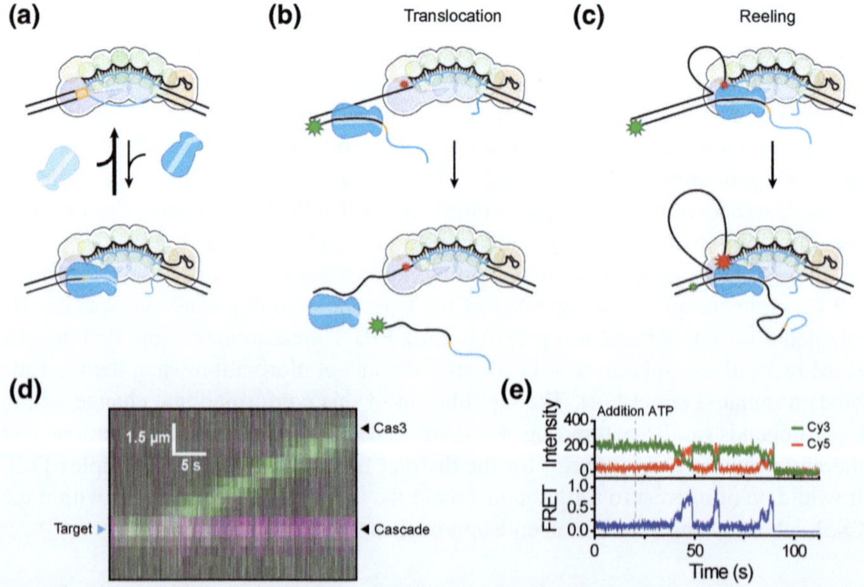

Fig. 9.7 DNA degradation in type I systems. **a** Schematic of recruitment of the Cas3 protein after R-loop formation by the Cascade effector complex. Once the R-loop is formed, Cascade licenses the DNA for degradation and recruits Cas3 to its Cse1 subunit. **b** Schematic of the translocation model, describing DNA unwinding by Cas3. In this model, Cas3 breaks its contacts with the Cascade complex while it unwinds the DNA. **c** Schematic of the reeling model, describing DNA unwinding by Cas3. When Cas3 reels the DNA, it remains tightly associated with the Cascade complex. As a result, loops are formed on the target strand. **d** Two colour kymographs of Cas3 translocation (green) away from the Cascade binding site (magenta) in DNA curtain assays. Blue arrow indicates the Cascade target site. Figure adapted from [31]. **e** Representative time trajectory of donor (green) and acceptor (red) fluorescence and corresponding FRET values (blue), displaying the repetitive DNA reeling by the Cas3 helicase

The degradation pattern observed in bulk biochemistry experiments can be explained by two different models. In the translocation model, Cas3 breaks its contacts with the Cascade complex when it unwinds and degrades the DNA. As a result, Cas3 translocates away from the Cascade binding site, leaving behind long single-stranded tracks of DNA (Fig. 9.7b) [31, 79, 80, 82, 84, 85]. Alternatively, Cas3 and Cascade remain tightly associated when Cas3 unwinds and degrades the DNA [31]. This mechanism, commonly referred to as reeling, has been observed for other helicases with a similar fold and results in the formation of DNA loops (Fig. 9.7c) [86–91].

Single-molecule fluorescence experiments have aided in understanding the mechanisms that underlie CRISPR interference. Initial DNA curtain assays showed that the majority (55%) of the Cas3 molecules remained stationary at the Cascade binding site without showing any dynamics, while the remaining 45% of the molecules would translocate away from Cascade binding site in a highly processive manner (Fig. 9.7d) [31]. Interestingly, a small fraction (14%) of these translocating molecule did not immediately break its contact with the Cascade complex, resulting in looped intermediates [31]. A recent DNA curtain paper further explored this rupture using force-dependent measurements on DNA curtains [36]. These measurements showed that the rupture between Cas3 and Cascade is highly dependent on the applied force [36].

The existence of the looped intermediate was further explored with a high spatio-temporal resolution using single-molecule FRET. To visualize reeling by Cas3 using single-molecule FRET, a donor and an acceptor fluorophore were placed on the DNA such that they could report on loop formation [35]. In these tension-free experiments, all Cas3 molecules remain tightly associated with the target-bound Cascade complex while reeling the DNA (Fig. 9.7c, e). Analysis of this reeling behaviour showed that Cas3 could go through multiple cycles of reeling on the same DNA substrate, allowing the helicase domain of Cas3 to repeatedly present ssDNA to its intrinsically inefficient nuclease domain (Fig. 9.7e) [35]. Characterization of the reeling showed that the reeling distance of Cas3 is finite, with an average distance of ~90 nucleotides. The short reeling distance of Cas3 explains why only a small fraction of the molecules showed this intermediate in diffraction-limited DNA curtain experiments [31, 35, 36]. Moreover, the repetitive behaviour of Cas3 provides an elegant mechanism to efficiently degrade the target DNA with an inefficient nuclease domain, while minimizing detrimental off-target degradation [35].

When Cas3 unwinds the invading DNA in the cell, it is likely to encounter DNA binding proteins and transcription factors that act as roadblocks on the DNA. To visualize the behaviour of Cas3 encountering a roadblock, the collisions between Cascade-Cas3 and an EcoRI restriction enzyme were monitored, using DNA curtain assays [36]. When the reeling population was analysed, it was shown that encounters with the restriction enzyme could block unwinding by Cas3. Cas3 was mostly observed to stall at the blockade, whereas some molecules would either dissociate or re-initiate another round of reeling. In contrast, the population of Cas3 molecules that translocated away from the Cascade binding site could remove the roadblock from the DNA [36]. This suggests two modes of DNA unwinding by Cas3 that drive distinct functions, namely DNA degradation through repetitive reeling and translocation to remove roadblocks.

Finally, to gain insights into the molecular mechanism of DNA unwinding by Cas3, the reeling events obtained through single-molecule FRET were analysed with a step-finding algorithm [35, 92]. These unwinding events were marked by a stepwise increase in FRET, that correspond to distinct steps of three base pairs [35]. Each of these steps consists of three hidden steps, indicating that Cas3 uses its helicase domain to break the dsDNA helix 1 nucleotide at a time [35]. After three successive 1-nucleotide steps, the DNA is released by the helicase generating a spring-loaded burst that moves the DNA by 3 base pairs [35]. This burst-like unwinding behaviour has been observed for nucleases [93, 94] and helicases with a RecA-like fold [87]. Given that the helicase domain of Cas3 is highly conserved [95], this spring-loaded unwinding likely reflects a general feature of Cas3 proteins.

While Cas3 uses a spring-loaded mechanism to reel the DNA in a forward direction, close inspection of the FRET events showed signs of backtracking [35]. This backtracking also occurred in three-base-pair steps, suggesting that the domain involved in holding the unwound nucleotides is responsible for constraining backtracking. In related helicases with a similar fold, the accessory domain (C-terminal domain) functions as a backstop, facilitating directional unwinding [87, 95, 96]. Apart from these short-range backtracking events, Cas3 also displayed long-range backtracking that returned the helicase to its initial FRET state [35]. These findings provide a mechanism for defining the unwinding distance of Cas3 that is required for efficient degradation of the invading DNA.

9.5.2 Type II Systems

Type II systems rely on a single effector complex for both target recognition and subsequent cleavage of the invading DNA. The Cas9 protein achieves cleavage of the target and non-target strand through its HNH and RuvC domains, respectively. DNA curtain and single-molecule FRET measurements revealed that Cas9 remains tightly bound to its DNA substrate when cleavage is triggered [29, 32, 33]. To release the DNA from the effector complex, harsh denaturing conditions were needed, suggesting that Cas9 is a single-turnover enzyme (Fig. 9.8a) [29, 32]. To probe the cleavage kinetics of Cas9 in real time, the base pairing beyond the target sequence was disrupted by truncating the PAM-distal flanking sequence on the non-target strand, which triggers the release of the non-target strand upon cleavage (Fig. 9.8b) [33]. By labelling the non-target strand and tracking the loss of fluorescence, it was shown that mismatches in the DNA result in slower cleavage kinetics (Fig. 9.8c), which is in agreement with bulk biochemistry data [29, 33, 73]. Taken together, these observations support the notion that HNH needs to license cleavage of the DNA [33].

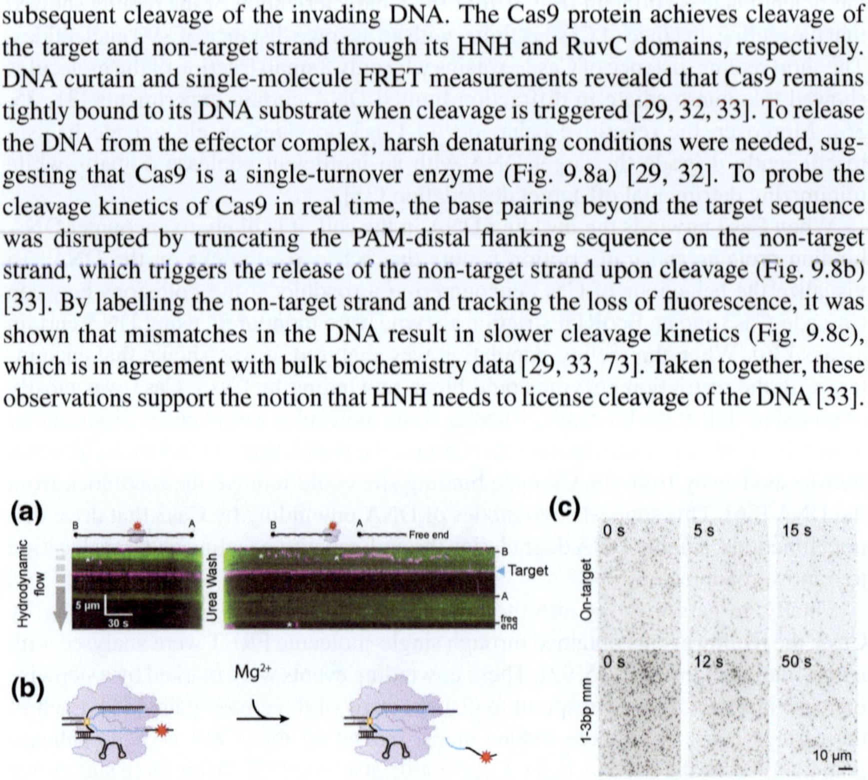

Fig. 9.8 DNA degradation in type II systems. **a** DNA curtain assay to visualize DNA cleavage by Cas9 effector complexes. The cleaved DNA was liberated from the Cas9 effector complex by flushing the chamber with 7 M urea. Asterisks denote quantum dots that are attached to the lipid bilayer. Figure adapted from [29]. **b** Single-molecule fluorescence assay to visualize DNA cleavage by Cas9 in real-time. **c** Images of surface immobilized Cas9 molecules bound to a partial duplex substrate. DNA cleavage by the Cas9 effector complex was induced by the addition of Mg^{2+} at $t = 0$

9.6 Outlook

The advent of single-molecule techniques has greatly advanced our understanding of the RNA-guided CRISPR-Cas immune systems. These techniques have provided tools to visualize molecular processes with short lifetimes of single RNA-guided proteins at a high spatial and temporal resolution. They have provided insights that extend beyond static high-resolution structures and bulk ensemble measurements, illuminating the dynamic nature of target search mechanisms, R-loop formation and DNA degradation in real time. Advances in the field of single-molecule biophysics promise to further deepen our understanding of CRISPR-Cas immunity, providing an unprecedented level of details in the molecular dynamics of these immune systems.

To date, most single-molecule studies have addressed fundamental questions that deepen our understanding of the biology behind the CRISPR immune system. However, recent efforts have focused on improving the specificity of CRISPR effector complexes for genome engineering purposes [34, 37]. These studies highlight that a comprehensive understanding of the underlying molecular dynamics of proteins can provide essential information for the rational design of proteins with enhanced functionalities. It would be of great interest to expand this line of research, focusing on the various stages of CRISPR immunity. Thus far, none of the single-molecule studies have focused on the CRISPR adaptation and crRNA maturation process (Fig. 9.2). A better understanding of the molecular dynamics of the spacer integration process could provide insights that may lead to new and enhanced tools to record cellular events on genomic loci [97, 98]. Moreover, single-molecule techniques could also aid in understanding the dynamics of crRNA maturation. For example, it has been shown that the Cas12 and Cas13 effector complexes (Fig. 9.2) process their own crRNA guides [99, 100]. A better understanding of this process could lead to enhanced tools for multiplexed genome engineering.

Acknowledgements This work was supported by the Netherlands Organization for Scientific Research [Vidi grant 864.14.002] and by a European Research Council Starting Grant under the European Union's Seventh Framework Programme [FP7/2007–2013/ERC grant 309509].

References

1. Mojica, F. J. M., Díez-Villaseñor, C., García-Martínez, J., & Soria, E. (2005). Intervening sequences of regularly spaced prokaryotic repeats derive from foreign genetic elements. *Journal of Molecular Evolution, 60,* 174–182.
2. Barrangou, R., et al. (2007). CRISPR provides acquired resistance against viruses in prokaryotes. *Science, 315,* 1709–1712.
3. Brouns, S. J. J., et al. (2008). Small CRISPR RNAs guide antiviral defense in prokaryotes. *Science, 340,* 216–219.
4. Marraffini, L. A., & Sontheimer, E. J. (2008). CRISPR interference limits horizontal gene transfer in staphylococci by targeting DNA. *Science, 322,* 1843–1845.

5. Abudayyeh, O. O., et al. (2016). C2c2 is a single-component programmable RNA-guided RNA-targeting CRISPR effector. *Science, 353*, aaf5573.
6. Wang, H., La Russa, M., & Qi, L. S. (2016). CRISPR/Cas9 in genome editing and beyond. *Annual Review of Biochemistry*. https://doi.org/10.1146/annurev-biochem-060815-014607.
7. Cox, D. B. T., et al. (2017). RNA editing with CRISPR-Cas13. *Science, 0180*, 1–15.
8. Mohanraju, P., et al. (2016). Diverse evolutionary roots and mechanistic variations of the CRISPR-Cas systems. *Science, 353*. https://doi.org/10.1126/science.aad5147.
9. Hsu, P. D., Lander, E. S., & Zhang, F. (2014). Development and applications of CRISPR-Cas9 for genome engineering. *Cell, 157,* 1262–1278.
10. Doudna, J. A., & Charpentier, E. (2014). The new frontier of genome engineering with CRISPR-Cas9. *Science, 346*. https://doi.org/10.1126/science.1258096.
11. Pennisi, E. (2013). The CRISPR craze. *Science, 341,* 833–836.
12. Makarova, K. S., et al. (2011). Evolution and classification of the CRISPR–Cas systems. *Nature Reviews Microbiology, 9,* 467–477.
13. Makarova, K. S., et al. (2015). An updated evolutionary classification of CRISPR-Cas systems. *Nature Reviews Microbiology, 13,* 722–736.
14. Shmakov, S., et al. (2015). Discovery and functional characterization of diverse class 2 CRISPR-Cas systems. *Molecular Cell, 60,* 385–397.
15. Konermann, S., et al. (2018). Transcriptome engineering with RNA-targeting type VI-D CRISPR effectors. *Cell, 173,* 665–676.e14.
16. Ray, S., Widom, J. R., & Walter, N. G. Life under the microscope: Single-molecule fluorescence highlights the RNA World. *Chemical Reviews*. https://doi.org/10.1021/acs.chemrev.7b00519 (in press).
17. Joo, C., Fareh, M., & Narry Kim, V. (2013). Bringing single-molecule spectroscopy to macromolecular protein complexes. *Trends in Biochemical Sciences, 38,* 30–37.
18. Jain, A., et al. (2011). Probing cellular protein complexes using single-molecule pull-down. *Nature, 473,* 484–488.
19. Fareh, M., et al. (2016). TRBP ensures efficient Dicer processing of precursor microRNA in RNA-crowded environments. *Nature Communications, 7,* 13694.
20. Kim, B., et al. (2015). TUT7 controls the fate of precursor microRNAs by using three different uridylation mechanisms. *The EMBO Journal, 34,* 1801–1815.
21. Miller, H., Zhou, Z., Shepherd, J., Wollman, A. J. M., & Leake, M. C. (2018). Reports on progress in physics single-molecule techniques in biophysics: A review of the progress in methods and applications. *Reports on Progress in Physics, 81,* 1–48.
22. Joo, C., Balci, H., Ishitsuka, Y., Buranachai, C., & Ha, T. (2008). Advances in single-molecule fluorescence methods for molecular biology. *Annual Review of Biochemistry, 77,* 51–76.
23. Cuculis, L., & Schroeder, C. M. (2017). A single-molecule view of genome editing proteins: Biophysical mechanisms for TALEs and CRISPR/Cas9. *Annual Review of Chemical and Biomolecular Engineering, 8,* 577–597.
24. Fareh, M., et al. (2015). Single-molecule pull-down for investigating protein-nucleic acid interactions. *Methods*.
25. Rutkauskas, M., Krivoy, A., Szczelkun, M., Rouillon, C., & Seidel, R. (2016). *Single-molecule insight into target recognition by CRISPR–Cas complexes* (1st ed.). Elsevier Inc. http://dx.doi.org/10.1016/bs.mie.2016.10.001.
26. Singh, D., & Ha, T. Understanding the molecular mechanisms of the CRISPR toolbox using single molecule approaches. *ACS Chemical Biology*. https://doi.org/10.1021/acschembio.7b00905 (in press).
27. Globyte, V., Kim, S. H., & Joo, C. Single-molecule view of small RNA–guided target search and recognition. *Annual Review of Biophysics*. https://doi.org/10.1146/annurev-biophys-070317-032923 (in press).
28. Roy, R., Hohng, S., & Ha, T. (2008). A practical guide to single-molecule FRET. *Nature Methods, 5,* 507–516.
29. Sternberg, S. H., Redding, S., Jinek, M., Greene, E. C., & Doudna, J. A. (2014). DNA interrogation by the CRISPR RNA-guided endonuclease Cas9. *Nature, 507,* 62–67.

30. Blosser, T. R., et al. (2015). Two distinct DNA binding modes guide dual roles of a CRISPR-Cas protein complex. *Molecular Cell, 58,* 60–70.
31. Redding, S., et al. (2015). Surveillance and processing of foreign DNA by the escherichia coli CRISPR-Cas system. *Cell, 163,* 854–865.
32. Singh, D., Sternberg, S. H., Fei, J., Doudna, J. A., & Ha, T. (2016). Real-time observation of DNA recognition and rejection by the RNA-guided endonuclease Cas9. *Nature Communications, 7,* 12778.
33. Dagdas, Y. S., Chen, J. S., Sternberg, S. H., Doudna, J. A., & Yildiz, A. (2017). A conformational checkpoint between DNA binding and cleavage by CRISPR-Cas9. *Science Advances, 3*. https://doi.org/10.1126/sciadv.aao0027.
34. Chen, J. S., et al. (2017). Enhanced proofreading governs CRISPR-Cas9 targeting accuracy. *Nature, 550,* 407–410.
35. Loeff, L., Brouns, S. J. J., & Joo, C. (2018). Repetitive DNA reeling by the Cascade-Cas3 complex in nucleotide unwinding steps. *Molecular Cell, 70,* 1–10.
36. Brown, M. W., et al. (2017). Assembly and translocation of a CRISPR-Cas primed acquisition complex. *BioRxiv*. https://doi.org/10.1101/208058.
37. Singh, D., et al. (2018). Mechanisms of improved specificity of engineered Cas9 s revealed by single-molecule FRET analysis. *Nature Structural & Molecular Biology, 25,* 347–354.
38. Singh, D., et al. (2018). Real-time observation of DNA target interrogation and product release by the RNA-guided endonuclease CRISPR Cpf1 (Cas12a). *Proceedings of the National Academy of Sciences of the United States of America, 1*, 201718686.
39. Szczelkun, M. D., et al. (2014). Direct observation of R-loop formation by single RNA-guided Cas9 and Cascade effector complexes. *Proceedings of the National Academy of Sciences of the United States of America, 111*, 9798–9803.
40. Rutkauskas, M., et al. (2015). Directional R-loop formation by the CRISPR-cas surveillance complex cascade provides efficient off-target site rejection. *Cell Reports, 10,* 1534–1543.
41. Chandradoss, S. D., Schirle, N. T., Szczepaniak, M., Macrae, I. J., & Joo, C. (2015). A dynamic search process underlies microRNA targeting. *Cell, 162,* 96–107.
42. Salomon, W. E., Jolly, S. M., Moore, M. J., Zamore, P. D., & Serebrov, V. (2015). Single-molecule imaging reveals that Argonaute reshapes the binding properties of its nucleic acid guides. *Cell, 162,* 84–95.
43. Jo, M. H., et al. (2015). Human Argonaute 2 has diverse reaction pathways on target RNAs. *Molecular Cell, 59,* 117–124.
44. Yao, C., Sasaki, H. M., Ueda, T., Tomari, Y., & Tadakuma, H. (2015). Single-molecule analysis of the target cleavage reaction by the Drosophila RNAi enzyme complex. *Molecular Cell, 59,* 125–132.
45. Mojica, F. J. M., Díez-Villaseñor, C., García-Martínez, J., & Almendros, C. (2009). Short motif sequences determine the targets of the prokaryotic CRISPR defence system. *Microbiology, 155,* 733–740.
46. Leenay, R. T., et al. (2015). Identifying and visualizing functional PAM diversity across CRISPR-Cas systems. *Molecular Cell, 62,* 137–147.
47. Levy, A., et al. (2015). CRISPR adaptation biases explain preference for acquisition of foreign DNA. *Nature, 520,* 505–510.
48. Jones, D. L., et al. (2017). Kinetics of dCas9 target search in *Escherichia coli*. *Science, 357,* 1420–1424.
49. Knight, S. C., et al. (2015). Dynamics of CRISPR-Cas9 genome interrogation in living cells. *Science, 350,* 823–826.
50. Lohman, T. M., & Ferrari, M. E. (1994). Escherichia Coli single—Stranded Dna-binding protein: Multiple DNA binding modes and cooperativities. *Annual Review of Biochemistry*.
51. von Hippel, P. H., & Berg, O. G. (1986). On the specificity of DNA-protein interactions. *Proceedings of the National Academy of Sciences of the United States of America, 83*, 1608–1612.
52. Hayes, R. P., et al. (2016). Structural basis for promiscuous PAM recognition in type I-E Cascade from E. coli. *Nature, 530*, 499–503.

53. Sashital, D. G., Wiedenheft, B., & Doudna, J. A. (2012). Mechanism of foreign DNA selection in a bacterial adaptive immune system. *Molecular Cell, 46,* 606–615.
54. Westra, E. R., et al. (2013). Type I-E CRISPR-Cas systems discriminate target from non-target DNA through base pairing-independent PAM recognition. *PLOS Genetics, 9.*
55. Globyte, V., Lee, S. H., Bae, T., Kim, J., & Joo, C. (2018). *CRISPR Cas9 searches for a protospacer adjacent motif by one-dimensional diffusion.* Kavli Institute of Nanoscience and Department of BioNanoScience, Delft University of Center for Genome Engineering, Institute for Basic Science, Seoul 08826, Republic o.
56. Anders, C., Niewoehner, O., Duerst, A., & Jinek, M. (2014). Structural basis of PAM-dependent target DNA recognition by the Cas9 endonuclease. *Nature, 513,* 569–573.
57. Jore, M. M., et al. (2011). Structural basis for CRISPR RNA-guided DNA recognition by Cascade. *Nature Structural & Molecular Biology, 18,* 529–536.
58. Xiao, Y., et al. (2017). Structure basis for directional R-loop formation and substrate handover mechanisms in type I CRISPR- Cas system article structure basis for directional R-loop formation and substrate handover mechanisms in type I CRISPR-Cas system. *Cell, 170,* 48–60.e11.
59. Jiang, F., et al. (2016). Structures of a CRISPR-Cas9 R-loop complex primed for DNA cleavage. *Science, 351,* 867–871.
60. Semenova, E., et al. (2011). Interference by clustered regularly interspaced short palindromic repeat (CRISPR) RNA is governed by a seed sequence. *Proceedings of the National Academy of Sciences of the United States of America, 108,* 10098–10103.
61. Fineran, P. C., et al. (2014). Degenerate target sites mediate rapid primed CRISPR adaptation. *Proceedings of the National Academy of Sciences of the United States of America,* E1629–E1638.
62. Jiang, W., Bikard, D., Cox, D., Zhang, F., & Marraffini, L. A. (2013). RNA-guided editing of bacterial genomes using CRISPR-Cas systems. *Nature Biotechnology, 31,* 233–239.
63. Jinek, M., et al. (2012). A programmable dual-RNA–Guided DNA endonuclease in adaptive bacterial immunity. *Science, 337,* 816–822.
64. Künne, T., Swarts, D. C., & Brouns, S. J. J. (2014). Planting the seed: Target recognition of short guide RNAs. *Trends in Microbiology, 22,* 74–83.
65. Klein, M., Eslami-Mossallam, B., Arroyo, D. G., & Depken, M. (2018). Hybridization kinetics explains CRISPR-Cas Off-targeting rules. *Cell Reports, 22,* 1413–1423.
66. Klein, M., Chandradoss, S. D., Depken, M., & Joo, C. (2016). Why Argonaute is needed to make microRNA target search fast and reliable Misha. *Seminars in Cell and Developmental Biology,* 1–9.
67. Schirle, N. T., Sheu-Gruttadauria, J., & MacRae, I. J. (2014). Structural basis for microRNA targeting. *Science, 346,* 608–613.
68. Klum, S. M., Chandradoss, S. D., Schirle, N. T., Joo, C., & MacRae, I. J. (2017). Helix-7 in Argonaute2 shapes the microRNA seed region for rapid target recognition. *The EMBO Journal, 37,* e201796474.
69. Wiedenheft, B., et al. (2011). Structures of the RNA-guided surveillance complex from a bacterial immune system. *Nature, 477,* 486–489.
70. Jackson, R. N., et al. (2014). Crystal structure of the CRISPR RNA-guided surveillance complex from Escherichia coli. *Science, 345,* 1473–1479.
71. Mulepati, S., Héroux, A., & Bailey, S. (2014). Crystal structure of a CRISPR RNA-guided surveillance complex bound to a ssDNA target. *Science, 345,* 1479–1484.
72. Xue, C., et al. (2016). Conformational control of cascade interference and priming activities in CRISPR immunity short article conformational control of cascade interference and priming activities in CRISPR immunity. *Molecular Cell, 64,* 1–9.
73. Sternberg, S. H., Lafrance, B., Kaplan, M., & Doudna, J. A. (2015). Conformational control of DNA target cleavage by CRISPR-Cas9. *Nature, 527,* 110–113.
74. Wu, X., et al. (2014). Genome-wide binding of the CRISPR endonuclease Cas9 in mammalian cells. *Nature Biotechnology, 32,* 670–676.

75. Kuscu, C., Arslan, S., Singh, R., Thorpe, J., & Adli, M. (2014). Genome-wide analysis reveals characteristics of off-target sites bound by the Cas9 endonuclease. *Nature Biotechnology, 32,* 677–683.
76. Zhao, H., et al. (2014). Crystal structure of the RNA-guided immune surveillance Cascade complex in Escherichia coli. *Nature, 515,* 147–150.
77. Swarts, D. C., Mosterd, C., van Passel, M. W. J., & Brouns, S. J. J. (2012). CRISPR interference directs strand specific spacer acquisition. *PLoS ONE, 7,* 1–7.
78. Datsenko, K. A., et al. (2012). Molecular memory of prior infections activates the CRISPR/Cas adaptive bacterial immunity system. *Nature Communications, 3,* 945.
79. Sinkunas, T., et al. (2011). Cas3 is a single-stranded DNA nuclease and ATP-dependent helicase in the CRISPR/Cas immune system. *The EMBO Journal, 30,* 1335–1342.
80. Westra, E. R., et al. (2012). CRISPR immunity relies on the consecutive binding and degradation of negatively supercoiled invader DNA by Cascade and Cas3. *Molecular Cell, 46,* 595–605.
81. Sinkunas, T., et al. (2013). In vitro reconstitution of Cascade-mediated CRISPR immunity in Streptococcus thermophilus. *The EMBO Journal, 32,* 385–394.
82. Mulepati, S., & Bailey, S. (2013). In vitro reconstitution of an Escherichia coli RNA-guided immune system reveals unidirectional, ATP-dependent degradation of DNA target. *Journal of Biological Chemistry, 288,* 22184–22192.
83. Huo, Y., et al. (2014). Structures of CRISPR Cas3 offer mechanistic insights into Cascade-activated DNA unwinding and degradation. *Nature Structural & Molecular Biology, 21,* 771–777.
84. Hochstrasser, M. L., et al. (2014). CasA mediates Cas3-catalyzed target degradation during CRISPR RNA-guided interference. *Proceedings of the National Academy of Sciences of the United States of America, 111,* 6618–6623.
85. Staals, R. H. J., et al. (2016). Interference dominates and amplifies spacer acquisition in a native CRISPR-Cas system. *Nature Communications, 23,* 127–135.
86. Park, J., et al. (2010). PcrA helicase dismantles RecA filaments by reeling in DNA in uniform steps. *Cell, 142,* 544–555.
87. Myong, S., Bruno, M. M., Pyle, A. M., & Ha, T. (2007). Spring-loaded mechanism of DNA unwinding by hepatitis C virus NS3 helicase. *Science*, 513–517.
88. Wu, W. Q., et al. (2017). Single-molecule studies reveal reciprocating of WRN helicase core along ssDNA during DNA unwinding. *Scientific Reports, 7,* 1–11.
89. Zhou, R., Zhang, J., Bochman, M. L., Zakian, V. A., & Ha, T. (2014). Periodic DNA patrolling underlies diverse functions of Pif1 on R-loops and G-rich DNA. *Elife, 3,* e02190.
90. Budhathoki, J. B., et al. (2016). A comparative study of G-quadruplex unfolding and DNA reeling activities of human RECQ5 helicase. *Biophysical Journal, 110,* 2585–2596.
91. Budhathoki, J. B., Stafford, E. J., Yodh, J. G., & Balci, H. (2015). ATP-dependent G-quadruplex unfolding by Bloom helicase exhibits low processivity. *Nucleic Acids Research, 43,* 5961–5970.
92. Kerssemakers, J. W. J., et al. (2006). Assembly dynamics of microtubules at molecular resolution. *Nature, 442,* 709–712.
93. Lee, G., Bratkowski, M. A., Ding, F., Ke, A., & Ha, T. (2012). Elastic coupling between RNA degradation and unwinding by an exoribonuclease. *Science, 336,* 1726–1729.
94. Lee, G., Yoo, J., Leslie, B. J., & Ha, T. (2011). Single-molecule analysis reveals three phases of DNA degradation by an exonuclease. *Nature Chemical Biology, 7,* 367–374.
95. Jackson, R. N., Lavin, M., Carter, J., & Wiedenheft, B. (2014). Fitting CRISPR-associated Cas3 into the Helicase Family Tree. *Current Opinion in Structural Biology, 24,* 106–114.
96. Fairman-Williams, M. E., Guenther, U. P., & Jankowsky, E. (2010). SF1 and SF2 helicases: Family matters. *Current Opinion in Structural Biology, 20,* 313–324.
97. Shipman, S. L., Nivala, J., Macklis, J. D., & Church, G. M. (2016). Molecular recordings by directed CRISPR spacer acquisition. *Science, 353.* https://doi.org/10.1126/science.aaf1175.
98. Sheth, R. U., Yim, S. S., Wu, F. L., & Wang, H. H. (2017). Multiplex recording of cellular events over time on CRISPR biological tape. *Science, 1461,* 1–9.

99. Liu, L., et al. (2017). The molecular architecture for RNA-guided RNA cleavage by Cas13a. *Cell, 170,* 714–726.e10.
100. Fonfara, I., Richter, H., BratoviÄ, M., Le Rhun, A., & Charpentier, E. (2016). The CRISPR-associated DNA-cleaving enzyme Cpf1 also processes precursor CRISPR RNA. *Nature, 532,* 517–521.

Chapter 10
Dynamics of MicroRNA Biogenesis

Mohamed Fareh

10.1 Introduction

RNA interference pathways have evolved in eukaryotes to regulate gene expression and suppress undesirable genetic elements such as viral nucleic acids and transposons. MicroRNAs (miRNAs) are a class of short non-coding RNAs (21–22 nucleotides) that regulate gene expression in nearly all biological processes. miRNAs associate with Argonaute (Ago) protein family and guide this nucleoprotein complex to recognize target RNAs via Watson-Crick RNA-RNA base-pairing [1]. The complementarity between the miRNA and the target RNA fosters a stable interaction required for the recruitment of additional protein effectors to mediate translational repression, mRNA deadenylation, decapping and ultimately RNA degradation [2]. The vast majority of miRNAs have to undergo a long biogenesis process before yielding functional mature miRNA. All began in the nucleus where the RNA polymerase II binds to miRNA-related promoters and transcribes primary precursor (pri-miRNA) that folds into a stem-loop structure with 5′ and 3′ flanking single-stranded (ss) regions. Canonical mature miRNAs are globally embedded in the stem region of pri-miRNAs, and their maturation requires processing by two endoribonuclease proteins (RNase III family) called Drosha and Dicer located in the nucleus and the cytoplasm, respectively [3] (Fig. 10.1).

Up to date, miRNA database called miRBase has catalogued at least 2585 miRNAs in human, 1899 in mice, 462 in *Drosophila melanogaster* and 435 in *Caenorhabditis elegans* that orchestrate gene expression in time and space, providing a rigorous control of the majority of cellular processes. miRNAs themselves are tightly regu-

M. Fareh (✉)
Peter MacCallum Cancer Centre, 305, Grattan St, Melbourne 3000, VIC, Australia
e-mail: Mohamed.Fareh@petermac.org

Sir Peter MacCallum Department of Oncology, University of Melbourne, Parkville, Melbourne 3010, VIC, Australia

C. Joo and D. Rueda (eds.), *Biophysics of RNA-Protein Interactions*,
Biological and Medical Physics, Biomedical Engineering,
https://doi.org/10.1007/978-1-4939-9726-8_10

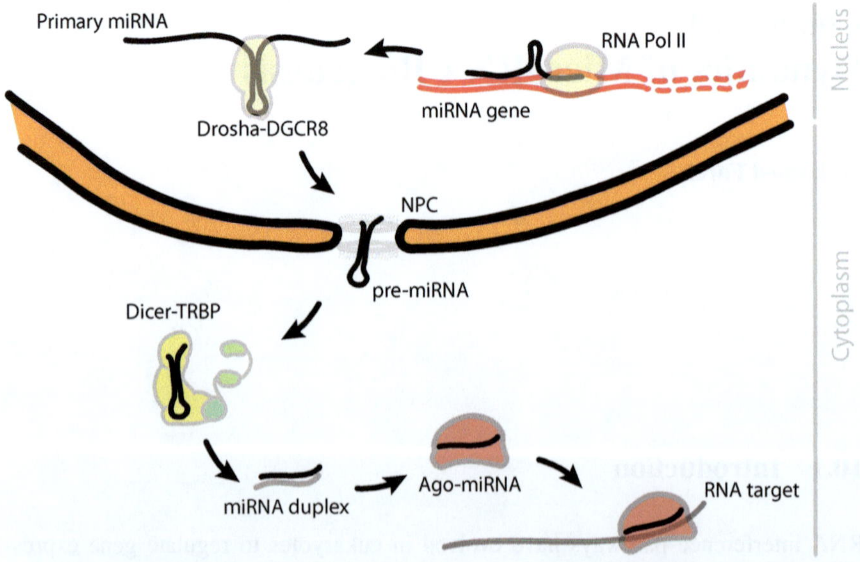

Fig. 10.1 microRNA biogenesis pathway. The majority of miRNA genes are transcribed by the RNA polymerase II. Promoters of miRNA genes are subject to transcriptional repression and activation by transcription factors and chromatin-remodelling enzymes. The primary miRNA (pri-miRNA) is typically a long transcript (1 Kb or longer) that contains a local stem-loop structure with 3′/5′ single-stranded overhangs. Drosha-DGCR8 complex (microprocessor) recognizes several features on the pri-miRNA and crops the lower part of the stem to generate a ~70-nucleotide precursor called pre-miRNA. Following Drosha processing, Exportin-5 and its cofactor RAN-GTP associate with the pre-miRNA and mediate the translocation through the nuclear pore complex (NPC). Pre-miRNA is released in the cytoplasm where Dicer-TRBP complex recognizes the hairpin structure of pre-miRNA and cleaves the terminal loop, creating a duplex miRNA composed of a guide and passenger strands. The passenger strand is ejected from the loading protein complex and undergoes rapid degradation, whereas the guide strand gets loaded into Argonaute (Ago) protein and mediates target recognition and translation repression

lated at the transcriptional and post-transcriptional levels, and their deregulation is frequently associated with human diseases such as cardiovascular disorders, obesity and cancer [4–7].

Lin-4 was the first microRNA discovered in *C. elegans* and appeared to be specific to worm [8, 9]. This short non-coding RNA plays an important role in controlling the timing of larval development. Soon thereafter, a second miRNA named let-7 was characterized in the same organism and turned out to be conserved in worms, flies, mammals and other eukaryotes, suggesting that miRNAs might have arisen through distinct events during the early stages of metazoan lineages evolution. In fact, miRNAs and miRNA-like RNAs have emerged independently in diverse eukaryotic lineages including plants, algae and fungi [10].

Further studies revealed that miRNAs are widespread in eukaryotes, and the number of miRNA genes soon exceeded the initial expectations. Advances in high-throughput sequencing technologies and computational analysis algorithms allowed scientists all over the world to predict hundreds of conserved miRNA genes in various species that were validated experimentally [11]. When looking at the conservation of miRNA, researchers found out that the so-called seed regions of miRNAs and their target sequences in the 3′ untranslated region (3′UTR) of mRNAs are well conserved, indicating that miRNAs have been under a high selection pressure throughout the evolution. In fact, microRNA losses are minor events compared to the emergence of new miRNAs sequences [11]. This high selection pressure led miRNAs to play key roles in regulating the majority of biological process. Indeed, knockout of miRNAs or mutations in the key biogenesis effectors are often associated with developmental defects and pathological disorders.

This chapter provides a general overview of miRNA lifecycle from the initial transcription in the nucleus, to the maturation by the endoribonucleases Drosha and Dicer, to cellular localization. This book chapter focuses on recent biophysical studies that have uncovered the molecular basis of miRNA biogenesis with high spatiotemporal resolution.

10.2 Genomic Architecture and Transcription Regulation of MicroRNA

MicroRNA genes are located in diverse genomic regions and are subject to transcriptional regulation by diverse mechanisms equivalent to those of coding genes. In human, miRNAs are often located in intragenic locus embedded within introns of coding or non-coding transcripts but also can be found occasionally within exonic regions [12]. Certain miRNA genes are also located in intergenic regions and their expression is driven by independent regulatory elements. MicroRNA genes are frequently organized in clusters containing several miRNA orthologs and are transcribed as polycistronic primary transcripts [13]. The miRNAs within those clusters often share sequence homology and have conserved ‘seed’ regions, suggesting gene duplication events that may have occurred during their evolution.

Computational and biochemical approaches demonstrated that the transcription of the vast majority of intragenic miRNAs is regulated by the RNA polymerase II (RNA pol II) together with their host gene, while intergenic miRNAs are controlled either by RNA pol II or RNA pol III [13–15]. The expression level of miRNAs is constantly modulated by RNA polymerases and their regulatory subunits in response to diverse intrinsic and extrinsic stimuli [3, 6, 12]. The promoters of miRNA genes host regulatory elements such as CpG islands, TATA box and initiation elements that are potent substrates for transcription factors and chromatin-remodelling enzymes. These regulators can enhance or repress the expression of miRNAs in a tissue-specific manner and under various environmental conditions. For example, transcription factors including TP53, MYC, and E2F1 have been shown to physically associate with miRNA promoters providing spatiotemporal regulation of their expression [16–20]. In addition to transcription factors, miRNA promoters are frequent targets of various chromatin-remodelling mechanisms such as DNA methylation and histone modifications that profoundly affect the accessibility and transcription activity of miRNA promoters [21, 22].

After transcription, the miRNA enters a long biogenesis process that starts in the nucleus and ends in the cytoplasm.

10.3 MicroRNA Processing by Drosha-DGCR8 Complex

In 2001, several groups reported the widespread of miRNA genes in eukaryotes, and the first clues of their genomic organization started to emerge, paving the way for a molecular understanding of their biogenesis pathways. The size difference between mature miRNAs and their precursors strongly indicated that this class of small non-coding RNAs has to undergo multiple maturation processes. A breakthrough study by Lee and co-workers demonstrated that upon transcription, the primary miRNA is subject to two successive maturation steps that are compartmentalized in the nucleus and cytoplasm, respectively [13]. A few years later, the endoribonuclease Drosha was reported to be the main enzyme that mediates the first cleavage of pri-miRNA in the nucleus [23–25].

The endoribonuclease Drosha recognizes and cleaves primary miRNAs (pri-miRNAs) harbouring a stem-loop structure and flanking single-stranded segments at the 3′ and 5′ ends [24]. Drosha is a ~160 KDa multi-domain endoribonuclease protein that belongs to the well-conserved RNase III family specialized in cleaving double-stranded (ds) RNA molecules. Drosha consists of N-terminal proline-rich and arginine/serine-rich domains, a central domain, two RNase III domains, and dsRNA-binding domain (dsRBD) at the C-terminal region [26–28]. The catalytic centre of Drosha contains two RNase III domains (RIIIDa and RIIIDb) that fold into an asymmetric intramolecular dimer. RNase III domains crop the lower part of the stem and the flanking single-stranded segments of pri-miRNA, generating a short hairpin-shaped RNA with 5′ phosphate group and two-nucleotide 3′ overhang called precursor miRNA (pre-miRNA) (Fig. 10.2).

The nucleus contains several thousands of RNA species that carry genetic information and important structural features. These RNA molecules often share structural modules with pri-miRNA and represent potential substrates of Drosha. The collateral cleavage of those RNAs would compromise their functions and subsequently alter the RNA homeostasis in the cell. The endonuclease Drosha evolved high-fidelity selection mechanisms to precisely recognize several features on canonical pri-miRNAs mainly in a sequence-independent manner and avoid collateral cleavage of other cellular RNAs.

Breakthrough biochemical, structural and biophysical studies allowed to decipher the molecular basis that Drosha uses when selecting for the canonical pri-miRNA molecules. Drosha relies on certain structural RNA features such as the terminal loop, the length of double-stranded structure in the stem region (33–35 base-par or bp) and the single-stranded flanking segments at the 3′/5′ ends to distinguish pri-miRNA substrates from other structured RNAs that are abundant in the nucleus [17, 27, 28]. Once Drosha recognizes a canonical pri-miRNA, it engages a productive interaction with this substrate to produce a second precursor miRNA called pre-miRNA.

10.3.1 Structure of Drosha

The cleavage positions of Drosha are precisely defined by the intermolecular distance between its catalytic site (RNase III domains) and other RNA-interacting domains within the core of Drosha, allowing this protein to act as a molecular 'ruler'. Drosha often cleaves pri-miRNA molecules at ~11 bp away from the basal junction between single-stranded and double-stranded RNA features (Fig. 10.2). The molecular basis governing such cleavage accuracy remained incomplete for more than a decade since the crystal structure of such big protein complex was challenging to obtain. Narry Kim laboratory has recently solved the atomic structure of Drosha at 3.2 Å resolution. The X-ray data revealed that the Bump helix and the surrounding structure at the base of the protein tightly interact with the lower stem region and lock the pri-miRNA substrate in a fixed position. Such interaction seems to constrain the basal loop to bifurcate into single-stranded (ss) RNA structure at exactly 11 bp away from the catalytic centre, namely RIIIDa and RIIIDb. Drosha then cleaves the pri-miRNA one helical turn from the basal junction (~11 bp) and two helical turns from the terminal loop (~22 bp). Surprisingly, the overall structure of Drosha is highly similar to another member of RNase III family called Dicer despite the poor sequence homology, suggesting that the two endoribonucleases may have evolved from the same ancestor [26].

The stem structure, the apical and basal junction and the 3′/5′ single-stranded flanking segments of pri-miRNA were shown to be essential features for substrate recognition and cleavage by Drosha [26, 27, 29]. Bartel and colleagues employed deep sequencing approach and molecular barcoding to identify additional features that are used by Drosha during substrate recognition. This study reported further functionally important pri-miRNA sequences such as UG motif at the interface between single-stranded and double-stranded segments (basal junction), UGUG motif at the beginning of the terminal loop (5′end), and CNNC motif at the 3′ end located ~17 nucleotides downstream of RNase IIIb cleavage sites [29, 30].

Altogether, these structural and functional studies indicate that Drosha relies on multiple RNA features and sequences to discriminate canonical pri-miRNAs from other cellular RNAs. Although these individual sequences and features contribute to the processing at different degrees, it appears that the basal modules (UG, basal junction, basal stem and CNNC) are the most determinant criteria in pri-miRNA recognition and processing.

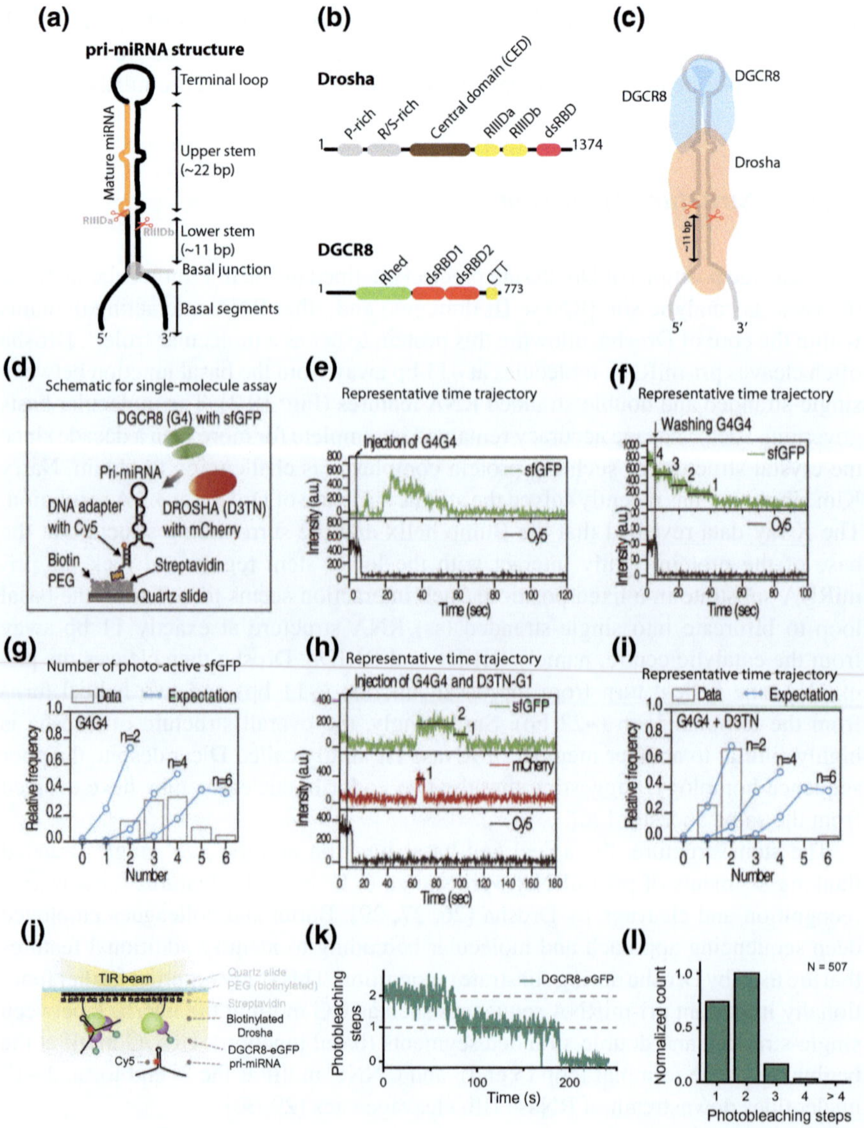
(a)
pri-miRNA structure
Terminal loop
Mature miRNA
Upper stem
(~22 bp)
RIIIDa
RIIIDb
Lower stem
(~11 bp)
Basal junction
Basal segments
5′
3′
(b)
Drosha
P-rich
R/S-rich
Central domain (CED)
RIIIDa
RIIIDb
dsRBD
1
1374
DGCR8
Rhed
dsRBD1
dsRBD2
CTT
773
(c)
DGCR8
DGCR8
Drosha
~11 bp
(d)
Schematic for single-molecule assay
DGCR8 (G4) with sfGFP
Pri-miRNA
DNA adapter with Cy5
DROSHA (D3TN) with mCherry
Biotin PEG
Streptavidin
Quartz slide
(e)
Representative time trajectory
Injection of G4G4
sfGFP
Cy5
Intensity (a.u.)
Time (sec)
(f)
Representative time trajectory
Washing G4G4
sfGFP
Cy5
Intensity (a.u.)
Time (sec)
(g)
Number of photo-active sfGFP
Data
Expectation
G4G4
n=2
n=4
n=6
Relative frequency
Number
(h)
Representative time trajectory
Injection of G4G4 and D3TN-G1
sfGFP
mCherry
Cy5
Intensity (a.u.)
Time (sec)
(i)
Representative time trajectory
Data
Expectation
G4G4 + D3TN
n=2
n=4
n=6
Relative frequency
Number
(j)
TIR beam
Quartz slide
PEG (biotinylated)
Streptavidin
Biotinylated Drosha
DGCR8-eGFP
Cy5
pri-miRNA
(k)
DGCR8-eGFP
Photobleaching steps
Time (s)
(l)
N = 507
Normalized count
Photobleaching steps

◀**Fig. 10.2 Primary miRNA processing by the Drosha-DGCR8 complex**. **a** Schematic of representative pri-miRNA structure. **b** Schematic of the Drosha and DGCR8 domains. **c** Schematic representation of the molecular architecture of the microprocessor complex composed by one Drosha and two DGCR8 bound to a pri-miRNA. **d** Schematic of single-molecule fluorescence assay used to count the number of DGCR8 molecules associated with Drosha within the microprocessor. **e, g** Representative time traces obtained through the binding of sfGFP-DGCR8 to surface-immobilized pri-miRNA (Cy5) in the absence of Drosha. The appearance of green (GFP) fluorescence reflects the binding of DGCR8 to surface-immobilized pri-miRNA. The photobleaching of the fluorophores is used to count the number of GFP-DGCR8 molecules. In the absence of Drosha, four-step photobleaching is dominant, indicating four DGCR8 can bind to a single pri-miRNA. **h, i** When Drosha is included in the assay, two steps of photobleaching become dominant, indicating that two DGCR8 are associated with Drosha and pri-miRNA (microprocessor). Adapted with permission from Nguyen et al., cell, 2015 [27]. **j** Single-molecule photobleaching assay to count the number of DGCR8 molecules associated with Drosha. Drosha was biotinylated in vivo and co-expressed with eGFP-DGCR8 in 293 HEK human cells. Drosha-DGCR8 are immobilized on the surface via streptavidin–biotin interaction in the presence of Cy5-labelled pri-miRNA. **k** The time trace reports the photobleaching of eGFP-DGCR8 molecules associated with surface-immobilized Drosha. **l** Quantification of the number of photobleaching steps observed in this single-molecule assay from N = 507 events. The data shows that two photobleaching steps are dominant, indicating that two DGCR8 molecules interact with one Drosha and pri-miRNA to form the microprocessor. Adapted with permission from Fareh et al., Methods, 2016 [36]

10.3.2 The Cofactor DGCR8

It has been repeatedly observed that in vivo, Drosha engages a stable interaction with a protein partner called DiGeorge syndrome critical region 8 (DGCR8) [31]. The two proteins define the core of a complex known as microprocessor that is indispensable for miRNA biogenesis in the animal kingdom. Mutations or deletions in the genomic region containing *DGCR8 gene* are associated with human genetic disorders [32], and experimental knockdown of DGCR8 in animal models leads to cellular dysfunction, demonstrating the importance of this cofactor in miRNA biogenesis and the maintenance of cell homeostasis [33–35].

DGCR8 is a protein of ~90 KDa that contains a nuclear localization signal (NLS) at its N-terminal region. DGCR8 forms a homodimer and bind to Drosha via a ~23 amino acids peptide at its C-terminal region called C-terminal tail (CTT). This protein–protein interaction appears to stabilize Drosha in vivo and regulate its RNA-binding activity and processing. In addition to the CTT domain, the C-terminal region contains a central RNA-binding heme (Rhed) and two dsRBDs that are responsible for the dimerization and the dsRNA-binding activity of DGCR8, respectively. Recent biochemical and structural studies from Narry Kim laboratory indicated that DGCR8 homodimer interacts with Drosha asymmetrically and defines the head of the microprocessor complex [26]. The digestion of pri-miRNA by the RNase A resulted in the disassembly of the microprocessor, indicating that Drosha and DGCR8 are likely to interact with each other and form a functional protein complex in a RNA-dependent manner [27].

Structural data showed that DGCR8 tends to form a homodimer when not associated with Drosha, and it remained unclear whether DGCR8 interacts with Drosha as a monomer or dimer. Nguyen and co-workers employed three-colour single-molecule fluorescence assays to define the stoichiometry of DGCR8 in a nucleoprotein complex together with Drosha and pri-miRNA [27]. Pri-miR-16-1 was immobilized on a passivated surface of imaging chambers through base-pairing with biotinylated DNA adaptor. Since the surface contains a monolayer of streptavidin, the biotinylated DNA was attached to the surface through near-covalent biotin–streptavidin interaction. The authors then introduced a mixture of GFP-labelled DGCR8 and mCherry-labelled Drosha into the microfluidic chamber and incubated them with a surface-immobilized pri-miRNA to promote the assembly of the microprocessor complex. This assay allowed for a real-time observation of Drosha complexed with DGCR8 and pri-miRNA by detecting the appearance of the fluorescence signal from spectrally different fluorophores. The colocalization of mCherry (Drosha), sfGFP (DGCR8) and Cy5 (pri-miRNA) reflected the assembly of the microprocessor at the single-molecule level. The photobleaching of the dye appears as a distinguishable decrease in the fluorescence signal in a stepwise manner offering the possibility of counting how many DGCR8 are associated with a single Drosha and pri-miRNA. In the absence of Drosha, the authors reported a wide range of photobleaching steps ranging from one to six, indicating that DGCR8 can bind to pri-miRNA with different stoichiometries, although a tetrameric form of DGCR8 seems to be the most frequently observed. When Drosha was included in the assay, two steps of photobleaching became dominant, indicating that DGCR8 form a dimer within the microprocessor complex. It is plausible that Drosha competes with DGCR8 binding sites on pri-miRNA, which results in the recruitment of a dimer DGCR8 instead of a tetramer when Drosha was included in the assay (Fig. 10.2). Additional sedimentation and gel filtration chromatography assays also confirmed that two DGCR8 proteins are likely to interact with Drosha in a pri-miRNA-dependent manner.

We also used different single-molecule fluorescence approach to probe the stoichiometry of DGCR8 within the microprocessor complex [36]. We engineered Drosha protein by adding an AP (acceptor peptide) tag to its N-terminus to allow for in vivo biotinylation during protein expression. DGCR8 fused to an eGFP was co-expressed with Drosha in 293 HEK human cells to promote the assembly of fluorescently labelled Drosha-DGCR8 complexes. Biotinylated Drosha-DGCR8 from crude cell extract was immobilized on the surface through biotin–streptavidin interaction. To determine the stoichiometry of DGCR8 within the microprocessor, eGFP was excited with a laser beam and the emitted fluorescence signal was recorded until the photobleaching of all the eGFP molecules. The number of photobleaching steps, defined as an apparent decrease in fluorescence intensity, reflects the number of DGCR8 molecules that are associated with a single Drosha protein. Our photobleaching data showed that ~46% (236 among 513 analysed molecules) of the microprocessor complexes are composed of one Drosha and two DGCR8 proteins (Fig. 10.2). Thus, it is in agreement with the model proposed by Nguyen and colleagues, in which the ~364 KDa microprocessor is a heterotrimeric protein complex composed of one Drosha and two DGCR8 molecules.

Biochemical and structural data have paved the way for a dynamic understanding of pri-miRNA recognition and processing by the microprocessor. Single-molecule fluorescence would be suitable to answer unsolved questions such as how the microprocessor coordinates the multiple RNA-binding domains in order to distinguish canonical pri-miRNA from other competing RNAs that are abundant in the nucleus? What are the kinetics of binding, cleavage and substrate releases? And to what extent the cofactor DGCR8 contribute to the substrate recognition and cleavage accuracy? Such question can be answered with high spatiotemporal resolution approaches such as single-molecule FRET and single-molecule fluorescence.

10.4 MicroRNA Transport Through the Nuclear Pore Complex

Following Drosha processing in the nucleus, pre-miRNAs are immediately exported to the cytoplasm through the nuclear pores. Nuclear pore complexes (NPC), incorporated into the nuclear envelope, regulate the bidirectional transport of macromolecules between the nucleus and the cytoplasm. The 100-nm-diameter NPC is perhaps the largest protein complex known in eukaryotic cells that allows the translocation of only biomolecules carrying import or export signals, although water and small metabolites are permitted to pass through freely. This mega-complex consists of ~1000 protein subunits and 30 distinct proteins called nucleoporins (NUPs) that fold into octagonal cylindrical shape (Fig. 10.3) [37–45]. The majority of RNA species including messenger RNAs (mRNAs), transfer RNAs (tRNAs), ribosomal RNAs, and miRNA precursors are exported to the cytoplasm through the NPC [46, 47]. A general model of nucleocytoplasmic transport was established through the analysis of the exchange of biomolecules between the nucleus and the cytoplasm. A well-conserved family of transporter proteins called nuclear transport receptors (importin-β family members or karyopherin), characterized by a α-superhelical structure, recognize a short peptide signal on a protein cargo—either a nuclear localization signal (NLS) or nuclear export signal (NES)—and mediate the cellular localization of the cargo through the NPC translocation [48–51]. The NPC should be viewed as a crowding environment where different macromolecules compete with each other during the translocation process. The basket of the nuclear pore contains disordered filaments rich in phenylalanine (F) and glycine (G) called FG-repeats [52, 53], which act as molecular barriers that prevent the translocation of macromolecules by simple diffusion. Only cargoes that are associated with transporters can specifically interact with the disordered filaments within the central channel of the pore and achieve selective translocation to the opposite compartment [47]. It is estimated that single NPC can accomplish the translocation of ~1000 macromolecules per second in both directions [47, 54, 55].

In addition to signal peptides on proteins, the importin-β family can also recognize specific structural motifs on RNAs called *cis*-acting localization elements (LEs) and mediate the transport either to the cytoplasm or to the nucleus. The transporters that import cargos to the nucleus are called importins, while the ones that export

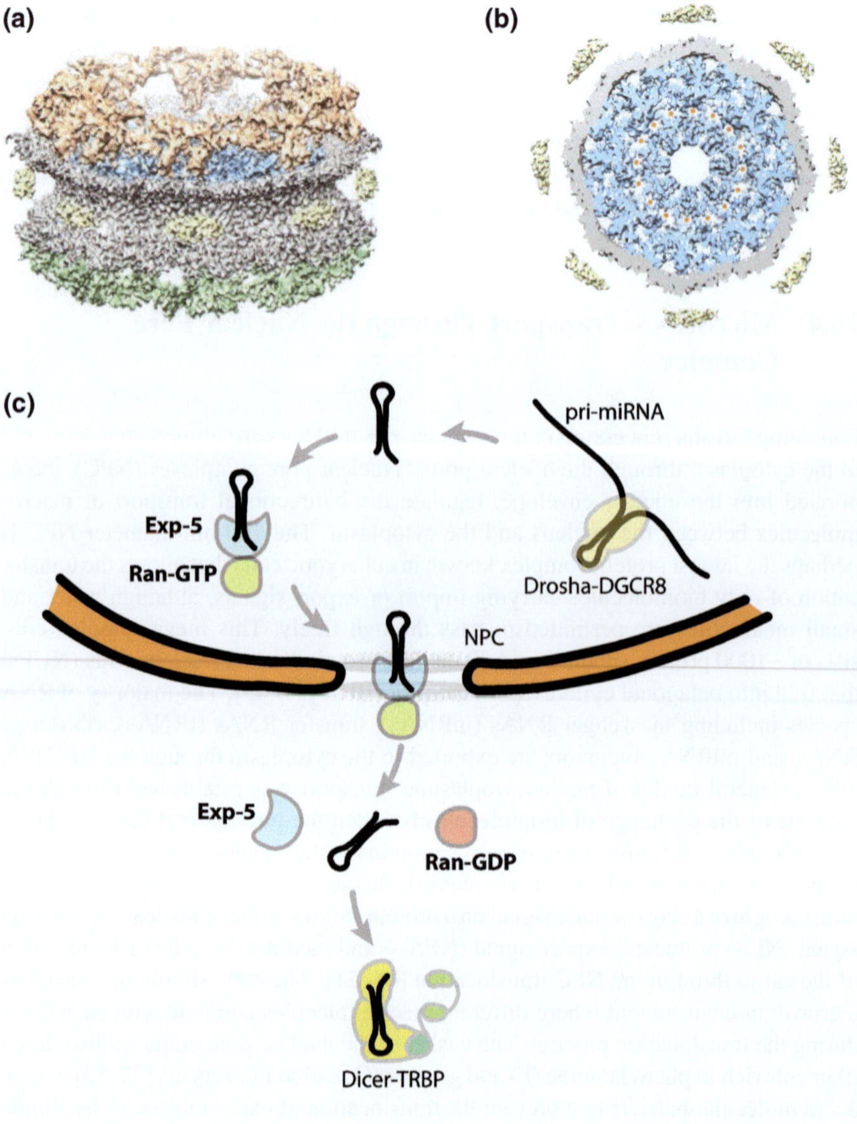
(a)
(b)
(c)
pri-miRNA
Exp-5
Ran-GTP
Drosha-DGCR8
NPC
Exp-5
Ran-GDP
Dicer-TRBP

◂**Fig. 10.3 Nuclear export of microRNA through the human NPC.** **a** Overview of the composite structure of the entire nuclear protein complex (NPC) that consists of around 1000 protein subunits organized in two parallel ring structures. Nucleoporins (Nups) preassemble into several distinct subcomplexes, which oligomerize to form an eightfold rotationally symmetric structure. The structural assignments in the outer rings are represented together with the assignments in the inner ring. **b** Top view of the nuclear pore complex (NPC). The central x–y-Sect. (10-nm thick) through the native NPC shows the organization of the central channel. Adapted with permission from Matthias Eibauer et al., Nature Communications, 2015 [45]. **c** The pri-miRNA is cleaved by the Drosha–DGCR8 complex to generate a pre-miRNA with a stem-loop structure containing two-nucleotide 3′ overhang. The pre-miRNA is specifically recognized by exportin-5 (Exp-5) and its cofactor RAN-GTP. This transporter complex actively mediates the transport of pre-miRNA to the cytoplasm through the nuclear pore complex (NPC). In the cytoplasm, the hydrolysis of GTP to GDP leads to the dissociation of the transporter complex and the release of pre-miRNA. The cytoplasmic Dicer-TRBP recognizes the hairpin structure of the pre-miRNA with 3′ overhang and cleaves the terminal loop to generate mature miRNA

cargos to the cytoplasm are named exportins. It is believed that the availability of those transporters is the major limiting factor in the nucleocytoplasmic transport. Importin-β family members are typically regulated by a small protein called Ran-GTPase [56, 57]. Ran is predominantly bound to a GTP molecule in the nucleus and GDP in the cytoplasm. Across the nuclear membrane, we can distinguish a gradient of Ran-GTP/Ran-GDP that is generated by two other cofactors: RanGEF (Ran-GDP-exchange factor) in the nucleus and RanGAP (Ran-GTPase activating protein) in the cytoplasm [47, 49]. These two major regulators create a driving force to ensure bidirectional nucleocytoplasmic transport of proteins and RNAs. The directionality is intimately linked to the release of the cargoes upon binding to specific cofactors that are available only in the correct side of the NPC. In the cytoplasm, importins bind cargo and mediate the NPC translocation before the release in the nucleus upon binding of RanGTP. In the other hand, exportins bind nuclear cargo containing NES together with RanGTP, translocate through the NPC and release the cargo in the cytoplasm upon the association with RanGAP, which stimulates the hydrolysis of GTP to GDP.

RNA molecules are major travellers through the NPC gate via the association with diverse protein transporters. Short RNA molecules such as miRNAs, tRNAs and small nuclear RNAs (snRNAs) follow a relatively simple transport route similar to the ones used by proteins, while mRNAs and large ribosomal RNAs use distinct pathways through the binding to other protein complexes [47, 58].

10.4.1 Nuclear Export of microRNA

In 2003, the nuclear export of miRNA was a very competitive field, with several laboratories attempting to identify the main exporters of miRNA [59–62]. Lund and colleagues investigated the molecular basis of miRNAs translocation through the NPC using *xenopus* oocytes as a model organism [59]. They found that pre-miRNAs are rapidly exported to the cytoplasm upon injection in the nucleus. Competition assays showed that pre-miRNA does use the same transporters as tRNA or mRNA. The export of pre-miRNAs was greatly reduced when the RanGTP was depleted, indicating that the export is likely to be mediated by a RanGTP-binding export receptor. While hunting the major transporters of pre-miRNA, the authors thought of exportin-5 (EXP-5) as a reliable candidate that mediate the export of small, structured and minihelix-containing RNAs such as tRNAs. Indeed, pre-miRNA co-immunoprecipitated with Exp-5, which demonstrated the physical association between the cargo and the transporter (Fig. 10.3). RanGTP was shown to increase the binding affinity of exportin-5 to pre-miRNA, and the depletion of Exp-5 using RNAi greatly impaired the efficiency of pre-miRNA export to the cytoplasm, demonstrating a direct and central role of Exp-5 in pre-miRNA export from the nucleus to the cytoplasm. Exp-5 relies on double-stranded motifs embedded within the secondary structure of the RNA cargo. Complementary studies further characterized the structural bases that govern the recognition pre-miRNA by Exp-5 [60–62]. At least 16 base-pairing in the stem region combined with 3′ overhang features are required to permit the binding and transport of pre-miRNA by Exp-5. On the other hand, the terminal loop seems to have a minor contribution to this recognition process (Fig. 10.3).

10.4.2 Nuclear Import of Mature miRNA

Although the majority of mature miRNAs are functionally active within subcellular compartments called P-bodies in the cytoplasm, recent reports have shown that miRNAs embedded within RISC complex can translocate back to the nucleus to regulate gene expression at the transcriptional and post-transcriptional levels [63–65]. However, the mechanisms mediating the relocation of miRNA and their protein effectors from the cytoplasm to the nucleus remain largely unresolved. Pitchiaya and co-workers used intracellular single-molecule, high-resolution localization and counting (iSHiRLoC) assays to probe the cellular localization and stability of various species of functional miRNAs [66]. To overcome technical challenges of single-molecule imaging in vivo, the authors used a flat cell line called U2OS (~2.5 to 5 μm) and a highly inclined laminal optical sheet (HILO) illumination that covers 3 μm of cellular depth. A single focal-plane in this approach allows illuminating an area containing approximately 50% of all miRNAs in the cell. They microinjected fluorescently labelled miRNAs either in the nucleus or in the cytoplasm and followed their cellular distribution and stability over time. Strikingly, the diffusion coefficient of single miRNAs allowed to distinguish a population associated with large nucleoprotein complexes with slow diffusion rate from another rapidly diffusing population that may represent free miRNAs species. The authors have shown that 9% of duplex miRNAs can translocate from the cytoplasm to the nucleus within 2 h of observation. These nuclear miRNAs assemble into low molecular weight complexes that are compositionally different from the cytoplasmic miRNA population. The nuclear localization and stability of guide miRNAs appeared to be highly dependent on the presence of Ago proteins and the base-pairing with their RNA targets [66].

Although several reports have shown nuclear localization of mature miRNAs and their protein effectors, it remains elusive the biological roles of nuclear miRNAs in mammals. Nevertheless, functional analyses of miRNAs in yeast, plants, nematodes and flies have shown that nuclear miRNAs can bind nascent RNA species and mediate RNA-induced transcriptional silencing (RITS), splicing, stability and chromatin remodelling [66].

10.4.3 Major Limitations for miRNA Tracking During the NPC Translocation

Overall, in vivo imaging of single miRNA and other short RNA species suffers from several technical challenges that impaired a dynamic view on their transport and cellular sub-localization. Unlike mRNA, miRNAs are relatively short RNA species and lack efficient fluorophore-labelling approaches that are suitable for in vivo single-molecule tracking. Long RNA species offer a better alternative to visualize the translocation through the NPC and obtain a dynamic understanding of their sub-cellular distribution at the single-molecule level, and perhaps, short and long RNA molecules may undergo relatively similar transport dynamics through NPC although the transporter proteins can differ between these RNA species.

Robert Singer laboratory pioneered fluorescence-based approaches to visualize RNA species throughout their journey from transcription to nuclear export and sub-cellular localization [55, 67]. Long RNA molecules were fused to a stem-loop structure encoded by bacteriophage genome. These stem-loop features engage in very strong interactions with MS2 bacteriophage coat protein [68], offering a great tool for the isolation and cellular localization of RNA of interest. MS2-GFP fusion allowed Singer laboratory to track the cellular localization and nuclear pore translocation of single mRNAs with high spatiotemporal resolution. Super-registration microscopy equipped with a fast camera allowed determining the speed of the NPC translocation. The researchers estimated that single mRNA translocates through the central channel of a nuclear pore within 5–20 ms [67]. Again this real-time observation confirmed that the size of the cargo, Ran-GTP concentration and especially the availability of transporter proteins are the limiting factors in this transport process. In fact, an enrichment in importin-β in the cytoplasmic buffer greatly enhanced the transport of mRNA mainly by decreasing the translocation dwell-time down to ~1 ms [69]. Given the high speed of translocation, approximately 200 RNA molecules are predicted to colocalize within the vicinity of a nuclear pore at a given time [67]. The high velocity of NPC translocation suggests that the rate of passage across the central channel alone is unlikely to be a rate-limiting step in this transport. Indeed, nucleoprotein complexes are reported to move bidirectionally through the NPC following a diffusion-based process [70]. Single-molecule FRET experiments showed that transporters with their cargoes can diffuse through the NPC and often change directionality within the central channel, while only 50% of the attempts are successful translocation events that would reach the opposite side of the NPC [71]. This implies that the transport is tightly regulated at the level of the cargo release since enhancer cofactors are exclusively located at one side of the nuclear pore. In this regard, miRNAs are expected to follow similar transport road to reach the cytoplasm. As the molecular crowding takes place in the central channel of the NPC, it is plausible that the abundance of other competitor RNA molecules bound to their transporters or transporters without cargoes might act as crowding agents and compete with the export and import of miRNAs. These remain very speculative due to the lack of experimental data, mainly because of the difficulty of tagging microRNA and other small RNA with specific tags (e.g. MS2, MCP) for in vivo tracking. Recent developments of small organic fluorophores and advances in microscopy techniques certainly paved the way to tackle these unsolved questions and obtain a dynamic view on microRNA translocation through NPC at the single-molecule level.

10.5 Pre-miRNA Recognition and Processing by Dicer-TRBP

Upon Drosha processing, the pre-miRNA is exported to the cytoplasm by exportin-5 and RAN-GTP. This complex binds to the stem and the 3′ overhang of a pre-miRNA and allows the export through the NPC. The hydrolysis of GTP initiates the disassembly of the transporter complex, leading to the release of pre-miRNA exclusively in the cytoplasm (Fig. 10.3). Once released, the cytoplasmic endoribonuclease Dicer recognizes the hairpin structure of pre-miRNA and cleaves the terminal loop to generate a duplex miRNA containing the guide and the passenger miRNA. Dicer is believed to play an essential role in the loading of the duplex miRNA to RISC complex and strand selection processes [72–75]. The thermodynamic stability of the duplex RNA ends defines its orientation within RISC complex and therefore dictates the fate of both RNA strands. The proximity of the guide strand to AGO fosters its loading into this effector protein, which protects the loaded miRNA from degradation by cellular nucleases. Conversely, the passenger strand is ejected from the RISC complex and undergoes rapid degradation [76, 77]. miRNA loaded into AGO guides the RISC complex to find mRNA target most likely through 1D and 3D target search mechanisms to initiate translational repression and/or mRNA decay (Fig. 10.1) (discussed in Chapter X) [78, 79].

Dicer homologues are widely conserved in fungi, plants and animal kingdoms and are considered as key actors in miRNA and siRNA processing [1, 80]. It is not surprising that Dicer deregulation impairs the steady state of miRNA production and cell homeostasis [81]. In fact, Dicer knockout in mice is lethal in the early stages of embryonic development [82], and conditional knockouts in a tissue-specific manner are often associated with severe defects in organs development [83–87], thus highlighting the central role of Dicer in microRNA production and maintaining cellular homeostasis.

10.5.1 Structure of Dicer

Human Dicer is a ~220 KDa multi-domain protein that belongs to the RNase III family and shares structure homology with Drosha [26, 88]. To date, the crystal structure of human Dicer has not been solved yet due to the large molecular weight of this protein and the complexity of purification. Nonetheless, the crystal structure of a Dicer homologue from *Giardia intestinalis* was solved in 2006, which was a cornerstone towards the understanding of the molecular architecture and function of this endoribonuclease [88]. Human Dicer contains multiple functional domains including a helicase domain (DExD/H-box helicase family), DUF883 domain (domain of unknown function), a PAZ domain, two RNase III domains, and a dsRBD at the C-terminus (Fig. 10.4) [45, 88, 89]. Electron microscopy allowed defining the global architecture of human Dicer associated with its cofactor TRBP. This protein complex adopts an L-shape form with the helicase domain in the base, the PAZ domain in the top and the pair of RNase III domains in the middle of Dicer's body defining its catalytic centre [90]. The two RNase III domains fold in an asymmetric manner and the shift between these two molecular scissors creates the 2-nucleotide overhang at the 3′ end of the duplex miRNA product, a hallmark of Dicer cleavage [91].

Upon the cleavage, human Dicer often generates miRNA with 21–23 nucleotides in length. This consistency in substrate cleavage is conferred to a molecular ruler property of Dicer that is defined by the intermolecular distance between the PAZ and RNase III pair. In fact, the 3′ 2-nucleotide overhang and 5′-phosphate ends of pre-miRNA are embedded within 3′ and 5′-phosphate pockets in the PAZ and platform domains, respectively [93, 94]. Those two pockets tightly hold the pre-miRNA termini, while the electrostatic interaction between the positively charged surface of Dicer's body and the negatively charged phosphate groups in the RNA backbone orients Dicer substrate towards the catalytic centre for the cleavage of the double-helix RNA at precise positions [88, 89, 91]. Thus, the length of miRNA duplex (21–23 nucleotides) is defined by the molecular distance between the 3′/5′ pockets and the pair of RNase III domains. Last, the helicase domain at the N-terminus of Dicer is also thought to regulate the substrate recognition and cleavage processes, mainly by sensing the terminal loop of a canonical pre-miRNA. Indeed, disturbing this interaction between the helicase domain and the terminal loop appears to compromise both cleavage efficiency and accuracy in vitro and in vivo [95, 96]. The Cryo-EM structure of hDicer-TRBP-Pre-miRNA ternary complex has been recently solved and suggests an important role of the N-terminal helicase domain towards the activation of the cleavage reaction [97]. The N-terminal DExD/H-box helicase domain appears to hold the terminal loop of pre-miRNA from accessing hDicer's processing centre in a pre-dicing state and may represent the limiting step in RNA processing. Sub-

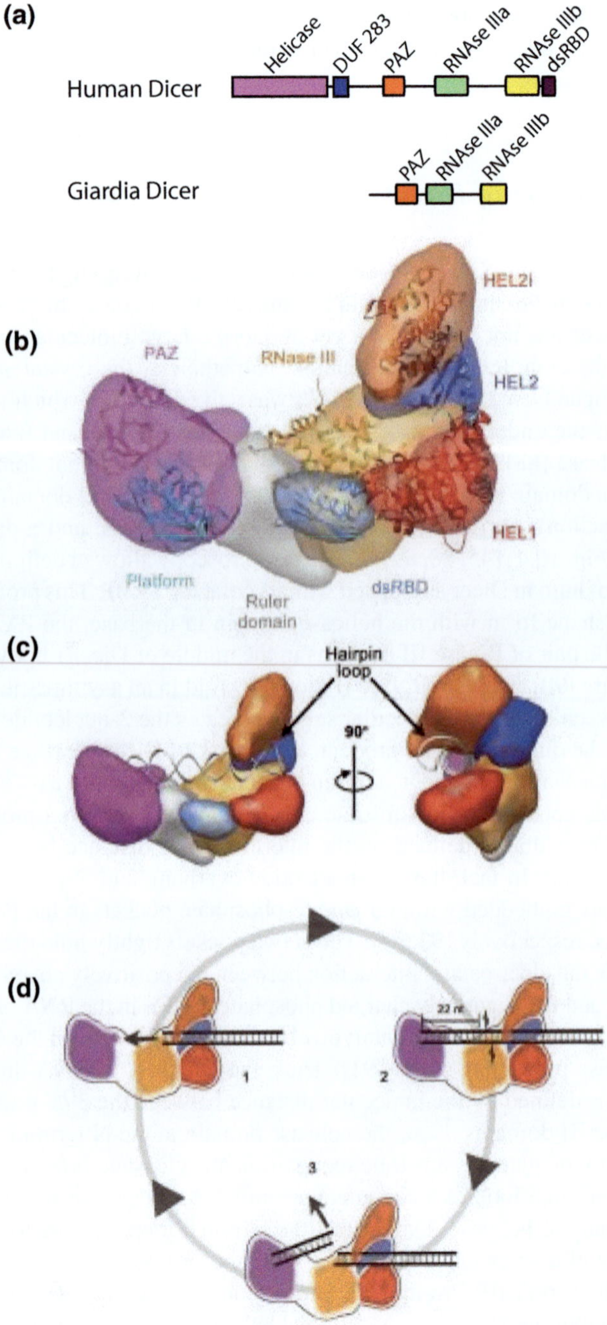
(a)
Helicase
DUF 283
PAZ
RNAse IIIa
RNAse IIIb
dsRBD
Human Dicer
PAZ
RNAse IIIa
RNAse IIIb
Giardia Dicer
(b)
PAZ
RNase III
HEL2i
HEL2
HEL1
Platform
dsRBD
Ruler domain
(c)
Hairpin loop
90°
(d)
22 nt
1
2
3

◀**Fig. 10.4 Architecture and Mechanism of Dicer.** **a** Schematic representation of the primary sequence of human and Giardia Dicers. *Giardia intestinalis* Dicer contains PAZ and tandem RNase III domains, but lacks the N-terminal DExD/H helicase, C-terminal double-stranded RNA-binding domain (dsRBD), and extended inter-domain regions associated with Dicer in higher eukaryotes. **b** Segmented map of human Dicer with crystal structures of homologous domains docked. **c** Model for pre-miRNA recognition. A pre-miRNA hairpin is modelled into the proposed binding channel of Dicer with the stem-loop fit in the RNA-binding cleft of the helicase. **d** Schematic for processive dicing: [1] The helicase translocates dsRNA into the nuclease core. [2] The PAZ domain (purple) recognizes the dsRNA end, positioning RNase III (orange) for cleavage. [3] The siRNA product is released while the dsRNA substrate remains bound to the helicase. Adapted with permission from Pick-Wei Lau et al., NSMB, 2012 [92]

sequent conformational rearrangement of the helicase domain presumably unlocks the RNA substrate, allowing the helical structure of the RNA to reach the RNase III pair for processing. This conformational change may mediate the transition from the pre-dicing state to the dicing state [97].

10.5.2 Dicer Cofactors

In addition to its own multi-domain structure, Dicer often interacts with other dsRNA-binding proteins (dsRBPs) that have been shown to enhance its cleavage efficiency and accuracy [98–101]. TRBP (TAR RNA-binding protein) is the most studied Dicer partner in the context of microRNA biogenesis. Dicer also interacts with other dsRNA-binding proteins such as PACT (protein activator of protein kinase R) [98] and the RNA editor ADAR 1 (*Adenosine Deaminase Acting on RNA 1*), which assists Dicer during miRNA/siRNA processing [102].

TRBP is a 39 KDa protein that possesses three consecutive dsRBDs interconnected with each other through long and flexible linkers [100, 103]. It is very common that dsRBPs contain several dsRBDs, which act on dsRNA cooperatively. Truncation or mutations in one of the dsRBDs have been shown to greatly impair the dsRNA-binding activity of this protein family [104, 105]. The first two dsRBDs of TRBP (dsRBD1 and dsRBD2) are located at the N-terminus and adopt a canonical $\alpha\beta\beta\beta\alpha$ topology responsible for the dsRNA-binding activity [103, 106], whereas the third domain (dsRBD3) lost its dsRNA-binding activity and often mediates the interaction with other protein partners including Dicer [100, 107]. Doudna and co-workers recently solved the crystal structure of the interface between Dicer and TRBP [100]. dsRBD3 of TRBP mediates the interaction with Dicer through its PBD domain (partner-binding domain) (Fig. 10.5). In this scenario, the remaining two dsRBDs of TRBP are free to interact with Dicer's RNA substrate through a classical recognition of two consecutive minor grooves of an A-form RNA helix (Fig. 10.6) [108–110]. These interactions between TRBP's dsRBD12 and the pre-miRNA have been demonstrated to participate in the strand selection process [75] and more importantly to fine-tune the cleavage accuracy of Dicer [100, 101]. Of note, it is important

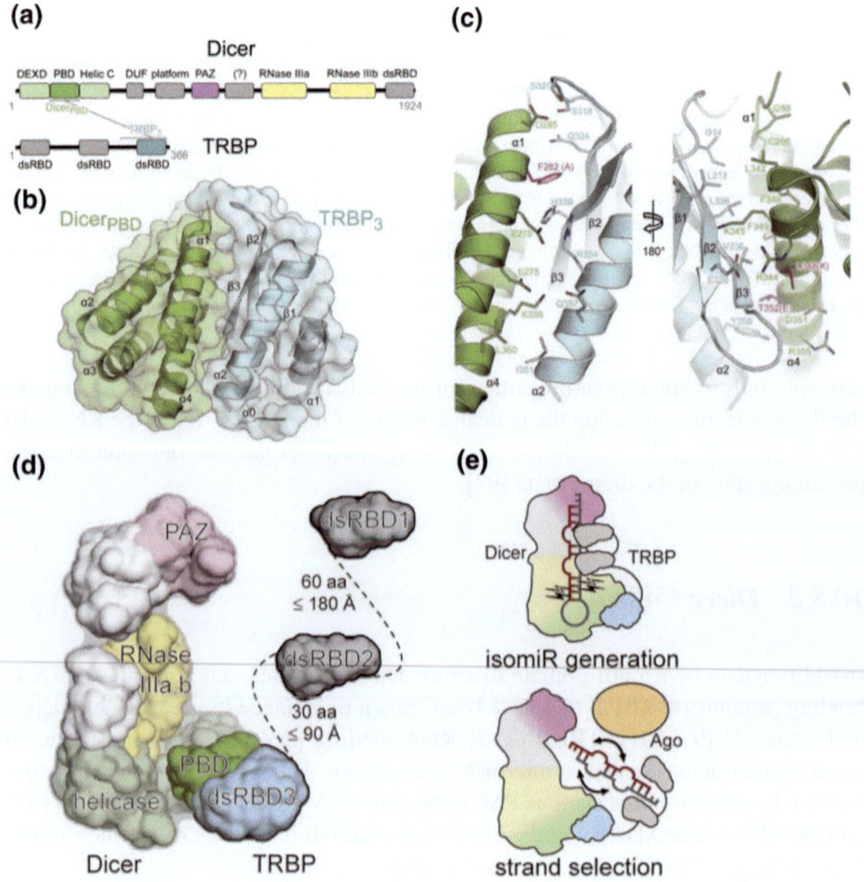

Fig. 10.5 Structure of the Dicer-TRBP Interface. a Cartoon representation of the primary sequence of Dicer and TRBP with brackets indicating the interacting domains. **b** Overlaid backbone cartoon and surface representations of the Dicer partner-binding domain (PBD) and the third dsRBD of TRBP. **c** Front and back views with interfacial residues shown. Dicer residues mutated to abrogate TRBP/PACT binding are shown in pink with resulting residues indicated in parentheses. **d** The human Dicer architecture (as determined by electron microscopy) is coloured according to functional domains (PAZ, pink; RNase IIIa/b, yellow; helicase, green), with the Dicer-TRBP interface structure determined in the present crystallographic work shown in dark green (Dicer-PBD) and cyan (TRBP3). NMR results suggest that the two N-terminal RNA-binding domains of an extended TRBP can readily access an RNA bound near the paired RNase III active sites of Dicer. **e** Models for how Dicer partner proteins contribute to isomiR formation (top) and strand selection fidelity during transfer of a Dicer product duplex to Argonaute (bottom). Adapted with permission from Wilson et al., Mol Cell, 2015 [100]

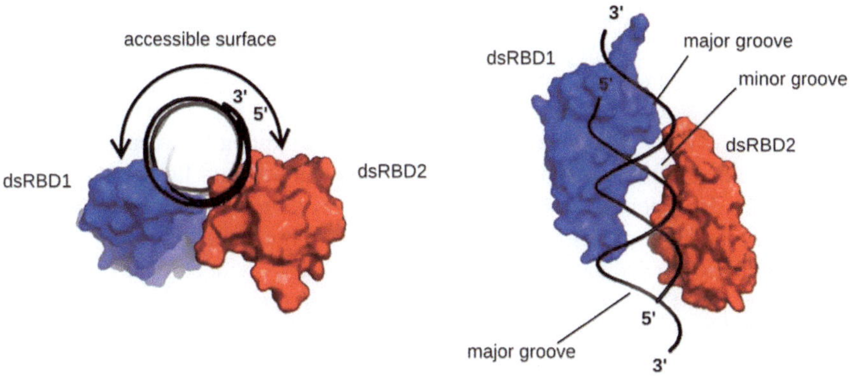

Fig. 10.6 Three-dimensional structures of TRBP's dsRBD1 and dsRBD2 in complex with dsRNA. Global view of the dsRBD12-dsRNA complex showing the half-cylinder RNA region left solvent-exposed, and potentially accessible to other proteins. dsRBD1 and dsRBD2 surfaces are represented in blue and red, respectively. dsRNA is represented as a black cartoon. Adapted with permission from Masliah et al., EMBO J, 2018 [108]

to regulate the cleavage accuracy of Dicer to avoid the production of various miRNA isoforms. Shifts in Dicer cleavage sites by only 1 nucleotide can revert the fate of the guide and passenger strands due to potential changes in the thermodynamic stability of the duplex ends. Furthermore, these shifts in cleavage sites can also change the whole 'seed' sequence and redefine the landscape of mRNA targets of a given miRNA. Both mechanisms listed above would lead to a disastrous off-target effect in case of inconsistency in Dicer cleavage sites. Likely, Dicer relies on its dsRBP cofactors to generate microRNA with a consistent length and a well-defined 'seed' region to avoid such collateral off-targeting. TRBP binding to a pre-miRNA seems to be rather cooperative than competitive in regard to Dicer binding, as supported by bulk cleavage assays containing the ternary complex [65, 108]. Thus, Dicer and TRBP within the same protein complex are likely to bind distinct motifs on pre-miRNA rather than competing for the same RNA features.

Biochemical and structural data revealed the main domains of Dicer and cofactors that orchestrate pre-miRNA cleavage. Such important information paved the way for a dynamic understanding of how Dicer and dsRBDs of its cofactors can coordinate the substrate recognition and cleavage processes. Single-molecule approaches are excellent tools to address such questions and get a real-time view of pre-miRNA recognition and cleavage.

10.5.3 Dynamic Binding of TRBP to dsRNA

Myong and co-workers employed various single-molecule approaches to understand how Dicer and TRBP interrogate an RNA substrate. First, they used smFRET (single-molecule Förster resonance energy transfer) to probe the molecular interactions between TRBP and various RNA molecules containing double-stranded motifs [111]. smFRET is an outstanding and widely used approach to probe protein–nucleic acid interactions at high spatiotemporal resolution [112–114]. The energy transfer (FRET) occurs through non-radiative dipole–dipole coupling between a donor fluorophore in an electronic-excited state and an adjacent acceptor fluorophore. FRET is a distance-dependent process and is suitable for revealing the molecular dynamics of nucleoprotein complexes that are in the range of 3–8 nanometre.

The authors labelled and surface-immobilized dsRNA with a donor fluorophore (Cy3, green) and TRBP with an acceptor fluorophore (Alexa 647, red). When the dye-labelled TRBP was introduced into the imaging chamber, rapid fluctuations between two FRET states (0.3 and 0.8) were observed, reflecting a dynamic movement of TRBP along the dsRNA (Fig. 10.7). This diffusion behaviour of TRBP appears to be strictly dependent on the presence of dsRNA motifs on the RNA investigated and was not observed with single-stranded RNA, RNA-DNA duplex, or very short dsRNA (shorter than 15 bp). TRBP requires dsRNA motif longer than 12 bp to engage a noticeable interaction with a given RNA molecule and dsRNA motifs longer than 15 bp to translocate from one end to another end of the dsRNA. Mutation analysis revealed that the diffusion behaviour of TRBP is dependent on both dsRBD1 and dsRBD2 since their truncation abrogated its ability to bind and move along the dsRNA, thus highlighting a cooperative interaction between the two dsRBDs at the N-terminus of TRBP. Conversely, dsRBD3 was dispensable and did not affect dsRNA-binding activity, neither the diffusion behaviour of TRBP [98]. This conclusion is in line with previous reports supporting a model in which the dsRBD3 had lost its RNA-binding activity throughout the evolution and retained protein–protein assembly as main activity [100, 107, 115, 116]. The authors confirmed the diffusion behaviour of TRBP by distinct single-molecule approaches such as three-colour smFRET and smPIFE (single-molecule protein-induced fluorescence enhancement). In the three-colour smFRET assay, two acceptor dyes (Cy5 and Cy7) were attached to opposite ends of dsRNA, and the translocation of Cy3-labelled TRBP (donor) was evident from the anti-correlation in the FRET intensity of Cy5 and Cy7 fluorophores. This observation corroborates that TRBP does diffuse along the dsRNA from one end to another.

smPIFE is also an interesting technique to track dynamic movements of proteins with 0–4 nanometre resolution, making it a complementary approach in order to fill the gap in the sensitivity of smFRET (3–8 nM) [117, 118]. This one-colour assay shows fluorescence enhancement when a protein comes to the vicinity of a dye covalently attached to a nucleic acid. The authors again observed a fluctuation in fluorescence intensity due to the introduction of an unlabelled TRBP into the imaging chamber, supporting the diffusion behaviour observed in smFRET assays. It is important to note that other dsRBPs with a tandem dsBBDs (e.g. PACT, R3D1-L, Staufen1) also exhibited similar diffusion behaviours, which highlight well-conserved and perhaps important biological functions related to the diffusion of dsRBPs in RNAi phenomenon and other biological processes [111, 119, 120].

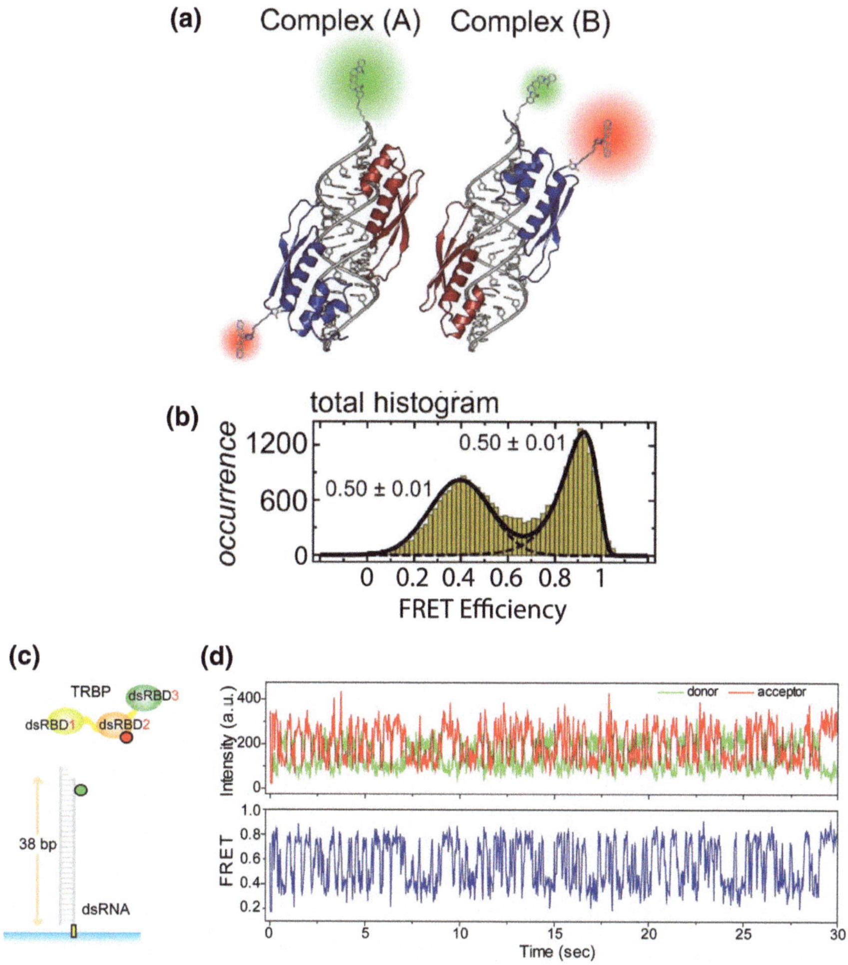

Fig. 10.7 TRBP's interaction with dsRNA at the single-molecule level. **a** Cartoons of labelled dsRBD12 of TRBP bound to Cy3B-labelled dsRNA. The different domain arrangements of dsRBD12 on dsRNA are characterized by different inter-dye distances, which were approximated as distances between their attachment points. **b** FRET efficiency histograms of dye-labelled dsRBD12 in complex with Cy3B-labelled dsRNA. Transfer efficiency histograms exhibit two subpopulations that are equally likely to occur. Errors associated with relative occurrences correspond to the standard deviation. This single-molecule FRET together with structural data support that dsRBD1 and dsRBD2 can bind a dsRNA in two symmetric orientations. Adapted with permission from Masliah et al., EMBO J, 2018 [108]. **c** Schematic representation of the single-molecule FRET assay in which Alexa 647-labelled TRBP (red) was added to an immobilized Cy3-labelled dsRNA (green), and their interaction was visualized by TIRF microscopy. **d** Time trajectory showing the nature of the interaction between TRBP and dsRNA. Repetitive FRET fluctuation was observed at the single-molecule level without TRBP dissociation from dsRNA, reflecting a repetitive distance change between TRBP and the end of dsRNA. Adapted with permission from Koh et al., PNAS, 2013 [111]

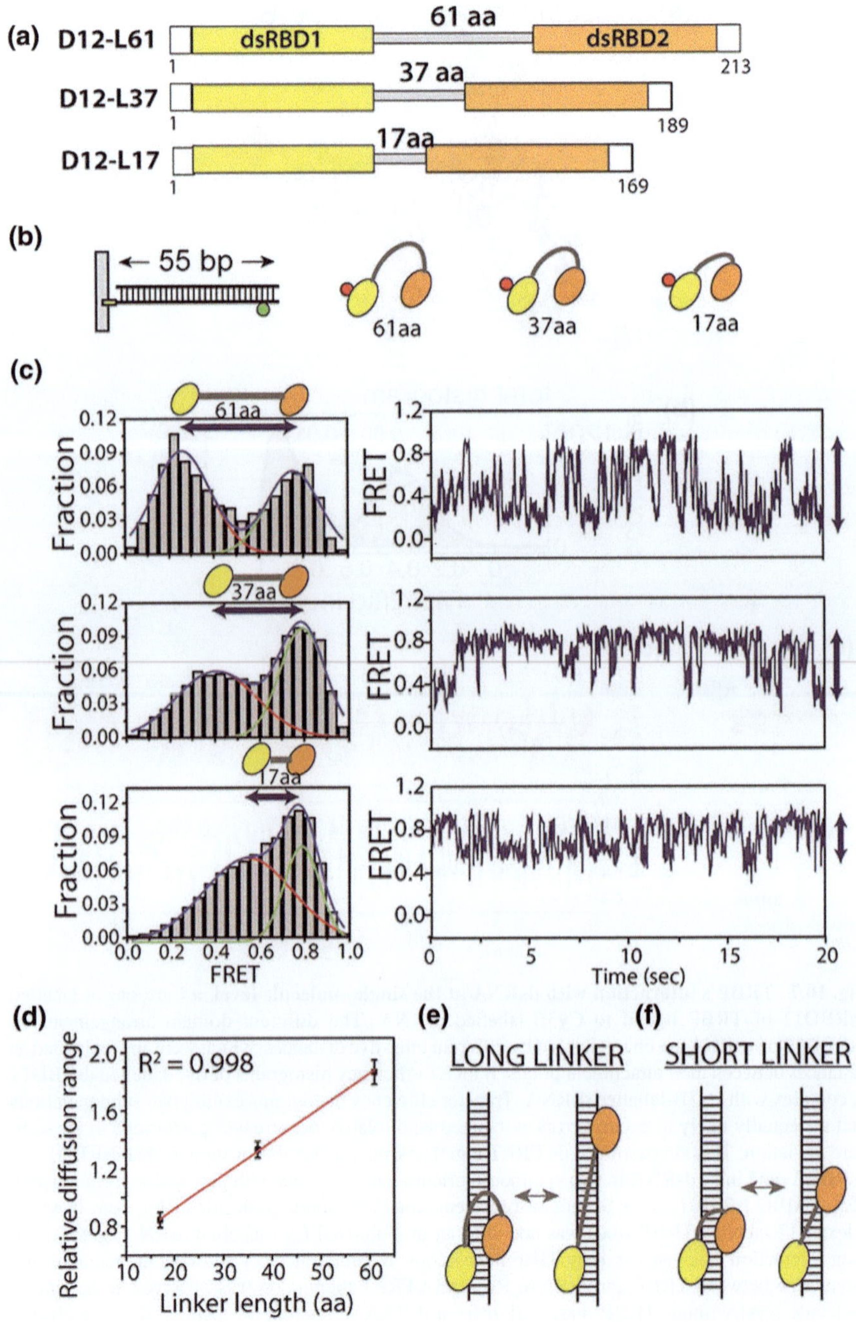
(a)
D12-L61
dsRBD1
61 aa
dsRBD2
1
213
D12-L37
37 aa
1
189
D12-L17
17aa
1
169
(b)
55 bp
61aa
37aa
17aa
(c)
61aa
37aa
17aa
Fraction
FRET
Time (sec)
(d)
R² = 0.998
Relative diffusion range
Linker length (aa)
(e)
LONG LINKER
(f)
SHORT LINKER

◄**Fig. 10.8 Length of the linker between dsRBD12 of TRBP controls the diffusion distance.** **a** Schematic representation of dsRBD1-2 variants used in this study. The length of the linker connecting dsRBD1 and dsRBD2 was reduced by mutagenesis. The linker in the constructs D12-L61, D12-L37 and D12-L17 contains 61, 37 and 17 amino acids, respectively. **b** Schematic representation of single-molecule FRET assay used to probe the linker length-dependence movement of dsRBD1-2 along dsRNA. **c** FRET histogram obtained from three measurements with varying linker lengths. **d** Linker length plotted against diffusion distance deduced from FRET values. **e**, **f** Schematic representation of D12 diffusion for long vs short linker length. The diffusion distance is highly correlated with the length of the linker connecting dsRBD1-2. A short linker constrains the movement of dsRBD1-2 along dsRNA. Adapted with permission from Koh et al., JACS, 2017 [120]

In a recent study, Myong and colleagues combined smFRET and smPIFE to understand how dsRBD1-2 cooperatively coordinates the movement of TRBP along the dsRNA. It remains unclear whether these two domains form a rigid structure that moves simultaneously on the dsRNA, or the structure of TRBP is flexible enough to allow for an independent movement of the two dsRBDs along the dsRNA substrate. Of note, structural data have shown that dsRBD1 and dsRBD2 are interconnected through a long and flexible linker, which likely support the second scenario [100]. The co-labelling of TRBP's dsRBD1 and dsRBD2 with two distinct dyes (Cy3 and Cy5) together with the use of smFRET was a key experiment to probe these two models. The binding to an unlabelled and surface-immobilized dsRNA was evident from the sudden appearance of the fluorescence on the imaged area. This binding exhibited a fluctuation in FRET efficiency between a low FRET state (open conformation) and a high FRET state (closed conformation). FRET histogram further displayed two broad peaks, consistent with the fluctuations in FRET that had been observed in the time trajectories and reflected a transition between a closed and open protein conformation (Fig. 10.8) [120]. These data strongly support the second model in which TRBP's dsRBD1 and dsRBD2 diffuse independently along the dsRNA. Given the dsRBD1 and dsRBD2 are interconnected through a linker, the freedom of independent movement of these two domains is expected to be restricted by the length of the linker. Indeed, when the length of the linker was shortened from 61 to 37 and 17 amino acids, the amplitude of the diffusion was greatly reduced as evidenced by shifts in the lower FRET peak from 0.25 to 0.4 and 0.55, respectively (Fig. 10.8). These data support that the length—and perhaps the flexibility—of the linker does constrain the distance on the dsRNA that dsRBD1 and dsRBD2 can explore independently [120].

It is unclear to what extent the dynamic movement of TRBP on dsRNA would be beneficial to the pre-miRNA processing by Dicer. In light of previous reports that showed an important role of TRBP in the regulation of Dicer cleavage accuracy, we may speculate that the diffusion along dsRNA might reflect attempts in which TRBP is positioning or transferring pre-miRNA to Dicer domains for accurate cleavage. The authors addressed this question and observed that when TRBP was associated with Dicer in a single protein complex, the translocation behaviour of TRBP decreased, and 40% of the binding became stable [60]. This might indicate that Dicer dictates the nature of the interaction by stabilizing the binding of dsRNA substrate within

Dicer-TRBP complex. The cease of TRBP diffusion could also reflect the transfer of pre-miRNA from TRBP to Dicer and a conformational change to a cleavage-competent state, where the 5′/3′ ends, terminal loop and stem of pre-miRNA are tightly held by the PAZ, helicase and RNase III domains, respectively. Such picture would fit a model supported by new biochemical and structural studies [100, 108].

10.5.4 Pre-miRNA Recognition in the Crowded Cellular Environment

It is intriguing that microRNA represents only a minor fraction of cellular RNA (0.01%), yet Dicer-TRBP complex succeeds in finding and cleaving pre-miRNA among approximatively 360,000 other competing RNAs in various cellular compartments (Fig. 10.9) [121]. Substrate recognition process is critical for the survival of the cell since collateral cleavage of structured RNA such as ribosomal RNA, tRNA or mRNA that carry important structural and functional information would have drastic consequences on the homeostasis of the cell. Therefore, Dicer-TRBP complex must

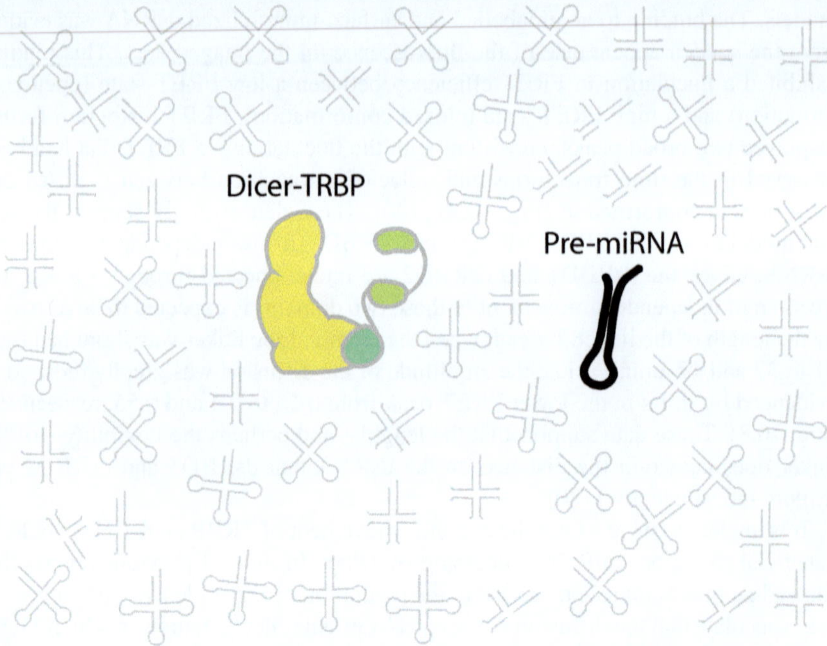

Fig. 10.9 Illustration of the RNA-crowded environment in the cell. Dicer-TRBP complex is constantly challenged by competitor structured RNAs that share structural homology with pre-miRNA. Dicer-TRBP complex has a very efficient and accurate substrate recognition mechanism to avoid cleaving structured RNAs carrying important structural and functional information

have an efficient and accurate substrate recognition process to unmask authentic pre-miRNA from other pre-miRNA-like cellular RNAs.

We sought to understand how Dicer and TRBP coordinate the substrate recognition process using biochemical and single-molecule fluorescence assays [122]. The advanced single-molecule techniques allow for a real-time tracking of the substrate selection process employed by Dicer-TRBP. To achieve this goal, it is crucial to purify homogeneous and functional protein assemblies containing Dicer and TRBP for the single-molecule analysis. The development of a unique pull-down assay to isolate functional Dicer-TRBP protein complexes from HEK 293T human cells was required, as this protein complex cannot be expressed in conventional bacterial expression systems. A gentle lysis, tandem purification and in vivo site-specific biotinylation of Dicer protein were key steps towards the purification of intact and functional protein complexes, and their immobilization on the surface of a microfluidic device for single-molecule imaging (Fig. 10.10) [36, 122]. We designed dye-labelled pre-miRNA molecules to record their interactions with surface-immobilized Dicer-TRBP with high temporal resolution (100–300 ms). The protein complexes were spatially separated and distinguishable within a diffraction limit, thus allowing tracking the performances of Dicer-TRBP proteins one at the time. This single-molecule assay showed that Dicer-TRBP was able to bind and process different species of pre-miRNAs harbouring the canonical stem-loop structure. When the RNA-binding activity of Dicer lacking its partner TRBP was tested, the number of binding events was reduced by one order of magnitude, thus confirming that Dicer's dsRBP cofactors enhance its RNA-binding activity [98, 99, 122].

Real-time observation of the recognition process by Dicer-TRBP in pre-steady-state condition exhibited two distinct binding behaviours. Half of the interactions were short-lived with an average lifetime of 1.5 s, while the second half of the population exhibited a more prolonged interaction with a lifetime of approximately 15 s (Fig. 10.11). The short binding observed here is unlikely to be productive and might reflect aborted tentative binding in which Dicer-TRBP failed to transfer the substrate towards the catalytic centre of the enzyme. An inverted orientation of the pre-miRNA during its initial recruitment by Dicer-TRBP is anticipated to yield non-productive interactions and could explain the short-lived binding events. Indeed, the encounter between Dicer-TRBP complex and pre-miRNA is expected to be stochastic since both the enzyme (Dicer-TRBP) and the substrate (pre-miRNA) freely diffuse within a given 3D area in the cytoplasm, and their encounter is mainly defined by random walks. The stretched and flexible structure of TRBP together with its high dsRNA-binding affinity renders this protein a potent candidate to trap and engage an initial contact with a nearby pre-miRNA before the transfer to Dicer. In fact, experimental data have shown that TRBP greatly increases the dsRNA-binding activity of Dicer [98, 99, 122]. The binding to TRBP is driven only by the interaction with the dsRNA region of pre-miRNA, given that TRBP is unable to recognize the termini of a structured RNA. Therefore, it is plausible that pre-miRNA gets loaded into TRBP with two distinct orientations. In fact, Masliah and co-workers used various biophysical approaches including NMR, EPR and single-molecule FRET to demonstrate that TRBP's dsRBD1 and dsRBD2 can swap their binding sites and associate with dsRNA in two distinct pseudo-symmetrical complexes, which is evidenced by the two FRET

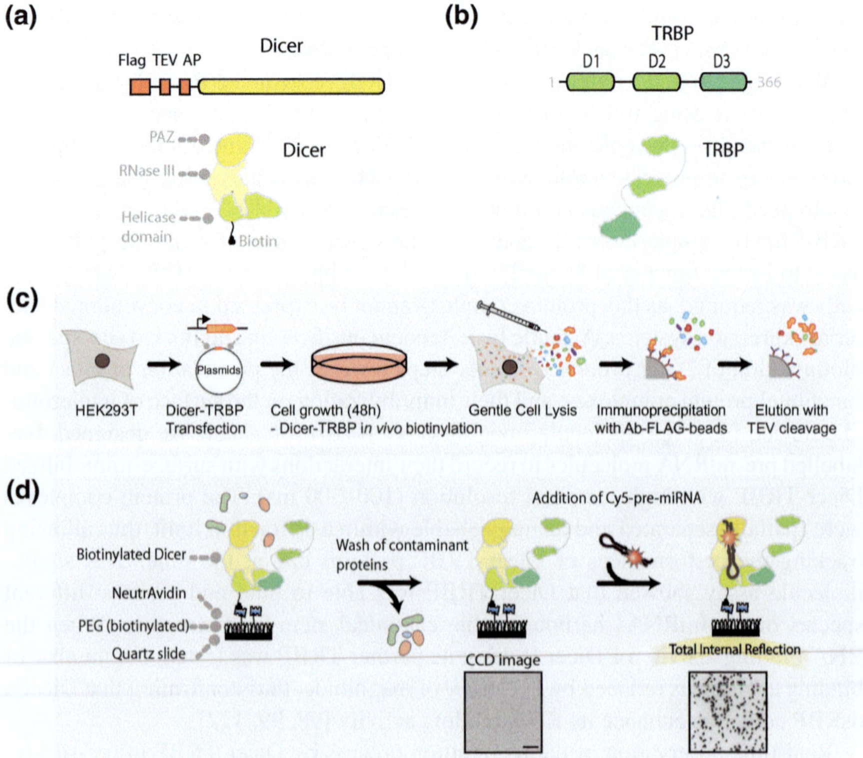

Fig. 10.10 Single-molecule pull-down assay to visualize the substrate recognition process employed by Dicer-TRBP complex. a Schematic representation of the engineered Dicer construct used in the single-molecule pull-down assay. FLAG, TEV and AP tags were used for immunoprecipitation, elution and in vivo biotinylation. **b** Schematic representation of TRBP used in this assay. **c** Workflow chart of the protein expression and purification for single-molecule pull-down assay. **d** Workflow chart for single-molecule fluorescence assay to capture pre-miRNA recognition by Dicer-TRBP with high spatiotemporal resolution. Adapted with permission from Fareh et al., Methods, 2016; Fareh et al., Nature Communications, 2016 [36, 122]

populations reported (Fig. 10.7) [108]. This observation supports that TRBP might help loading a dsRNA into Dicer enzyme in two distinct orientations. In contrast to TRBP, Dicer's PAZ and helicase domains have the capability to discriminate between 5′/3′ ends and the terminal loop, and consequently, are able to distinguish a well oriented from misoriented pre-miRNA. Dicer-TRBP complex is, therefore, expected to reject a misoriented RNA substrate since the PAZ domain and the helicase domain cannot engage stable interactions with their canonical substrates, namely 5′/3′ ends and terminal loop. It is likely that the quickly aborted binding events observed in pre-steady-state single-molecule assay reflect a misoriented substrate being recruited by TRBP and rejected by Dicer after the initial probing (Fig. 10.11).

10.5.4.1 TRBP-PAZ Coordinate the Substrate Recognition

In an RNA-crowded environment such as the cytoplasm, random collisions between Dicer and abundant RNA species constantly challenge the fidelity and the turnover of this enzyme. PAZ domain contains a 3′ pocket to recognize the 3′ 2-nucleotide overhang on a pre-miRNA and has been proposed to be an important player in the substrate recognition process. Conversely, the recognition of dsRNA motifs on pre-miRNA by dsRBD1-2 domains of TRBP appear to be crucial for the substrate recognition process, given their truncation compromises the dsRNA-binding affinity of the protein complex [122]. For a decade, it remained elusive how TRBP and the PAZ domain coordinate the substrate recognition process due to the lack of high-resolution data.

Blocking the interaction between PAZ and the 3′ end of a pre-miRNA was a key approach to comprehend how and when the PAZ domain gets involved in this substrate recognition. In this assay, the recognition of 3′ 2-nucleotide overhang by the PAZ domain was impaired through the attachment a small chemical group (biotin) to the 3′ end of pre-miRNA. This modified pre-miRNA was labelled with Cy3 dye (green) and introduced into a microfluidic chamber containing surface-immobilized Dicer-TRBP. As a control, an equal amount of Cy5-labelled pre-miRNA with an unmodified 3′ end was simultaneously introduced in the same imaging chamber while recording the interactions in a two-colour assay (Fig. 10.12). The analysis of time trajectories obtained from this assay allowed to visualize single Dicer-TRBP at the work and unrevealed how this enzyme can discriminate between two RNA substrates sharing extensive structural features. Strikingly, approximatively 85% of the interactions with pre-miRNA containing blocked 3′ end (green) were short-lived and got rejected after less than 1 s of probing, while a large fraction of the encounters with the canonical pre-miRNA achieved a stable and productive binding. The ~1-s of interaction indicates that Dicer-TRBP complex does interact with double-stranded RNA molecules in a termini-independent manner. This initial interaction reflects the timeframe of substrate sensing used by Dicer-TRBP. In fact, cleavage bulk assay demonstrated that the short interaction with pre-miRNA containing a modified 3′ end is a non-productive binding, as this substrate could not get cleaved into mature miRNA. TRBP itself is unable to recognize the 3′ end of a pre-miRNA. However, the PAZ domain can sense the 3′ end of a freshly recruited dsRNA and dictates the nature of the interaction. Canonical pre-miRNA that are well-oriented will be fully transferred to Dicer's body for cleavage within 15 s timeframe, while non-canonical and misoriented pre-miRNAs are rapidly rejected to avoid compromising the turnover of this enzyme. This rapid rejection implements Dicer-TRBP with a time-efficient and accurate substrate recognition mechanism to avoid non-productive engagement with non-canonical and misoriented pre-miRNA (Fig. 10.12).

Surprisingly, Dicer lacking its partner TRBP failed in rejecting non-canonical pre-miRNA and remained associated with those non-optimal substrates for a long time. This non-productive long binding does compromise Dicer turnover in a RNA-crowded environment as demonstrated by bulk cleavage assays containing an excess of competitor RNA species. These data highlight that TRBP and the PAZ domain

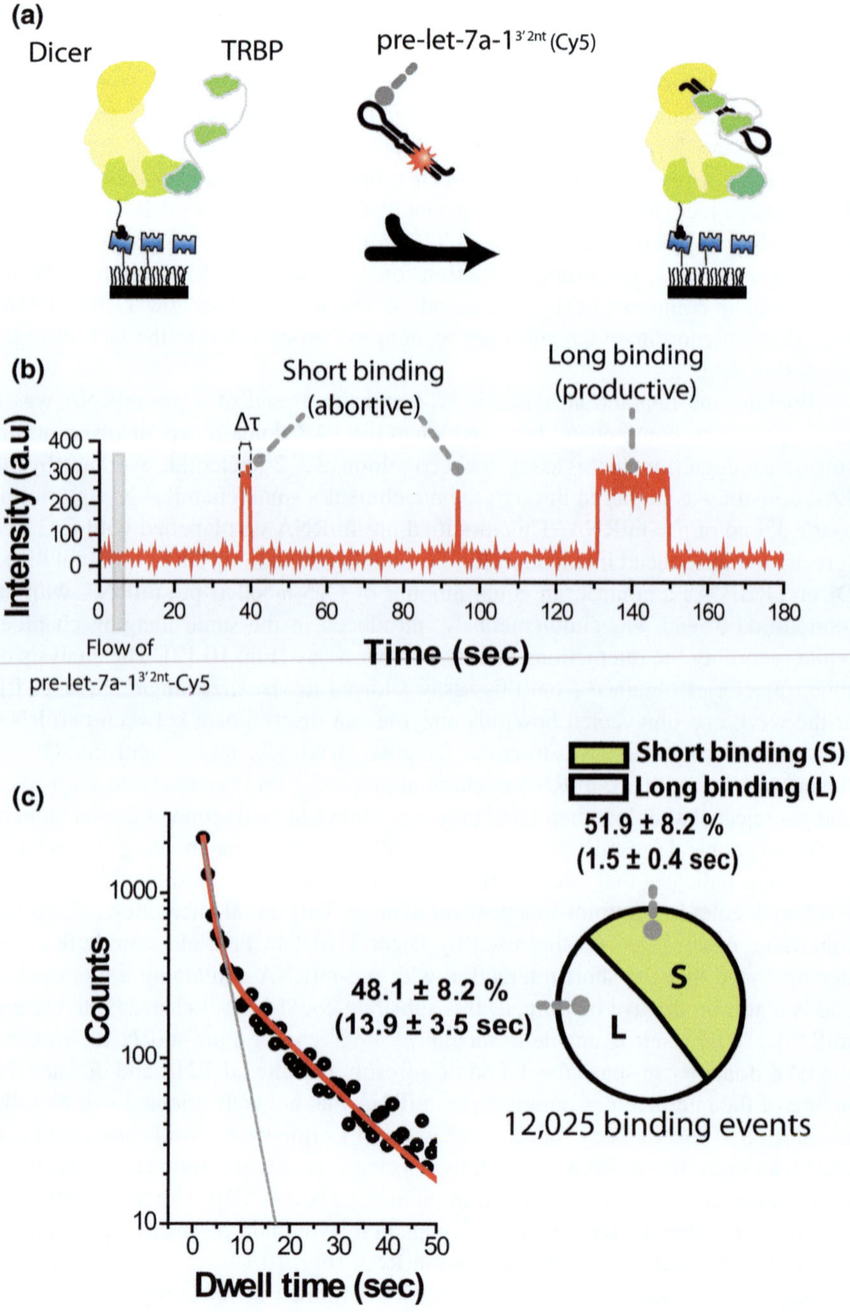
(a)
Dicer
TRBP
pre-let-7a-1 3′2nt (Cy5)
(b)
Short binding (abortive)
Long binding (productive)
Δτ
Intensity (a.u)
Time (sec)
Flow of pre-let-7a-1 3′2nt-Cy5
(c)
Short binding (S)
Long binding (L)
51.9 ± 8.2 % (1.5 ± 0.4 sec)
48.1 ± 8.2 % (13.9 ± 3.5 sec)
Counts
Dwell time (sec)
S
L
12,025 binding events

◂**Fig. 10.11 Real-time observation of pre-miRNA recognition by the Dicer-TRBP complex.** **a** Schematic representation of a single-molecule assay to capture pre-miRNA recognition by the Dicer-TRBP complex in real time. **b** Representative time trace (a time resolution 300 ms) exhibiting recognition of multiple Cy5-labelled pre-let-7a-13′ 2nt by a single Dicer-TRBP complex. The dwell-time ($\Delta\tau$) is the time between docking and dissociation. Cy5-labelled pre-let-7a-13′ 2nt was added at time 5 s. **c** Dwell-time histogram derived from binding events recorded in a pre-steady-state condition. The distribution was fitted with a double exponential decay (red line). The dashed grey line is a fit to a single exponential decay. The pie chart displays the ratio between short binding ($\Delta\tau$short $= 1.5 \pm 0.4$ s, green) and long binding ($\Delta\tau$long $= 13.9 \pm 3.5$ s, white). The error is the SD of four independent measurements. The two populations of binding reported reflect distinct binding modes of Dicer-TRBP to pre-miRNA. The short binding may represent a quick release of a misoriented substrate after the initial probing, while the long binding reflects sensing, cleavage and product release Adapted with permission from Fareh et al., Nature Communications, 2016 [122]

are required for the rapid rejection mechanism that allows Dicer-TRBP to possess a high fidelity and turnover during the substrate recognition process. As a model, we propose that TRBP acts as a gatekeeper precluding Dicer from engaging long and non-productive bindings with non-canonical pre-miRNA (Fig. 10.13) [122].

10.5.4.2 Energy Landscape of Substrate Recognition

The energy landscape of Dicer–RNA interactions provides a comprehensive understanding of this substrate recognition mode (Fig. 10.13). The short binding mode ((i) in the landscape) reflects the entry of RNA to Dicer, which leads to a long binding mode (ii) and consequently to the cleavage-competent state of the enzyme if the pre-miRNA is well oriented (iii). Non-canonical pre-miRNA exhibited a short binding mode. Dicer-TRBP can shortly interact with any dsRNA, yet only canonical pre-miRNA bypasses this entry checkpoint (i) and reaches the long binding mode (ii). When Dicer was challenged with three orders of magnitude excess of competitor RNAs, the long and short binding modes were equally influenced, indicating that these two binding modes are likely to be on the same reaction pathway.

The 3′ end of RNA is first recognized by the PAZ domain of Dicer in the short binding mode (i) within approximately one second. If the RNA has the canonical 3′ 2-nt overhang, it gets transferred to a more stable binding mode (ii). TRBP lowers the energy barrier between the free RNA state and the first binding state (i). The high RNA-binding affinity of TRBP also deepens the energy level of (i), which makes RNA difficult to go over the barrier between (i) and (ii). It is likely that these two alternations in the energy landscape make RNA more readily associated with Dicer and also prevents other cellular RNA from falling into (ii) as reflected by the rapid rejection mechanism. Other RNA-processing enzymes that partners with dsRBP are anticipated to follow similar recognition process to discriminate real substrates from other competitor RNAs.

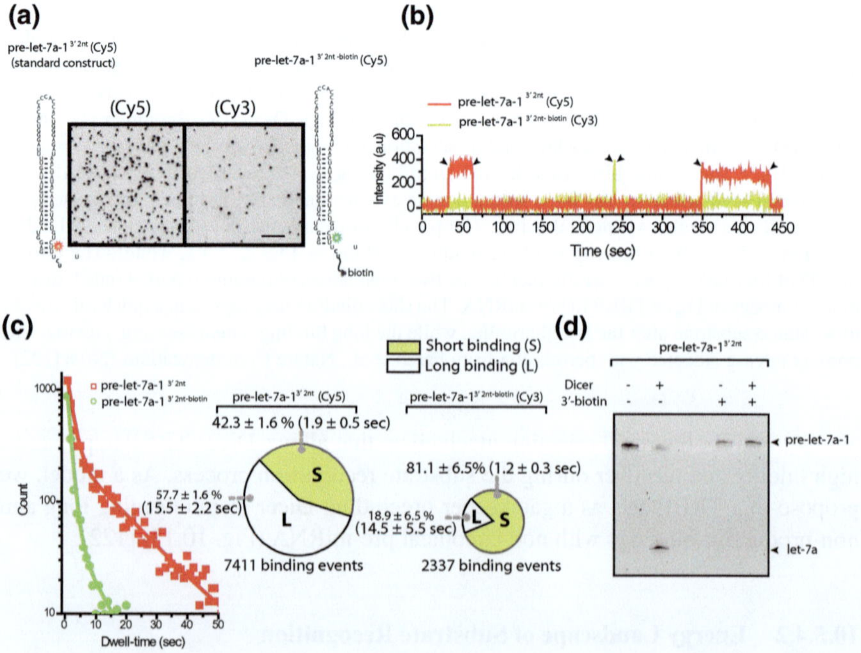

Fig. 10.12 TRBP and PAZ domain coordinate the substrate selection process. a Two-colour competition assay to track the recognition of two different substrates by Dicer-TRBP. The standard pre-let-7a-1$^{3'\ 2nt}$ was labelled with Cy5. A biotin group was attached to the 3′ overhang of Cy3-labelled pre-let-7a-1 (pre-let-7a-1$^{3'\ 2nt-biotin}$). The CCD images show docking of standard pre-let-7a-1$^{3'\ 2nt}$ (left) and pre-let-7a-1$^{3'\ 2nt-biotin}$ (right). Scale bar, 5 μm. **b** Representative time trajectory (time resolution 300 ms) showing long binding of two standard pre-let-7a-1$^{3'\ 2nt}$ substrates (Cy5, red) and short binding of one pre-let-7a-1$^{3'\ 2nt-biotin}$ substrate (Cy3, green) to a single Dicer-TRBP complex. **c** Dwell-time histograms derived from binding of standard pre-let-7a-1$^{3'\ 2nt}$ (red) and pre-let-7a-1$^{3'\ 2nt-biotin}$ (green). The distributions were fitted with a double exponential decay. The pie chart in the left displays the percentage of short binding ($\Delta\tau$short = 1.9 ± 0.5 s, green) and long binding ($\Delta\tau$long = 15.5 ± 2.2 s, white) obtained with Cy5-labelled standard pre-let-7a-1$^{3'\ 2nt}$. The pie chart in the right displays the percentage of short binding ($\Delta\tau$short = 1.2 ± 0.3 s, green) and long binding ($\Delta\tau$long = 14.5 ± 5.5 s, white) obtained with Cy3-labelled pre-let-7a-1$^{3'\ 2nt-biotin}$. The size of the pie charts is proportional to the total number of binding events. The error is the SD of four independent measurements. These data show that a pre-miRNA with a non-canonical 3′end is rapidly rejected by Dicer-TRBP, while pre-miRNA with canonical 3′end achieves long and productive binding. **d** In vitro cleavage of standard pre-let-7a-1$^{3'\ 2nt}$ (left) and pre-let-7a-1$^{3'\ 2nt-biotin}$ (right) by wild-type Dicer-TRBP. The top arrow indicates pre-let-7a-1$^{3'\ 2nt}$, and the bottom arrow indicates a cleaved product (mature let-7a). This bulk cleavage experiment demonstrates that the short binding is not productive, but reflects the probing time of an RNA substrate by TRBP-PAZ domain. Adapted with permission from Fareh et al., Nature Communications, 2016 [122]

In line with this model, recent cryo-EM structure of hDicer-TRBP-pre-miRNA ternary complex has revealed that this nucleoprotein complex adopts two distinct conformations: a pre-dicing state where the RNA substrate closely interacts with PAZ, the helicase and TRBP's dsRBD12 but positioned away from the catalytic centre of hDicer. Conformational changes are believed to dictate the transition to a dicing competent state where the helical structure of pre-miRNA gets transferred to the proximity of hDicer catalytic centre for the cleavage [97].

10.6 Concluding Remarks

This chapter has discussed the journey of microRNA from the initial transcription in the nucleus by RNA polymerases, to the processing by Drosha and the export through the nuclear pore complex, to the final maturation step in the cytoplasm by the endoribonuclease Dicer, which generates a functionally active guide microRNA. Recent structural and single-molecule fluorescence studies revealed the molecular basis governing the biogenesis of microRNA with high spatial and temporal resolution. These milestone investigations have revealed hidden steps and intermediate conformations that are difficult to obtain by conventional biochemical approaches.

In the coming years, it is inevitable that single-molecule approaches will move forward and conquer a more physiologically relevant environment such as the cell and its organelles, where the dynamic interactions between microRNAs and their processing enzymes are controlled by various intracellular and extracellular stimuli. Obviously, it is more challenging to obtain a dynamic view on microRNA biogenesis in vivo due to several technical challenges including difficulties in imaging single molecules in the crowded cellular environment (diffraction limit) and the lack of reliable targeted RNA labelling strategies. On the other hand, recent advances in optical microscopy, RNA/protein labelling strategies, and computational tools will undoubtedly allow for in vivo tracking of single microRNA in the near future. For instance, the revolution of super-resolution microscopy has allowed researchers all over the world to overcome the diffraction limit of conventional fluorescence imaging and track single protein complexes performing their functions in vivo [123–125]. Breakthroughs in RNA labelling approaches such as the use of specific RNA tags, organic dyes and RNA aptamers are rapidly emerging, allowing visualizing an RNA molecule throughout its life cycle [67, 126, 127]. These milestones in technical development are expected to make in vivo single-molecule tracking of RNA a routine technique available for biological and biomedical sciences.

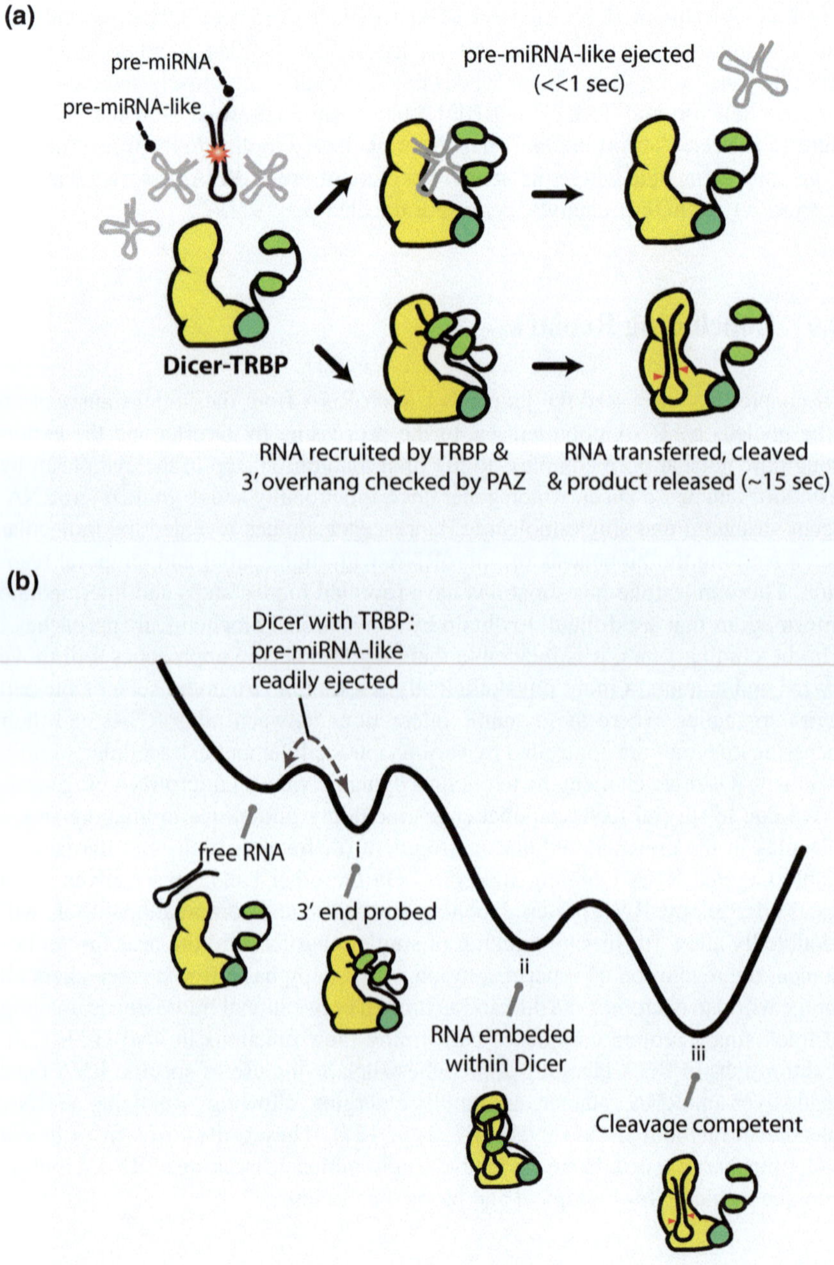
(a)
pre-miRNA
pre-miRNA-like
Dicer-TRBP
pre-miRNA-like ejected
(<<1 sec)
RNA recruited by TRBP &
3′ overhang checked by PAZ
RNA transferred, cleaved
& product released (~15 sec)
(b)
Dicer with TRBP:
pre-miRNA-like
readily ejected
free RNA
i
3′ end probed
ii
RNA embeded
within Dicer
iii
Cleavage competent

◀**Fig. 10.13 Model and the free energy landscape of substrate recognition by Dicer-TRBP. a** The double-stranded RNA-binding domains (dsRBD1-2) of TRBP recruit pre-miRNA by interacting with the stem region of the RNA. TRBP relocates the dsRNA into Dicer, where the PAZ domain verifies the length of 3′ overhang. If the dsRNA molecule possesses all required pre-miRNA features, the RNA becomes stably associated and cleaved within 15 s. If it does not, it is ejected far faster than 1 s. Dicer lacking TRBP partner fails in rejecting non-canonical pre-miRNA substrates rapidly, which compromises the enzyme turnover in an RNA-crowded environment. **b** Free energy landscape of substrate recognition. (i) is a state in which the 3′ end of RNA is recognized by TRBP-PAZ. (ii) is a state in which RNA is embedded within Dicer. (iii) is the cleavage-competent state of Dicer. The energy barrier between free RNA and (i) is lowered by TRBP. The energy level of (i) is deepened by TRBP to prevent long binding of non-canonical pre-miRNA into Dicer body and subsequent cleavage. Cellular RNA such as tRNA is trapped in (i) and readily ejected out of PAZ-TRPB, which enhance the processing activity of Dicer in an RNA-crowded cellular environment. Conversely, canonical pre-miRNA overcomes the energy barrier in (i) and is cleaved upon the transition from a pre-miRNA bound state (ii) to a cleavage-competent state (iii). After the cleavage, the product is released to allow the enzyme to probe a new dsRNA substrate. Adapted with permission from Fareh et al., Nature Communications, 2016 [122]

Acknowledgements I thank Chirlmin Joo and Joe Trapani laboratory members for their help, support, and the insightful discussions. This work was supported by the Fondation pour la Recherche Medicale [SPE20120523964 to Mohamed Fareh]; and the Zero Childhood Cancer initiative in Australia led by Prof. Michelle Haber (executive director, Children's Cancer Institute, Sydney, Australia) and Prof. Joe Trapani (head of Cancer Immunotherapy program, Peter MacCallum Cancer Centre, Melbourne, Australia).

References

1. Ha, M., & Kim, V. N. (2014). Regulation of microRNA biogenesis. *Nature Reviews Molecular Cell Biology, 15,* 509–524.
2. Huntzinger, E., & Izaurralde, E. (2011). Gene silencing by microRNAs: Contributions of translational repression and mRNA decay. *Nature Reviews Genetics, 12,* 99–110.
3. Krol, J., Loedige, I., & Filipowicz, W. (2010). The widespread regulation of microRNA biogenesis, function and decay. *Nature Reviews Genetics, 11,* 597–610.
4. Calin, G. A., & Croce, C. M. (2006). MicroRNA signatures in human cancers. *Nature Reviews Cancer, 6,* 857–866.
5. Nicoloso, M. S., Spizzo, R., Shimizu, M., Rossi, S., & Calin, G. A. (2009). MicroRNAs–the micro steering wheel of tumour metastases. *Nature Reviews Cancer, 9,* 293–302.
6. Fareh, M., Turchi, L., Virolle, V., Debruyne, D., Almairac, F., de-la-Forest Divonne, et al. (2012). The miR 302-367 cluster drastically affects self-renewal and infiltration properties of glioma-initiating cells through CXCR4 repression and consequent disruption of the SHH-GLI-NANOG network. *Cell Death and Differentiation, 19,* 232–244.
7. Fareh, M., Almairac, F., Turchi, L., Burel-Vandenbos, F., Paquis, P., Fontaine, D., et al. (2017). Cell-based therapy using miR-302-367 expressing cells represses glioblastoma growth. *Cell Death and Disease, 8,* e2713.
8. Lee, R. C., Feinbaum, R. L., & Ambros, V. (1993). The *C. elegans* heterochronic gene lin-4 encodes small RNAs with antisense complementarity to lin-14. *Cell, 75,* 843–854.

9. Wightman, B., Ha, I., & Ruvkun, G. (1993). Posttranscriptional regulation of the heterochronic gene lin-14 by lin-4 mediates temporal pattern formation in *C. elegans*. *Cell, 75,* 855–862.
10. Bartel, D. P. (2018). Metazoan MicroRNAs. *Cell, 173,* 20–51.
11. Berezikov, E. (2011). Evolution of microRNA diversity and regulation in animals. *Nature Reviews Genetics, 12,* 846–860.
12. Kim, V. N., & Nam, J. W. (2006). Genomics of microRNA. *Trends in genetics: TIG, 22,* 165–173.
13. Lee, Y., Jeon, K., Lee, J. T., Kim, S., & Kim, V. N. (2002). MicroRNA maturation: Stepwise processing and subcellular localization. *The EMBO Journal, 21,* 4663–4670.
14. Cai, X., Hagedorn, C. H., & Cullen, B. R. (2004). Human microRNAs are processed from capped, polyadenylated transcripts that can also function as mRNAs. *RNA, 10,* 1957–1966.
15. Lee, Y., Kim, M., Han, J., Yeom, K. H., Lee, S., Baek, S. H., et al. (2004). MicroRNA genes are transcribed by RNA polymerase II. *The EMBO Journal, 23,* 4051–4060.
16. Li, K., Li, Z., Zhao, N., Xu, Y., Liu, Y., Zhou, Y., et al. (2013). Functional analysis of microRNA and transcription factor synergistic regulatory network based on identifying regulatory motifs in non-small cell lung cancer. *BMC Systems Biology, 7,* 122.
17. O'Donnell, K. A., Wentzel, E. A., Zeller, K. I., Dang, C. V., & Mendell, J. T. (2005). c-Myc-regulated microRNAs modulate E2F1 expression. *Nature, 435,* 839–843.
18. Ma, L., Young, J., Prabhala, H., Pan, E., Mestdagh, P., Muth, D., et al. (2010). miR-9, a MYC/MYCN-activated microRNA, regulates E-cadherin and cancer metastasis. *Nature Cell Biology, 12,* 247–256.
19. Chang, T. C., Yu, D., Lee, Y. S., Wentzel, E. A., Arking, D. E., West, K. M., et al. (2008). Widespread microRNA repression by Myc contributes to tumorigenesis. *Nature Genetics, 40,* 43–50.
20. Mestdagh, P., Fredlund, E., Pattyn, F., Schulte, J. H., Muth, D., Vermeulen, J., et al. (2010). MYCN/c-MYC-induced microRNAs repress coding gene networks associated with poor outcome in MYCN/c-MYC-activated tumors. *Oncogene, 29,* 1394–1404.
21. Barros-Silva, D., Costa-Pinheiro, P., Duarte, H., Sousa, E. J., Evangelista, A. F., Graca, I., et al. (2018). MicroRNA-27a-5p regulation by promoter methylation and MYC signaling in prostate carcinogenesis. *Cell Death and Disease, 9,* 167.
22. Song, S. J., Poliseno, L., Song, M. S., Ala, U., Webster, K., Ng, C., et al. (2013). MicroRNA-antagonism regulates breast cancer stemness and metastasis via TET-family-dependent chromatin remodeling. *Cell, 154,* 311–324.
23. Gregory, R. I., Yan, K. P., Amuthan, G., Chendrimada, T., Doratotaj, B., Cooch, N., et al. (2004). The Microprocessor complex mediates the genesis of microRNAs. *Nature, 432,* 235–240.
24. Lee, Y., Ahn, C., Han, J., Choi, H., Kim, J., Yim, J., et al. (2003). The nuclear RNase III Drosha initiates microRNA processing. *Nature, 425,* 415–419.
25. Han, J., Lee, Y., Yeom, K. H., Kim, Y. K., Jin, H., & Kim, V. N. (2004). The Drosha-DGCR25 complex in primary microRNA processing. *Genes & Development, 18,* 3016–3027.
26. Kwon, S. C., Nguyen, T. A., Choi, Y. G., Jo, M. H., Hohng, S., Kim, V. N., et al. (2016). Structure of human DROSHA. *Cell, 164,* 81–90.
27. Nguyen, T. A., Jo, M. H., Choi, Y. G., Park, J., Kwon, S. C., Hohng, S., et al. (2015). Functional anatomy of the human microprocessor. *Cell, 161,* 1374–1387.
28. Han, J., Lee, Y., Yeom, K. H., Nam, J. W., Heo, I., Rhee, J. K., et al. (2006). Molecular basis for the recognition of primary microRNAs by the Drosha-DGCR28 complex. *Cell, 125,* 887–901.
29. Fang, W., & Bartel, D. P. (2015). The menu of features that define primary MicroRNAs and enable De Novo design of MicroRNA genes. *Molecular Cell, 60,* 131–145.
30. Auyeung, V. C., Ulitsky, I., McGeary, S. E., & Bartel, D. P. (2013). Beyond secondary structure: Primary-sequence determinants license pri-miRNA hairpins for processing. *Cell, 152,* 844–858.

31. Denli, A. M., Tops, B. B., Plasterk, R. H., Ketting, R. F., & Hannon, G. J. (2004). Processing of primary microRNAs by the microprocessor complex. *Nature, 432,* 231–235.
32. Sellier, C., Hwang, V. J., Dandekar, R., Durbin-Johnson, B., Charlet-Berguerand, N., Ander, B. P., etal. (2014). Decreased DGCR8 expression and miRNA dysregulation in individuals with 22q11.2 deletion syndrome. *PloS one, 9,* e103884.
33. Bartel, D. P. (2004). MicroRNAs: Genomics, biogenesis, mechanism, and function. *Cell, 116,* 281–297.
34. Stark, K. L., Xu, B., Bagchi, A., Lai, W. S., Liu, H., Hsu, R., et al. (2008). Altered brain microRNA biogenesis contributes to phenotypic deficits in a 22q11-deletion mouse model. *Nature Genetics, 40,* 751–760.
35. Fenelon, K., Mukai, J., Xu, B., Hsu, P. K., Drew, L. J., Karayiorgou, M., et al. (2011). Deficiency of Dgcr8, a gene disrupted by the 22q11.2 microdeletion, results in altered short-term plasticity in the prefrontal cortex. *Proceedings of the National Academy of Sciences of the United States of America, 108,* 4447–4452.
36. Fareh, M., Loeff, L., Szczepaniak, M., Haagsma, A. C., Yeom, K. H., & Joo, C. (2016). Single-molecule pull-down for investigating protein-nucleic acid interactions. *Methods, 105,* 99–108.
37. Doye, V., & Hurt, E. (1997). From nucleoporins to nuclear pore complexes. *Current Opinion in Cell Biology, 9,* 401–411.
38. Beck, M., & Hurt, E. (2017). The nuclear pore complex: Understanding its function through structural insight. *Nature reviews. Molecular cell biology, 18,* 73–89.
39. Raices, M., & D'Angelo, M. A. (2012). Nuclear pore complex composition: A new regulator of tissue-specific and developmental functions. *Nature Reviews Molecular Cell Biology, 13,* 687–699.
40. Grossman, E., Medalia, O., & Zwerger, M. (2012). Functional architecture of the nuclear pore complex. *Annual Review of Biophysics, 41,* 557–584.
41. Finlay, D. R., Meier, E., Bradley, P., Horecka, J., & Forbes, D. J. (1991). A complex of nuclear pore proteins required for pore function. *The Journal of Cell Biology, 114,* 169–183.
42. Hinshaw, J. E., Carragher, B. O., & Milligan, R. A. (1992). Architecture and design of the nuclear pore complex. *Cell, 69,* 1133–1141.
43. Akey, C. W., & Radermacher, M. (1993). Architecture of the Xenopus nuclear pore complex revealed by three-dimensional cryo-electron microscopy. *The Journal of Cell Biology, 122,* 1–19.
44. Kosinski, J., Mosalaganti, S., von Appen, A., Teimer, R., DiGuilio, A. L., Wan, W., et al. (2016). Molecular architecture of the inner ring scaffold of the human nuclear pore complex. *Science, 352,* 363–365.
45. Eibauer, M., Pellanda, M., Turgay, Y., Dubrovsky, A., Wild, A., & Medalia, O. (2015). Structure and gating of the nuclear pore complex. *Nature communications, 6,* 7532.
46. Okamura, M., Inose, H., & Masuda, S. (2015). RNA export through the NPC in eukaryotes. *Genes, 6,* 124–149.
47. Kohler, A., & Hurt, E. (2007). Exporting RNA from the nucleus to the cytoplasm. *Nature Reviews Molecular Cell Biology, 8,* 761–773.
48. Mattaj, I. W., & Englmeier, L. (1998). Nucleocytoplasmic transport: The soluble phase. *Annual Review of Biochemistry, 67,* 265–306.
49. Gorlich, D., & Kutay, U. (1999). Transport between the cell nucleus and the cytoplasm. *Annual Review of Cell and Developmental Biology, 15,* 607–660.
50. Izaurralde, E., & Adam, S. (1998). Transport of macromolecules between the nucleus and the cytoplasm. *RNA, 4,* 351–364.
51. Pemberton, L. F., Blobel, G., & Rosenblum, J. S. (1998). Transport routes through the nuclear pore complex. *Current Opinion in cell Biology, 10,* 392–399.
52. Hurt, E. C. (1988). A novel nucleoskeletal-like protein located at the nuclear periphery is required for the life cycle *of Saccharomyces cerevisiae. The EMBO Journal, 7,* 4323–4334.
53. Doye, V., & Hurt, E. C. (1995). Genetic approaches to nuclear pore structure and function. *Trends in Genetics: TIG, 11,* 235–241.

54. Lowe, A. R., Siegel, J. J., Kalab, P., Siu, M., Weis, K., & Liphardt, J. T. (2010). Selectivity mechanism of the nuclear pore complex characterized by single cargo tracking. *Nature, 467,* 600–603.
55. Grunwald, D., Singer, R. H., & Rout, M. (2011). Nuclear export dynamics of RNA-protein complexes. *Nature, 475,* 333–341.
56. Moore, M. S., & Blobel, G. (1993). The GTP-binding protein Ran/TC4 is required for protein import into the nucleus. *Nature, 365,* 661–663.
57. Moroianu, J. (1999). Nuclear import and export pathways. *Journal of Cellular Biochemistry,* (Suppl 32–33), 76–83.
58. Rodriguez, M. S., Dargemont, C., & Stutz, F. (2004). Nuclear export of RNA. *Biology of the Cell, 96,* 639–655.
59. Lund, E., Guttinger, S., Calado, A., Dahlberg, J. E., & Kutay, U. (2004). Nuclear export of microRNA precursors. *Science, 303,* 95–98.
60. Yi, R., Qin, Y., Macara, I. G., & Cullen, B. R. (2003). Exportin-5 mediates the nuclear export of pre-microRNAs and short hairpin RNAs. *Genes & Development, 17,* 3011–3016.
61. Zeng, Y., & Cullen, B. R. (2004). Structural requirements for pre-microRNA binding and nuclear export by Exportin 5. *Nucleic Acids Research, 32,* 4776–4785.
62. Bohnsack, M. T., Czaplinski, K., & Gorlich, D. (2004). Exportin 5 is a RanGTP-dependent dsRNA-binding protein that mediates nuclear export of pre-miRNAs. *RNA, 10,* 185–191.
63. Gagnon, K. T., Li, L., Chu, Y., Janowski, B. A., & Corey, D. R. (2014). RNAi factors are present and active in human cell nuclei. *Cell Reports, 6,* 211–221.
64. Khudayberdiev, S. A., Zampa, F., Rajman, M., & Schratt, G. (2013). A comprehensive characterization of the nuclear microRNA repertoire of post-mitotic neurons. *Frontiers in Molecular Neuroscience, 6,* 43.
65. Liao, J. Y., Ma, L. M., Guo, Y. H., Zhang, Y. C., Zhou, H., Shao, P., et al. (2010). Deep sequencing of human nuclear and cytoplasmic small RNAs reveals an unexpectedly complex subcellular distribution of miRNAs and tRNA 3′ trailers. *PloS One, 5,* e10563.
66. Pitchiaya, S., Heinicke, L. A., Park, J. I., Cameron, E. L., & Walter, N. G. (2017). Resolving subcellular miRNA trafficking and turnover at single-molecule resolution. *Cell Reports, 19,* 630–642.
67. Grunwald, D., & Singer, R. H. (2010). In vivo imaging of labelled endogenous beta-actin mRNA during nucleocytoplasmic transport. *Nature, 467,* 604–607.
68. Stockley, P. G., Stonehouse, N. J., Murray, J. B., Goodman, S. T., Talbot, S. J., Adams, C. J., et al. (1995). Probing sequence-specific RNA recognition by the bacteriophage MS2 coat protein. *Nucleic Acids Research, 23,* 2512–2518.
69. Yang, W., & Musser, S. M. (2006). Nuclear import time and transport efficiency depend on importin beta concentration. *The Journal of Cell Biology, 174,* 951–961.
70. Yang, W., Gelles, J., & Musser, S. M. (2004). Imaging of single-molecule translocation through nuclear pore complexes. *Proceedings of the National Academy of Sciences of the United States of America, 101,* 12887–12892.
71. Sun, C., Yang, W., Tu, L. C., & Musser, S. M. (2008). Single-molecule measurements of importin alpha/cargo complex dissociation at the nuclear pore. *Proceedings of the National Academy of Sciences of the United States of America, 105,* 8613–8618.
72. Noland, C. L., & Doudna, J. A. (2013). Multiple sensors ensure guide strand selection in human RNAi pathways. *RNA, 19,* 639–648.
73. Tants, J. N., Fesser, S., Kern, T., Stehle, R., Geerlof, A., Wunderlich, C., et al. (2017). Molecular basis for asymmetry sensing of siRNAs by the Drosophila Loqs-PD/Dcr-2 complex in RNA interference. *Nucleic Acids Research, 45,* 12536–12550.
74. Meijer, H. A., Smith, E. M., & Bushell, M. (2014). Regulation of miRNA strand selection: Follow the leader? *Biochemical Society Transactions, 42,* 1135–1140.
75. Noland, C. L., Ma, E., & Doudna, J. A. (2011). siRNA repositioning for guide strand selection by human Dicer complexes. *Molecular Cell, 43,* 110–121.
76. Filipowicz, W., Bhattacharyya, S. N., & Sonenberg, N. (2008). Mechanisms of post-transcriptional regulation by microRNAs: Are the answers in sight? *Nature Reviews Genetics, 9,* 102–114.

77. Guo, L., & Lu, Z. (2010). The fate of miRNA* strand through evolutionary analysis: Implication for degradation as merely carrier strand or potential regulatory molecule? *PloS One, 5,* e11387.
78. Chandradoss, S. D., Schirle, N. T., Szczepaniak, M., MacRae, I. J., & Joo, C. (2015). A dynamic search process underlies MicroRNA targeting. *Cell, 162,* 96–107.
79. Klein, M., Chandradoss, S. D., Depken, M., & Joo, C. (2017). Why Argonaute is needed to make microRNA target search fast and reliable. *Seminars in Cell & Developmental Biology, 65,* 20–28.
80. Grishok, A., Pasquinelli, A. E., Conte, D., Li, N., Parrish, S., Ha, I., et al. (2001). Genes and mechanisms related to RNA interference regulate expression of the small temporal RNAs that control *C. elegans* developmental timing. *Cell, 106,* 23–34.
81. Kim, Y. K., Kim, B., & Kim, V. N. (2016). Re-evaluation of the roles of DROSHA, Export in 5, and DICER in microRNA biogenesis. *Proceedings of the National Academy of Sciences of the United States of America, 113,* E1881–E1889.
82. Bernstein, E., Kim, S. Y., Carmell, M. A., Murchison, E. P., Alcorn, H., Li, M. Z., et al. (2003). Dicer is essential for mouse development. *Nature Genetics, 35,* 215–217.
83. Krill, K. T., Gurdziel, K., Heaton, J. H., Simon, D. P., & Hammer, G. D. (2013). Dicer deficiency reveals microRNAs predicted to control gene expression in the developing adrenal cortex. *Molecular Endocrinology, 27,* 754–768.
84. Mori, M. A., Thomou, T., Boucher, J., Lee, K. Y., Lallukka, S., Kim, J. K., et al. (2014). Altered miRNA processing disrupts brown/white adipocyte determination and associates with lipodystrophy. *The Journal of Clinical Investigation, 124,* 3339–3351.
85. Mudhasani, R., Zhu, Z., Hutvagner, G., Eischen, C. M., Lyle, S., Hall, L. L., et al. (2008). Loss of miRNA biogenesis induces p19Arf-p53 signaling and senescence in primary cells. *The Journal of Cell Biology, 181,* 1055–1063.
86. Soukup, G. A., Fritzsch, B., Pierce, M. L., Weston, M. D., Jahan, I., McManus, M. T., et al. (2009). Residual microRNA expression dictates the extent of inner ear development in conditional Dicer knockout mice. *Developmental Biology, 328,* 328–341.
87. Chen, J. F., Murchison, E. P., Tang, R., Callis, T. E., Tatsuguchi, M., Deng, Z., et al. (2008). Targeted deletion of Dicer in the heart leads to dilated cardiomyopathy and heart failure. *Proceedings of the National Academy of Sciences of the United States of America, 105,* 2111–2116.
88. Macrae, I. J., Zhou, K., Li, F., Repic, A., Brooks, A. N., Cande, W. Z., et al. (2006). Structural basis for double-stranded RNA processing by Dicer. *Science, 311,* 195–198.
89. MacRae, I. J., Zhou, K., & Doudna, J. A. (2007). Structural determinants of RNA recognition and cleavage by Dicer. *Nature Structural & Molecular Biology, 14,* 934–940.
90. Lau, P. W., Potter, C. S., Carragher, B., & MacRae, I. J. (2009). Structure of the human Dicer-TRBP complex by electron microscopy. *Structure, 17,* 1326–1332.
91. Zhang, H., Kolb, F. A., Jaskiewicz, L., Westhof, E., & Filipowicz, W. (2004). Single processing center models for human Dicer and bacterial RNase III. *Cell, 118,* 57–68.
92. Lau, P. W., Guiley, K. Z., De, N., Potter, C. S., Carragher, B., & MacRae, I. J. (2012). The molecular architecture of human Dicer. *Nature Structural & Molecular Biology, 19,* 436–440.
93. Park, J. E., Heo, I., Tian, Y., Simanshu, D. K., Chang, H., Jee, D., et al. (2011). Dicer recognizes the 5′ end of RNA for efficient and accurate processing. *Nature, 475,* 201–205.
94. Tian, Y., Simanshu, D. K., Ma, J. B., Park, J. E., Heo, I., Kim, V. N., et al. (2014). A phosphate-binding pocket within the platform-PAZ-connector helix cassette of human Dicer. *Molecular Cell, 53,* 606–616.
95. Tsutsumi, A., Kawamata, T., Izumi, N., Seitz, H., & Tomari, Y. (2011). Recognition of the pre-miRNA structure by Drosophila Dicer-1. *Nature Structural & Molecular Biology, 18,* 1153–1158.
96. Gu, S., Jin, L., Zhang, Y., Huang, Y., Zhang, F., Valdmanis, P. N., et al. (2012). The loop position of shRNAs and pre-miRNAs is critical for the accuracy of dicer processing in vivo. *Cell, 151,* 900–911.

97. Liu, Z., Wang, J., Cheng, H., Ke, X., Sun, L., Zhang, Q. C., et al. (2018). Cryo-EM structure of human Dicer and its complexes with a Pre-miRNA substrate. *Cell, 173,* 1549–1550.
98. Lee, H. Y., Zhou, K., Smith, A. M., Noland, C. L., & Doudna, J. A. (2013). Differential roles of human Dicer-binding proteins TRBP and PACT in small RNA processing. *Nucleic Acids Research, 41,* 6568–6576.
99. Chakravarthy, S., Sternberg, S. H., Kellenberger, C. A., & Doudna, J. A. (2010). Substrate-specific kinetics of Dicer-catalyzed RNA processing. *Journal of Molecular Biology, 404,* 392–402.
100. Wilson, R. C., Tambe, A., Kidwell, M. A., Noland, C. L., Schneider, C. P., & Doudna, J. A. (2015). Dicer-TRBP complex formation ensures accurate mammalian microRNA biogenesis. *Molecular Cell, 57,* 397–407.
101. Kim, Y., Yeo, J., Lee, J. H., Cho, J., Seo, D., Kim, J. S., et al. (2014). Deletion of human tarbp2 reveals cellular microRNA targets and cell-cycle function of TRBP. *Cell Reports, 9,* 1061–1074.
102. Ota, H., Sakurai, M., Gupta, R., Valente, L., Wulff, B. E., Ariyoshi, K., et al. (2013). ADAR1 forms a complex with Dicer to promote microRNA processing and RNA-induced gene silencing. *Cell, 153,* 575–589.
103. Yamashita, S., Nagata, T., Kawazoe, M., Takemoto, C., Kigawa, T., Guntert, P., et al. (2011). Structures of the first and second double-stranded RNA-binding domains of human TAR RNA-binding protein. *Protein Science: A Publication of the Protein Society, 20,* 118–130.
104. Schmedt, C., Green, S. R., Manche, L., Taylor, D. R., Ma, Y., & Mathews, M. B. (1995). Functional characterization of the RNA-binding domain and motif of the double-stranded RNA-dependent protein kinase DAI (PKR). *Journal of Molecular Biology, 249,* 29–44.
105. Krovat, B. C., & Jantsch, M. F. (1996). Comparative mutational analysis of the double-stranded RNA binding domains of Xenopus laevis RNA-binding protein A. *The Journal of Biological Chemistry, 271,* 28112–28119.
106. Takahashi, T., Miyakawa, T., Zenno, S., Nishi, K., Tanokura, M., & Ui-Tei, K. (2013). Distinguishable in vitro binding mode of monomeric TRBP and dimeric PACT with siRNA. *PloS One, 8,* e63434.
107. Daniels, S. M., & Gatignol, A. (2012). The multiple functions of TRBP, at the hub of cell responses to viruses, stress, and cancer. *Microbiology and Molecular Biology Reviews: MMBR, 76,* 652–666. https://mmbr.asm.org/content/76/3/652.abstract; https://www.ncbi.nlm.nih.gov/pmc/articles/PMC3429622/
108. Masliah, G., Maris, C., Konig, S. L., Yulikov, M., Aeschimann, F., Malinowska, A. L., et al. (2018). Structural basis of siRNA recognition by TRBP double-stranded RNA binding domains. *The EMBO Journal, 37,* e97089. https://doi.org/10.15252/embj.201797089.
109. Ryter, J. M., & Schultz, S. C. (1998). Molecular basis of double-stranded RNA-protein interactions: Structure of a dsRNA-binding domain complexed with dsRNA. *The EMBO Journal, 17,* 7505–7513.
110. Masliah, G., Barraud, P., & Allain, F. H. (2013). RNA recognition by double-stranded RNA binding domains: A matter of shape and sequence. *Cellular and Molecular Life Sciences: CMLS, 70,* 1875–1895.
111. Koh, H. R., Kidwell, M. A., Ragunathan, K., Doudna, J. A., & Myong, S. (2013). ATP-independent diffusion of double-stranded RNA binding proteins. *Proceedings of the National Academy of Sciences of the United States of America, 110,* 151–156.
112. Joo, C., McKinney, S. A., Nakamura, M., Rasnik, I., Myong, S., & Ha, T. (2006). Real-time observation of RecA filament dynamics with single monomer resolution. *Cell, 126,* 515–527.
113. Joo, C., Balci, H., Ishitsuka, Y., Buranachai, C., & Ha, T. (2008). Advances in single-molecule fluorescence methods for molecular biology. *Annual Review of Biochemistry, 77,* 51–76.
114. Abbondanzieri, E. A., Bokinsky, G., Rausch, J. W., Zhang, J. X., Le Grice, S. F., & Zhuang, X. (2008). Dynamic binding orientations direct activity of HIV reverse transcriptase. *Nature, 453,* 184–189.
115. Lee, Y., Hur, I., Park, S. Y., Kim, Y. K., Suh, M. R., & Kim, V. N. (2006). The role of PACT in the RNA silencing pathway. *The EMBO Journal, 25,* 522–532.

116. Benoit, M. P., Imbert, L., Palencia, A., Perard, J., Ebel, C., Boisbouvier, J., et al. (2013). The RNA-binding region of human TRBP interacts with microRNA precursors through two independent domains. *Nucleic Acids Research, 41,* 4241–4252.
117. Hwang, H., & Myong, S. (2014). Protein induced fluorescence enhancement (PIFE) for probing protein-nucleic acid interactions. *Chemical Society Reviews, 43,* 1221–1229.
118. Myong, S., Cui, S., Cornish, P. V., Kirchhofer, A., Gack, M. U., Jung, J. U., et al. (2009). Cytosolic viral sensor RIG-I is a 5′-triphosphate-dependent translocase on double-stranded RNA. *Science, 323,* 1070–1074.
119. Wang, X., Vukovic, L., Koh, H. R., Schulten, K., & Myong, S. (2015). Dynamic profiling of double-stranded RNA binding proteins. *Nucleic Acids Research, 43,* 7566–7576.
120. Koh, H. R., Kidwell, M. A., Doudna, J., & Myong, S. (2017). RNA scanning of a molecular machine with a built-in ruler. *Journal of the American Chemical Society, 139,* 262–268.
121. Peltier, H. J., & Latham, G. J. (2008). Normalization of microRNA expression levels in quantitative RT-PCR assays: Identification of suitable reference RNA targets in normal and cancerous human solid tissues. *RNA, 14,* 844–852.
122. Fareh, M., Yeom, K. H., Haagsma, A. C., Chauhan, S., Heo, I., & Joo, C. (2016). TRBP ensures efficient Dicer processing of precursor microRNA in RNA-crowded environments. *Nature Communications, 7,* 13694.
123. Willig, K. I., Kellner, R. R., Medda, R., Hein, B., Jakobs, S., & Hell, S. W. (2006). Nanoscale resolution in GFP-based microscopy. *Nature Methods, 3,* 721–723.
124. Rust, M. J., Bates, M., & Zhuang, X. (2006). Sub-diffraction-limit imaging by stochastic optical reconstruction microscopy (STORM). *Nature Methods, 3,* 793–795.
125. Betzig, E., Patterson, G. H., Sougrat, R., Lindwasser, O. W., Olenych, S., Bonifacino, J. S., et al. (2006). Imaging intracellular fluorescent proteins at nanometer resolution. *Science, 313,* 1642–1645.
126. Dean, K. M., & Palmer, A. E. (2014). Advances in fluorescence labeling strategies for dynamic cellular imaging. *Nature Chemical Biology, 10,* 512–523.
127. Paige, J. S., Nguyen-Duc, T., Song, W., & Jaffrey, S. R. (2012). Fluorescence imaging of cellular metabolites with RNA. *Science, 335,* 1194.

MIX
Papier aus verantwortungsvollen Quellen
Paper from responsible sources
FSC® C105338

If you have any concerns about our products,
you can contact us on
ProductSafety@springernature.com

In case Publisher is established outside the EU,
the EU authorized representative is:
Springer Nature Customer Service Center GmbH
Europaplatz 3, 69115 Heidelberg, Germany

Printed by Libri Plureos GmbH
in Hamburg, Germany